Chevrolet Volt

Development Story of the Pioneering Electrified Vehicle

Other SAE books of interest

Hybrid-Powered Vehicles, Second Edition
By John M. German
(Product Code: T-125)

Electric and Hybrid-Electric Vehicles
by Ronald K. Jurgen,
(Product Code: PT-143.SET)

Advanced Hybrid Powertrains for Commercial Vehicles
By Haoran Hu, Rudy Smaling, and Simon J. Basely
(Product Code: R-396)

Chevrolet Volt
Development Story of the Pioneering Electrified Vehicle

By Lindsay Brooke

Published by
SAE International
400 Commonwealth Drive
Warrendale, PA 15096-0001 USA

Phone: (724) 776-4861
Fax: (724) 776-5760
www.sae.org

PT-149

400 Commonwealth Drive
Warrendale, PA 15096-0001 USA

E-mail: **CustomerService@sae.org**
Phone: **877-606-7323** *(inside USA and Canada)*
724-776-4970 *(outside USA)*
Fax: **724-776-1615**

ISBN 978-0-7680-4765-3
Library of Congress Catalog Number 2011922124
SAE Order No. PT-149

To purchase bulk quantities,
please contact:

SAE Customer Service
E-mail: **CustomerService@sae.org**
Phone: **877-606-7323** *(inside USA and Canada)*
724-776-4970 *(outside USA)*
Fax: **724-776-1615**

Visit the SAE Bookstore at
http://store.sae.org

Table of Contents

SAE International vehicle ELECTRIFICATION SERIES

Table of Contents

SAE International
vehicle ELECTRIFICATION SERIES

Preface

GM on the Critical Path

Four years of Volt development

When veteran GM engineer Jon Lauckner sketched out his idea for a new type of electrified propulsion system for his boss, Bob Lutz, in 2007, he reckoned there were two major hurdles in the way of the idea actually reaching production.

The first hurdle was simply getting the idea for an "extended range" electric vehicle approved. But Lutz, then GM's Vice Chairman for Product Development, was immediately convinced this was something the automaker had to do. The second hurdle was far more daunting. For Lauckner's idea to work as conceived, it needed a high power/high energy automotive battery that did not exist.

Lutz likened the program to a "moon shot," because of the high level of invention and critical-path engineering required to meet the aggressive 2010 production target. Lauckner remained confident that the issue of developing a suitable battery would be solved.

"We have the best technical organization in the industry," he told me about a year into the Volt's development. "We're going to execute this program and deliver an exceptional new vehicle as planned."

With volume production of the 2011 Chevrolet Volt underway at GM's Detroit-Hamtramck assembly plant, the development team and its strategic suppliers have delivered on Lauckner's promises. Volt's overall performance, battery range, NVH attenuation, driver interfaces, and build quality exceeded my high expectations during recent drives. Currently the car has no peer in its approach to "green" mobility.

My four years of reporting on this milestone vehicle inspired the special Volt digital magazine which launched on SAE International's new Vehicle Electrification web portal (www.evsae.com) in November 2010. In covering the program's development I filled a dozen notebooks and compiled hundreds of hours of interview recordings with the engineers who made Volt happen. This SAE technical publication is the logical follow-up to the online magazine.

With GM set to publish 12 technical papers related to Volt at the 2011 SAE World Congress, it made sense to combine them with the Volt digital magazine and other Volt-related content previously published in *Automotive Engineering International* magazine. Taken together, the compendium provides the most comprehensive insight into Volt's genesis, engineering, and development.

Competitors will find many things of interest when they tear down a Volt for analysis. Its innovative powertrain brings a number of high-volume, industry-first technology applications. Among them are the liquid-cooled Li-ion battery pack; smart-phone driver interface for remote cabin conditioning and battery-charge control; SAE J1772 charge coupler; and the use of Behr's novel "chiller" heat exchanger that uses both A/C refrigerant and glycol-based coolant to help cool the battery in extreme operating conditions.

The more than 200 patents GM has filed related to the Volt program will prove useful as more competitors enter the electrified-vehicle space. Particularly important is the intellectual property related to power controls and the functionality of the new 4ET50 electrified transaxle.

My key takeaway about Volt as a product is that it represents the advent of the industry's turn toward vehicles that are primarily electronics platforms, rather than mechanical ones. By the time the car had entered production, it had undergone more than 20 major vehicle control software calibrations (known as VESCOMs), plus many more v.1, v.2, etc., iterations. No wonder GM is investing so heavily in new resources in the rapidly growing electrical/electronics engineering arena.

But the highlight of covering Volt for me was getting to know the high-caliber people who made it happen. Its execution reflects the dedication and focus of everyone involved.

Lindsay Brooke
Senior Editor, SAE International
March 2011

CHAPTER ONE:

Why Volt?

Engineers will debate whether Volt is technically a series-type hybrid or an extended-range EV (E-REV as GM prefers to call it). Lear Corp. helped develop the car's offboard 120-V and 240-V charging set. GM continues to evaluate charging technologies (Coulomb Technologies charging station shown).

After 48 months' development, the 2011 Chevrolet Volt has entered series production. The pioneering "E-REV" is as important to the mobility industry as it is to GM.

Why Volt?

A production-spec Volt undergoes aerodynamic testing by technician Nina Tortosa at GM's Warren, MI, wind tunnel in 2009.

Mention the words "fuel injector" to anyone who's reasonably savvy about vehicles, and they'll likely know something about the device that controls the metering of liquid fuel into an engine's cylinders and helps cause the vehicle to drive faster or slower.

But mention the term "IGBT" and the response may elicit a lot of blank stares. IGBTs, the acronym for insulated gate bipolar transistors, are the fuel injectors of electrically driven vehicles. They control the supply of battery electrical energy into the electric motors that drive the wheels. IGBTs are valves for electrons, much like a fuel injector is a valve for liquid fuels.

Comparing these two simple and essential components shows how fundamentally the auto industry is going to change, as it moves from the petroleum-based model that has sustained it for more than 100 years to the electrified model that many experts believe will literally propel it for the next 100 years.

Vehicle electrification—the industry's shift to hybrids, plug-in hybrids, and electric vehicles (EVs)—is bringing more than a different vocabulary. It is bringing new technologies, patents, core competencies, and skill sets for engineers.

Electrification is attracting investment capital that has helped create new companies aimed at disrupting the status quo. It has helped jump-start a U.S. battery industry (aided by billions in federal and state subsidies), invigorated development of more efficient electric motors, and forced traditional automakers and suppliers to rethink their own product-development strategies, R&D paths, and industry alliances. And it has prodded them toward new relationships with the energy sector and government.

The 2011 Chevrolet Volt is in the vanguard of this shift. What literally began in 2006 as a napkin sketch by a General Motors executive engineer will soon be the world's first production extended-range electric passenger vehicle (E-REV).

Jon Lauckner, President of the new GM Ventures group, is the "godfather" of Volt.

The Volt's long-anticipated launch—the concept debuted at the 2007 North American International Auto Show in Detroit—kicks off a new age of vehicle propulsion offering significantly cleaner tailpipe emissions than the engines used in today's conventional cars and light trucks. Nearly 500 hybrid and EV programs of all types are in the works worldwide through 2015, according to forecaster IHS Global Insight.

Following Volt closely into showrooms are Nissan's battery-electric Leaf, also slated for 2011, as well as Volt's European-market cousin, the 2012 Opel Ampera, and a flotilla of new hybrids and EVs from Toyota, Honda, Ford, VW, BMW, Tesla, Fisker, Hyundai, Peugeot, Renault, BYD, Mercedes, and others.

Many of the OEMs' technology roadmaps are linked to collaborative ventures with battery and power-electronics suppliers and systems integrators. They're teaming with university researchers and creating new EV-focused curricula with engineering schools. They're also supported by global engineering and testing specialists including AVL, FEV, Ricardo, and IAV. These companies have invested millions to develop their own hybrid and EV prototypes (including E-REVs) and related powertrain systems.

Even Ferrari and Porsche have committed to some degree of powertrain electrification. Company leaders say this will enable them to remain viable within the increasingly stringent global CO_2-emissions environment.

SAE International, too, is deeply involved in creating new standards, including the pioneering J1772 charge-coupler standard introduced earlier this year. More standards related to vehicle electrification are on the way. SAE also is working with a growing list of authors to publish technical papers covering all facets of hybrid and EV development. (See table, page 4.)

"With Volt and our next-gen programs that fol-

Volt Sparks SAE Technical Papers

During the vehicle's development, GM engineers hinted to *AEI* that they would be chronicling Volt's development with a flurry of technical papers in 2011. The following titles are included in this book:

Title	Author	Paper No.
Aerodynamic Development of the 2011 Chevrolet Volt	Nina Tortosa	2011-01-0168
The Voltec 4ET50 Electric Drive System	Khwaja Rahman	2011-01-0355
High Voltage Battery Tray Design Optimization	Kristel Coronado	2011-01-0671
Optimizing 12 Volt Start - Stop for Conventional Powertrains	Darrell Robinette	2011-01-0699
Voltec Charging System EMC Requirements and Test Methodologies	Vipul M. Patel	2011-01-0742
Developing A Portable Charge Cord for the Chevy Volt	Tony Argote	2011-01-0878
The GM Voltec 4ET50 Multi-Mode Electric Transaxle	Michael A. Miller	2011-01-0887
High Voltage Power Allocation Management of Hybrid/Electric Vehicles	James D. Marus	2011-01-1022
High Voltage Contactor Control Function	Trista Schieffer	2011-01-1266
Voltec Battery Design and Manufacturing	Robert Parrish	2011-01-1360
Voltec Battery System for Electric Vehicles with Extended Range	Roland Matthe	2011-01-1373
Co-development of Chevy Volt Tire Properties for Electric Range and Performance Balance	Dean C. Degazio	2011-01-0096
The CO2 Benefits of Electrification - E-REVs, PHEVs, and Charging Scenarios	Peter J. Savagian	2009-01-1311
The Electrification of the Automobile: From Conventional Hybrid, to Plug-in Hybrids, to Extended-Range Electric Vehicles	E.D. Tate	2008-01-0458

low, we're talking about transforming the vehicle and the industry—and that's not just the auto industry because it includes fuels, the charging infrastructure, and energy producers as well," said Jon Lauckner, President of GM Ventures.

Lauckner is considered 'godfather' of the Volt. It was his napkin sketch over cocktails that convinced a skeptical GM Vice Chairman Bob Lutz of the E-REV concept.

"We can hypothesize whether E-REVs will be a permanent thing, or whether we all move toward pure electrics," he observed. "But the reason Volt is so significant is it represents the tipping point between the centrality of the internal-combustion engine (ICE) as the prime mover of vehicles, and electric propulsion systems. It's the tipping point between the petroleum and electric models."

Others agree on the vehicle's importance to its maker (which sees it as a new technical foundation for the company, rather than as a single model) and its significance to the industry. "Volt is perhaps the auto industry's most notable technological advancement of the past 50 years," said Dr. David Cole, Chairman Emeritus of the Center of Automotive Research.

Added the *Wall St. Journal*: "[Volt is] GM's most important model in decades—and possibly the key to its survival." Indeed, the program remained fully funded and supported by management, all through GM's financial turmoil and bankruptcy. It even outlasted four key engineers and three CEOs.

A "bridge" to an EV future

What makes the Volt different than the various hybrid vehicles currently on sale, and battery-electric cars like Leaf, is its "extended range" capability. This is made possible by a new type of powertrain developed by GM. The system, known as "Voltec," provides 25 to 50 mi (40 to 80 km) of zero emission electric-only operation, thanks to its fairly large 16 kW·h lithium-ion battery pack. When battery power is depleted, an onboard generator powered by a small gasoline engine engages automatically to provide up to 310 additional miles (500 km) of range.

The generator's role is to sustain a minimum state of battery charge while the car returns to a location where it can be charged via a 110/120-V or 220/240-V ac electrical outlet—the cleanest and

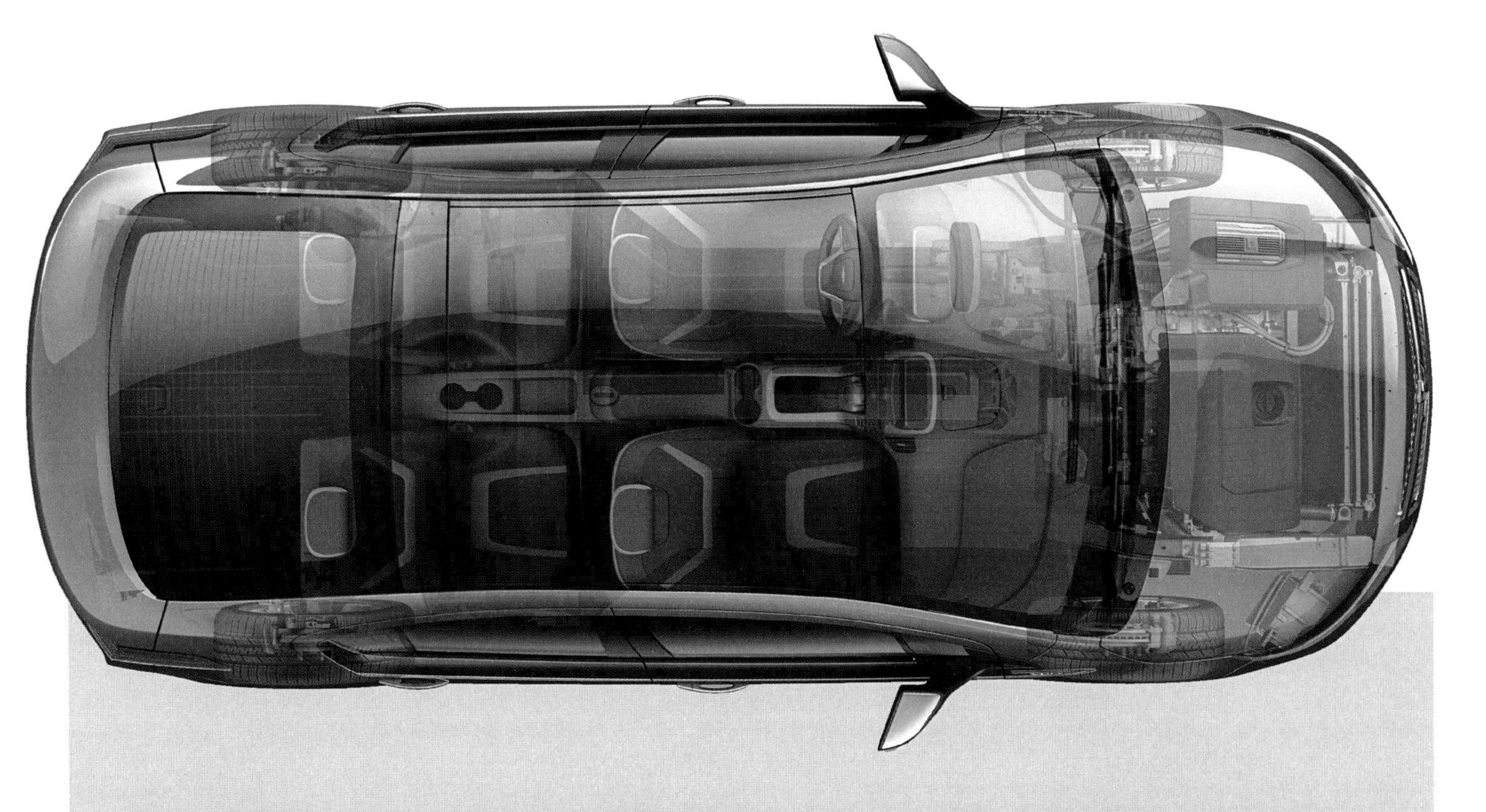

Volt's 16-kW·h lithium-ion battery pack weighs 375 lb (170 kg) and serves as a structural member along the car's center line. The 1.4-L engine is designed to run on ethanol as well as gasoline and features sophisticated control algorithms for its starting regimen, including periodic starts to circulate fuel and oil when the car has been driven in EV mode for extended periods.

most economical source of energy. Volt's regenerative braking system also provides a modest amount of energy back to the battery.

Studies by GM and other automakers show that more than 70% of American and European car owners drive 40 mi (64 km) or less in their typical daily commuting. In such a duty cycle, Volt would never burn gasoline, making it essentially an electric car. Indeed, GM engineers expect Volt's 1.4-L gasoline engine will serve mainly in an occasional support role for many owners. (The engine is programmed to start every 45 days, regardless of use. The brief running time is designed to circulate lubricating oil, check the emissions-control system, and burn off some gasoline lest the unused fuel become stale.)

The gasoline-powered generator gives Volt a sort of umbilical cord to the existing liquid-fuel infrastructure. It also allowed the car's development team to balance battery size and capacity with the size and power of the ICE. This strategy produced a car with an attractive balance of efficiency and performance. It also kept systems cost down. And perhaps most importantly, it eliminated the dreaded "range anxiety" that will limit the appeal of pure electric vehicles until widespread public charging is available.

"In this role the ICE is the bridge to pure electrics, until we can get more capability from, and lower costs related to the batteries," Lauckner explained. "We realize that having two systems on board to create electricity is less efficient than settling on one. And we also know that sometime in the future we'll move away from petroleum as the centrality of our whole propulsion strategy."

At that point, would a piston engine even be necessary to generate electricity on board as efficiently as possible? Because it serves only as a stationary generator, the power unit wouldn't need to be capable of the high revs required by a conventional car. (Volt's engine is governed to 4800 rpm.) Nor would it need to run all the time. Indeed, engineers at GM and its competitors and suppliers are investigating super-optimized alternatives, opening up new avenues for mechanical systems and combustion science, and even bringing new life to proven engine types.

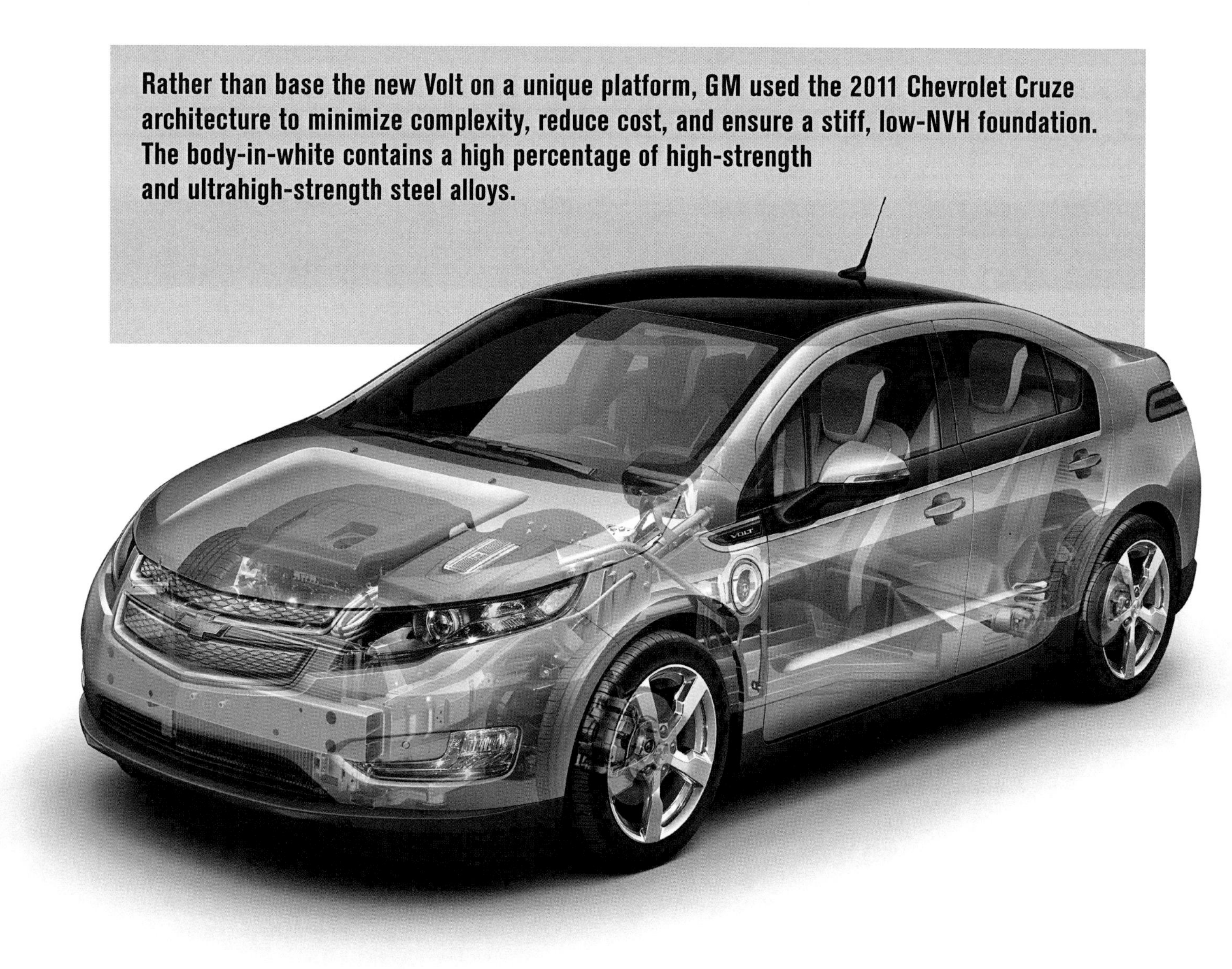

Rather than base the new Volt on a unique platform, GM used the 2011 Chevrolet Cruze architecture to minimize complexity, reduce cost, and ensure a stiff, low-NVH foundation. The body-in-white contains a high percentage of high-strength and ultrahigh-strength steel alloys.

"People are talking about specialized twin- and three-cylinder engines, miniature gas turbines, Sterling cycle engines, Wankel rotaries, and fuel cells," Lauckner noted. "For future-generation E-REVs, we can consider a lot of things that can create electricity—including some we don't even understand at the moment. Because when we step across that line from petroleum centrality into where it's all about electricity, all of a sudden the game changes very quickly."

There has been much debate over exactly how to categorize the 2011 Volt, given its novel E-REV powertrain. It's not a classic EV. Neither is it technically a series-type hybrid, as defined by the traditional example of a diesel-electric locomotive. With the Voltec powertrain, GM has combined the operating characteristics of EV, series-HEV, and plug-in hybrid types to create a unique (and patented) propulsion system. It offers some unique attributes that are discussed elsewhere in this special digital SAE publication.

A "call to arms" for suppliers

Vehicle electrification skeptics often point out the industry's risk in investing billions in new technologies. Many of the technologies, like lithium-based batteries, are still unproven in mass automotive use and offer no certain return on investment in the short term. The skeptics also question a greater push toward vehicle electrification when U.S. gasoline prices remain relatively low. On these points the skeptics are correct, to a degree.

Experts expect a steady global march to zero-emissions laws.

But industry leaders recognize that whether it's CAFE in the U.S. or all-encompassing CO_2 laws in Europe, the regulators are going to be "cranking it down in terms of CO_2 emissions and cranking it up in terms of higher mpg," as one engineer observed.

For the major global regions, including emergent China and India, it will be a slow-but-steady march to zero-emissions laws, experts believe.

"They're going to keep banging away and, despite all the creativity and cleverness we engineers can muster, we are not going to be able to keep the ICE as the prime mover. We just can't. The Second Law of Thermodynamics just grinds down on it," Lauckner asserted.

The advent of Volt/Ampera and their 2011-2015 hybrid and EV competitors is also a "call to arms" for suppliers who have entered this space.

"We need a supply base with both the product and process capability to design components and systems to be smaller, lighter, more efficient, reduced bill of materials—in quantity and with lower cost," noted Tony Posawatz, Volt's Vehicle Line Director. "This is where automotive experience can play a key role, because manufacturing critical propulsion-system items to automotive durability and reliability specs in 200,000-unit annual volumes is very different from making stuff for research."

Skeptics point to the most expensive piece of the car—Volt's battery pack, a hefty 375-lb (170-kg) T-shaped module estimated to cost GM $10,000 per unit. (GM executives won't reveal actual cost but say it is lower.) Industry analysts reckon, however, that production scale and evolving technologies will steadily lower the break-even point. The experts note that the first commercial mobile phones were the size of bricks and cost $3000. Smart and steady engineering usually wins for the consumer.

"It's the typical case of having a pretty steep learning curve whenever you bring a new technology to the market," explained CAR's Dr. Cole. "It never happens immediately. No battery manufacturer is going to put up a facility to build a million lithium car batteries in the first year.

"Rather, they'll aim their first-year production volume at 20,000 to 40,000 units. They'll incorporate their learnings into the next-generation batteries," he said. "Pretty quickly, by perhaps the third generation, they'll begin to boost scale and become cost-competitive."

Cole believes by the third generation, GM and its battery-cell partner LG Chem will have reduced the cost of the Volt packs by about half.

In his new GM Ventures job, Lauckner is charged with sniffing out promising new-technology partners and sharing GM's home-grown technologies with promising collaborators. To him, the electrification journey started by the famous EV-1 in the early 1990s and re-ignited by Volt in 2011 is running faster than many observers believe.

"If you look back five years, the number of companies on the OEM and supplier level who had any level of competency in electrified vehicles would be close to zero. Count the number of companies who are on board today, and the number of battery companies. Not just U.S. or European companies; I'm talking about everybody everywhere. It's incredible growth," he said.

"Combine money and brainpower and you're very likely to get results. This thing is going to happen—it's not a question of 'if' but 'when.' The companies that step out and harness money and intellectual commitment are the companies that are going to win,"
he said.

And the entire auto industry will be chasing them.

The Electrification of the Automobile: From Conventional Hybrid, to Plug-in Hybrids, to Extended-Range Electric Vehicles

2008-01-0458
Published
04/14/2008

E. D. Tate, Michael O. Harpster and Peter J. Savagian
General Motors Corporation

doi:10.4271/2008-01-0458

ABSTRACT

A key element of General Motors' Advanced Propulsion Technology Strategy is the electrification of the automobile. The objectives of this strategy are reduced fuel consumption, reduced emissions and increased energy security/diversification. The introduction of hybrid vehicles was one of the first steps as a result of this strategy. To determine future opportunities and direction, an extensive study was completed to better understand the ability of Plug-in Hybrid Electric Vehicles (PHEV) and Extended-Range Electric Vehicles (E-REV) to address societal challenges. The study evaluated real world representative driving datasets to understand actual vehicle usage. Vehicle simulations were conducted to evaluate the merits of PHEV and E-REV configurations.

As derivatives of conventional full hybrids, PHEVs have the potential to deliver a significant reduction in petroleum usage. However, the fuel consumption benefits are limited by the underlying constraints of the base hybrid systems and vehicles. Even with incremental electric power and speed improvements, the PHEV's lack of full-performance, all-electric capability requires engine operation under everyday speed and/or load conditions, regardless of available battery energy. This creates emissions concerns and can severely limit the actual all-electric driving range in the real world.

The E-REV is principally an Electric Vehicle (EV) with full vehicle performance available as an EV. Significantly, it overcomes the historical EV re-charge time limitations by adding a fuel-powered electric generator to extend driving range. Actual all-electric driving can regularly be experienced throughout the working energy range of the vehicle's battery without fear of being stranded. The E-REV offers the opportunity for petroleum independence, and a dramatic reduction in emissions for many drivers.

An E-REV traction drive and battery system needs to be specifically designed for the task. The systems are significantly more capable and larger than those designed for PHEVs. An E-REV is typically also architected to accommodate packaging of these systems while retaining performance and utility. The compelling benefits of the E-REV drive GM to address these challenges.

The study results indicate that both the PHEVs and the E-REVs can play a role in addressing future needs. The study shows that in the real world the PHEV is quite likely to run with blended operation, but the E-REV is very likely to remain in EV mode for most drivers.

GM is currently developing both PHEV and E-REV vehicles. The Saturn VUE Green Line PHEV is being developed as a derivative of the conventional 2-Mode Hybrid. The Chevrolet Volt E-REV is also under development with full performance, all-electric capability, but without practical range limitations.

INTRODUCTION

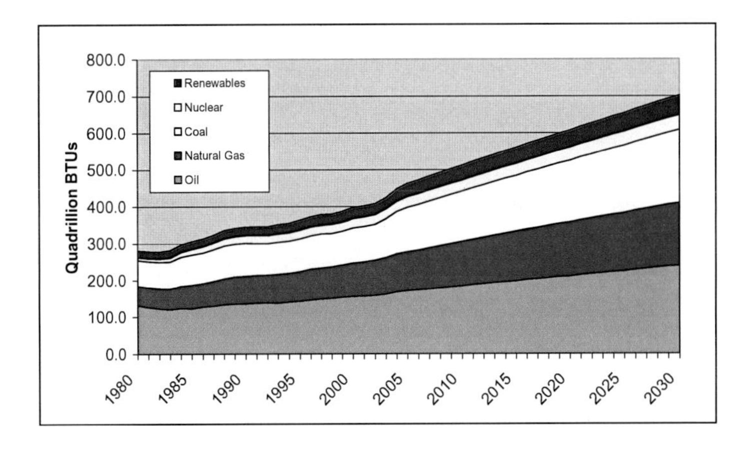

Figure 1 – Projected worldwide energy sources

ENERGY OUTLOOK - Worldwide energy production is projected to grow at an annual rate of over 2% providing for an expanding population and industrial development, despite increasing efficiencies in consumption. Figure 1 shows that fuels, primarily petroleum oil is projected to grow at a similar rate, even in scenarios where the fuel remains at relatively high historical costs. 2030 World oil consumption is accordingly projected at 210 quadrillion Btu (118 million barrels annually), an increase of over 30% compared to 2004. The portion of oil used for transportation is growing and is projected to use 68% of liquid fuel energy over the period 2004 - 2030. [1]

Significant concerns have been raised about the security of oil supply and initiatives have been outlined to diversify energy in transportation including initiatives proposed by the US Administration and the Department of Energy [2]. These initiatives include the development of a vehicle that plugs-in and derives a great deal of its utility using energy from the electric power grid. Recent enthusiasm in PHEVs and E-REVs, in part, stems from these concerns.

General Motors' Advanced Propulsion Technology Strategy is to remove automobiles from the environmental dialogue. The strategy calls for reduced consumption, reduced emissions, and diversification of energy sources. Continued improvements in base vehicle and powertrain efficiencies figure prominently in GM's plans, as does an aggressive rollout of ethanol-blended fuels.

Another key element of the strategy is to allow automobiles to shift significant portions of their required energy from petroleum to other sources. Figure 2 shows a network of the various energy sources, energy pathways, and possible on-vehicle energy storage media. Higher power motors, higher energy on-board electrical storage, and systems that allow for driving without a combustion engine enable vehicles that can use non-petroleum energy sources for transportation. We call the increase in electrical content and magnitude onto the vehicle "electrification".

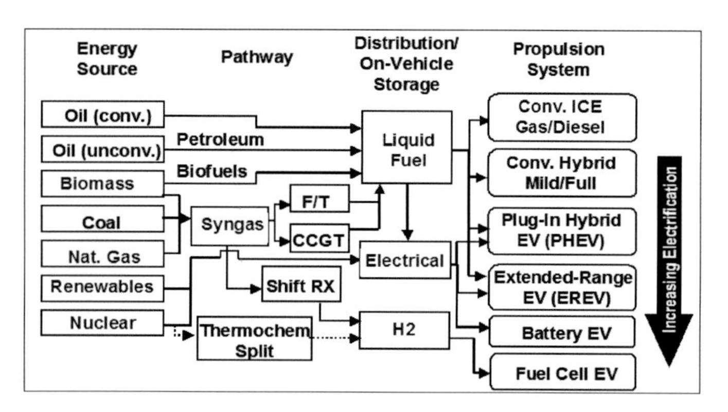

Figure 2 – Energy sources, paths, on-vehicle storage and vehicle propulsion systems

Diverse energy sources figure into the future of the worlds total energy bill. Yet today, automobiles rely almost exclusively on liquid fuels as the on-vehicle storage medium. Note that most other sources can be, and are already, used as part of the electric grid as shown Figure 3.

Electric grid power is a natural candidate for transportation energy distribution with on-vehicle storage. Worldwide grid electricity is expected to approximately double in the next two decades, outstripping the growth in total energy consumption. Improved generating efficiency from new plants means that electrical generation will continue to use approximately 40% of the world energy sources [1].

If electric energy can be effectively stored and integrated to propel automobiles, the full range of energy sources could be tapped for future automotive needs. Furthermore, future improvements in the efficiency and environmental impact of electric power generation will be directly realized by the PHEVs and E-REVs on the road at that time.

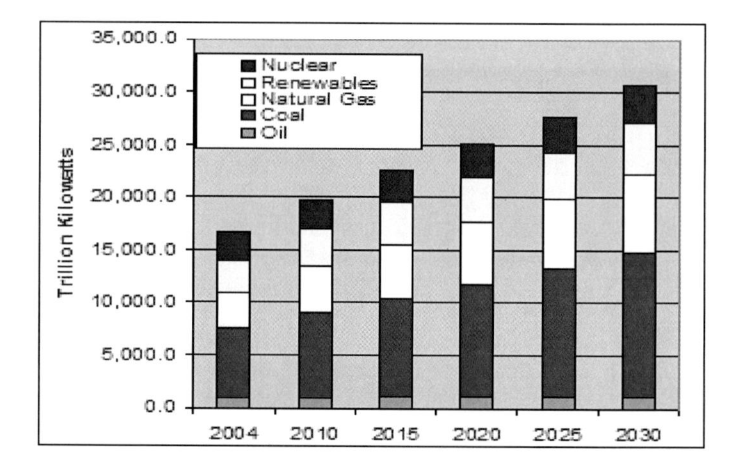

Figure 3 – Projected worldwide electric power capacity

ELECTRIFICATION - Several technical and commercial challenges are associated with charging, storing and making use of electric energy to propel automobiles. The term "electrification" means development and integration of systems and components that enable electric energy to be used for transportation. Challenges of electrification include providing automotive levels of reliability and durability, package density, acceptable noise, vibration and harshness, and automotive levels of cost in a set of new components and control algorithms.

Electrification has its roots in modern times with introduction of the first modern electric vehicle, the 1996 GM EV1. The EV1 was a full performance Battery Electric Vehicle (BEV) with a complete set of vehicle electrification technologies [3]. They included the first modern:

- Vehicle charging interface

- Off-board charging system
- Battery pack and charge control system
- Electric isolation system
- Electric drive and traction system
- Electric accessory power system
- Electric steering system
- Electric air conditioning system
- Electrically actuated brakes
- Regenerative braking systems

The EV1 also included an efficient vehicle package and structure designed at the outset to accommodate a battery pack with sufficient energy and power for full performance as an electric vehicle.

Unfortunately, the market experience with the EV1 indicated that additional improvements in BEVs were needed. Some EV1 drivers gave the term "range anxiety" to their continual concern and fear of becoming stranded with a discharged battery in a limited range vehicle. Improvements in on-board energy storage and charging time were necessary for more widespread deployment of BEVs. Most EV-enabling electric components and systems have found utility in the meanwhile when used in mild and full Hybrids Electric Vehicles (HEVs).

TRANSITIONAL ELECTRIC VEHICLES - Electrification has continued in the industry with the development of conventional HEVs. These vehicles do not provide full performance on electric power alone, and therefore the power and energy level required for the systems is reduced when compared with full performance BEVs. In addition, while conventional hybrids (both mild and full) improve vehicle efficiency and directly reduce petroleum consumption and CO_2 emissions, all the energy they consume is from an on-board liquid medium. In HEVs, electric storage and power on board are only used to better utilize the liquid fuel. Therefore, HEVs offer no additional energy pathways for diverse sources, including CO_2 neutral renewables.

However, hybrids are being produced at volumes now sufficient to develop and improve the reliability and cost effectiveness of electrification subsystems. Considering a long-term goal of greater energy diversification, hybrids can be considered the first of transitional electric vehicles. Figure 4 is a table illustrating the features of transitional electric vehicle types, including HEVs, Plug-In Hybrid Electric Vehicles (PHEVs) and Extended-Range Electric Vehicles (E-REVs). Each is progressively more electrified and we expect plays a progressively larger role toward shifting a portion of the transportation energy burden toward other sources and away from petroleum.

Vehicle Type	Electric Power	Onboard Electric Storage	Grid Conn.	Electric Driving
Mild HEV	low	low	no	no
Full HEV	med	low	no	very limited
PHEV	med	med	yes	limited
E-REV	high	high	yes	Full

Figure 4 - Features of Transitional Electric Vehicles

MAIN SECTION

DEFINITIONS – Before beginning a detailed discussion on the benefits of varying degrees of electrification it is necessary to define the key systems.

Hybrid - A hybrid is defined by SAE [4] as: "A vehicle with two or more energy storage systems both of which must provide propulsion power – either together or independently." In practice, hybrid vehicles typically require both sources to provide full vehicle capability. The engine is also typically the larger of the two propulsion sources, being sized to provide most of the power during high power vehicle events. The motor is typically the smaller of the two propulsion sources, being sized to maximize the amount of energy that can be captured during braking and for limited low speed EV operation.

Plug-in Hybrid (PHEV) - A PHEV has been defined by SAE [4] as: "A hybrid vehicle with the ability to store and use off-board electrical energy in the RESS (rechargeable energy storage system)." These systems are in effect an incremental improvement over the Hybrid with the addition of a large battery with greater energy storage capability, a charger, and modified controls for battery energy management and utilization.

There are two types of PHEVs operating strategies. These operating strategies require the definition of a schedule for discussion. The EPA urban, referred to as the urban, schedule is a common reference for PHEVs and will be used for this discussion.

The first type has an operating strategy which is very similar to the conventional hybrid. The engine use is required for most accelerations and speeds. The characteristics of this mode of operation are shown in Figure 5. Here you can see that the engine starts almost from the start and engine power is used throughout driving to supplement battery power usage. For PHEVs that use conventional hybrids as the starting point, this mode of operation is typical due to the operational speed and electrical power capabilities of the underlying hybrid systems. We call such a hybrid a conversion PHEV.

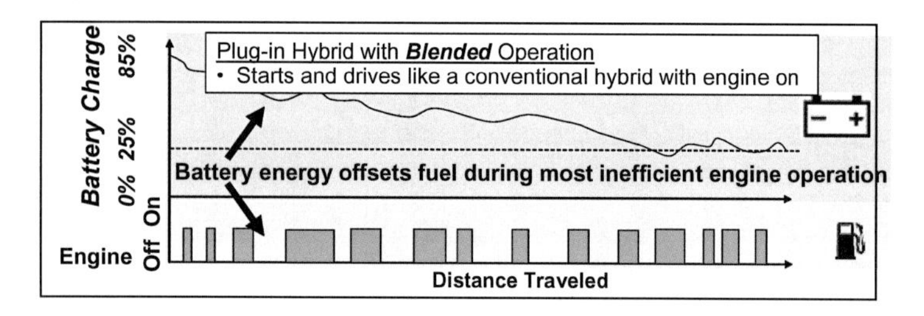

Figure 5 – A *Conversion* PHEV using a *blended* operating strategy

The second type of PHEV operation strategy is referred to as *Initial EV*. This type of system requires that the battery, motors, thermal systems, power electronics and system configuration be set to allow electric-only operation over the complete power and speed range of a cycle, in this case the urban schedule. The characteristics of this mode of operation are shown in Figure 6. In this figure you see that while the battery is depleting the stored energy the vehicle operates electric-only, without any engine operation. Once the battery is depleted, the vehicle operates like a conventional hybrid.

In real world operation, anytime during operation that the driver requests more power or speed than the motor and battery can provide, the engine is required to start and the controls will default to a *Blended* mode of operation to maintain emissions capability. In order to delay engine start in an Initial EV PHEV, a full hybrid system may have to be additionally modified to raise the traction power capability, and to extend the speed at which the vehicle operates electrically. We will examine the benefits of such modifications sufficient to drive the urban driving schedule, and call it an "urban-capable PHEV".

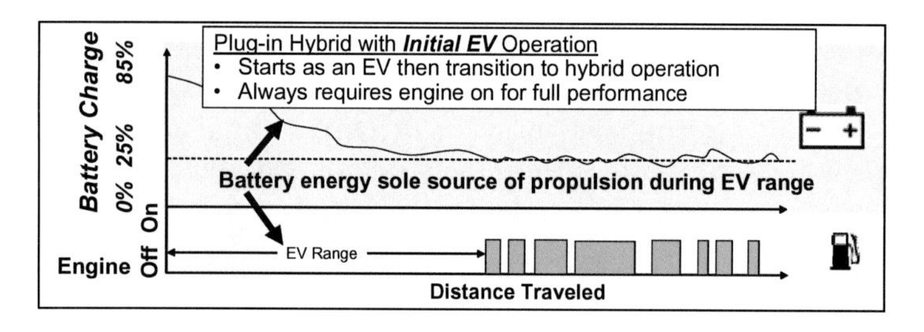

Figure 6 – A *Urban-Capable* PHEV using an *Initial EV* operating strategy

Extended-Range Electric Vehicle (E-REV) - SAE has not established a definition for an E-REV. The authors propose the following definition: "A vehicle that functions as a full-performance battery electric vehicle when energy is available from an onboard RESS and having an auxiliary energy supply that is only engaged when the RESS energy is not available." This is very similar to a vehicle envisioned in the 2007 ARB Expert panel report [5], when they described,

> " ....This type of vehicle can operate as a full performance ZEV, during the time the ICE is not operating, and can avoid the cold start emission problem discussed above and it requires a relatively large energy battery and a large full performance, electric drive propulsions system, similar to FPBEV or FCEV...."

The E-REV is unique from a PHEV in that the vehicle, battery and propulsion system are sized such that the engine never is required for operation of the vehicle when energy is available from the battery. The definition of this type of vehicle does not require the specification of a operating cycle (urban schedule in the PHEV discussion). As a full-performance battery electric vehicle, the battery, motor, and power electronics must be sized for the full capability of the vehicle. The vehicle must also be architected to allow packaging of the large E-REV battery which has a greater size due to the full EV requirement. The characteristics an E-REV are shown Figure 7. The operation of an E-REV looks similar to that of an Initial EV PHEV; however an E-REV must maintain this mode of operation on all operating schedules when energy is available from the battery. An E-REV does not need to start the engine for speed or power demands from the driver and therefore does not need to transition to a *Blended* operation strategy when battery energy is available, unlike the *Initial EV* PHEV.

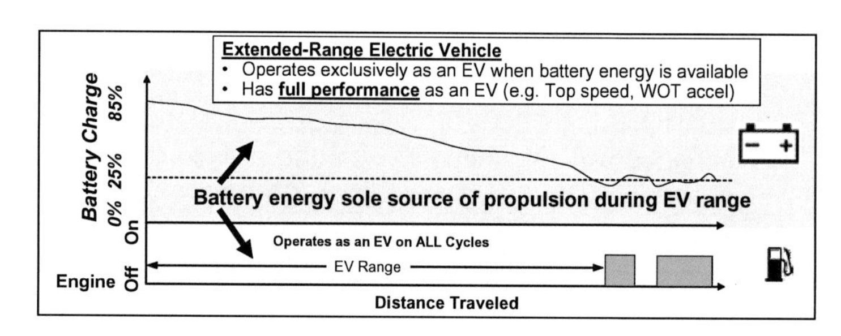

Figure 7–Characteristics of E-REV operation

A better understanding of the different systems is gained by examining Figures 8. This figure shows that as greater electric-only operation is required, there is a need to increase motor size and overall electric propulsion capability. Hybrid and PHEVs are able to blend electric and engine power to propel the vehicle and therefore require less total onboard power than the E-REV. The E-REV requires full electric propulsion capability and as such, the additional power capability of the total onboard power exceeds what is required to propel the vehicle. The E-REV does allow significant engine downsizing in that the engine power is not required to meet peak vehicle power demands and can be sized to meet only continuous power demand.

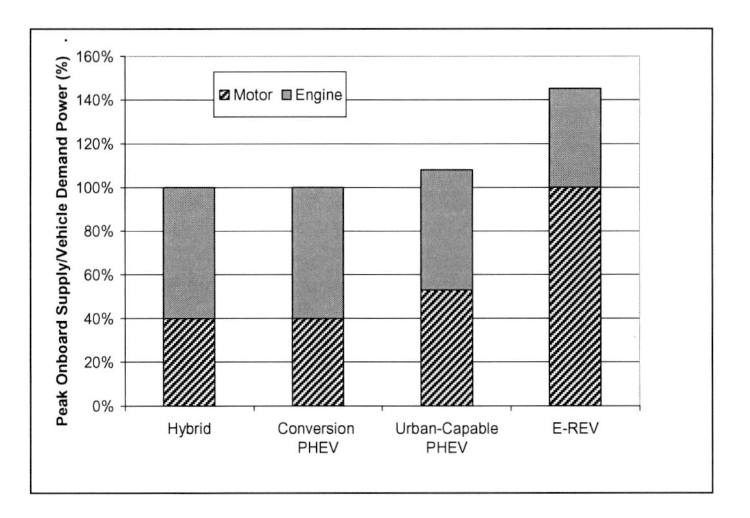

Figure 8 – Peak motor and engine power as a percentage of the peak vehicle power demand

ELECTRIC RANGE BENEFIT

There are three benefits to vehicle electrification: reduced petroleum consumption, reduced emissions, and energy diversification. Energy security comes from the ability to use multiple energy sources via electric pathways and on-vehicle storage, and a net reduction in fuel usage. Since energy security is so closely related to fuel usage reduction, we only consider fuel usage reduction in our study. To analytically evaluate these benefits three sources of data were used: the 2001 National Personal Transportation Survey, time series data from the Southern California Association of Governments (SCAG) Regional Travel Survey, and the Environmental Protection Agency's (EPA) testing cycles.

Vehicle Usage & Powertrain Constraints Discussion

For E-REV's and PHEV's, there are three primary constraints that affect vehicle performance. Those constraints are the total energy stored in the batteries, battery power limits, and electric motor or mechanical speed limits. Any one of these limits can significantly reduce the system's EV operation and subsequent opportunity to displace petroleum. At one extreme, the lowest level of performance for a PHEV is obtained by converting an existing HEV into a PHEV only by increasing on-board electrical energy storage and adding a charger, resulting in the conversion PHEV. This type of PHEV has blended-type operation and typically has speed, power, and energy constraints which cause the engine to start very soon after the vehicle moves. At the other extreme is the E-REV. An E-REV does not start the engine until all useable on-board electrical energy has been used. Between these two extremes are solutions which are able to operate like EVs to different degrees.

To evaluate the benefits of a non-conventional powertrain, the typical approach has been to consider the EPA's testing cycles. These define a speed the vehicle must follow versus time during a test. Three of the most commonly considered cycles are the urban, EPA highway, referred to as the highway cycle, and US06 test cycles. Using these cycles as a reference in designing powertrains naturally leads to systems which are designed to maximize performance against those cycles. One proposed level of PHEV performance is the ability to follow the urban and highway cycles while operating as an EV until the useable on-board electrical energy is exhausted. A transition from EV operation to Blended PHEV or HEV behavior occurs when either wheel power or wheel speed exceeds the level required to follow the urban cycle. This is the type of system previously defined as an 'urban-capable PHEV.'

In addition to EPA cycles, other potential sources of design requirements exist. At a very high level, the National Personal Transportation Survey has information on travel times and distances. This information can provide insights into the potential fuel economy and emissions benefits of different PHEV and E-REV vehicle types. Additionally, some regions have collected detailed driving information on vehicle usage. The SCAG Regional Travel Survey (RTS) [6] is one example with data from the southern California area. With permission from SCAG, The National Renewable Energy Lab (NREL) released a copy of the RTS data set. The data was made anonymous by NREL to remove all personally identifiable information. It was reduced to 621 traces of speed versus time, and ignition key state versus time. Each trace corresponded to one vehicle instrumented in the RTS. In the raw form provided, the data set had problems with missing points. The data was processed to interpolate and fill in for missing data. The statistical characteristics of this data for distance and trip times are different than the NPTS data because it is a sub-sample of the population of vehicles from a specific geographic region.

One use of the RTS dataset is to evaluate vehicle power usage for 'real-world' drivers compared to existing federal driving schedules. To normalize the data to account for distance driven, the 'power intensity' for driving is calculated. Power intensity is defined as energy used by a vehicle divided by distance driven. Figure 9 shows the power intensity of each sample vehicle and the daily distance driven. Comparing the RTS samples to the federal schedules, we note that greater than 94% of vehicles operate at a power intensity higher than occurs in the urban and highway schedules.

These results imply that having an electric motor and battery power capable of driving the urban schedule may not satisfy the vast majority of drivers' needs. To understand this fully, the *peak* power demands and vehicle's constraints must be evaluated. As will be shown later, even with sufficient energy on–board the engine will need to operate frequently in order to drive an urban-capable vehicle in a blended fashion. On the other hand, if a vehicle has the ability to run the US06 schedule on electric power alone, that vehicle can satisfy the majority of vehicle usage in the RTS without using the engine.

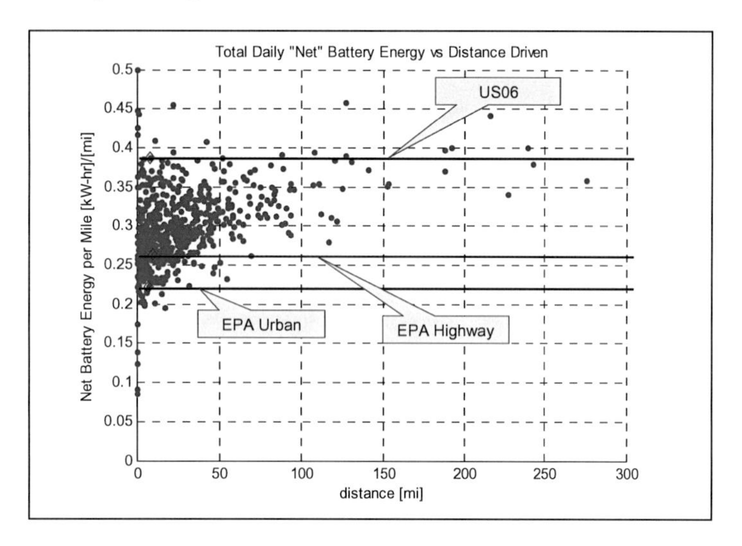

Figure 9a - RTS vehicle power intensity versus daily distance driven

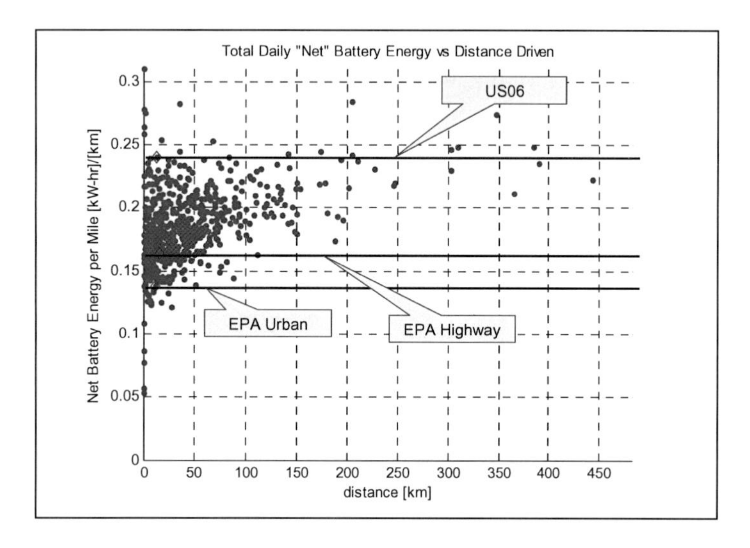

Figure 9b (metric) - RTS vehicle power intensity versus daily distance driven

The RTS data provides some interesting insights into how vehicles are used in southern California. To see how this usage compares to the existing EPA cycles, the cumulative distribution of energy to accelerate and decelerate a typical mid-sized vehicle versus the instantaneous wheel power was evaluated. This is calculated by sampling the data set at a one hertz rate. The speed and instantaneous acceleration along with a simple vehicle model are used to calculate the instantaneous power at the wheel. The energy is approximated by assuming the power level remains constant for the sample. This form of summarizing the data identifies the amount of energy that cannot be generated or absorbed by a power limited electrical traction system. The bounds on the RTS data are plotted in figure 10. The bound for 10%, 50%, and 90% of the sample are identified. The energy distribution for the EPA highway, urban and US06 cycles are also plotted in this figure.

For example, in figures 10 and 11, consider a typical mid-sized vehicle with a powertrain that is limited to 40kW at the wheels for motoring and regeneration. For the urban and highway cycles, only a trivial amount of energy can not be delivered or absorbed. However, for the US06 cycle, almost twenty percent of the energy is delivered at power levels that exceed the electric capability of the vehicle. This power must be delivered by the engine. Additionally, almost twenty percent of braking energy occurs at power levels beyond the electrical limits of the system, limiting the energy recovery.

The distribution of wheel energy from the RTS data shows that the bulk of the vehicle usage is somewhere between the urban and US06 in energy distribution versus wheel power. The urban cycles represent the lower limits on vehicle energy distribution and the US06 represents a reasonable upper limit. Consider the previous example with a 40 kW power limit at the wheel. Since the y-axis of the chart is the cumulative distribution of energy, the amount of energy available which higher power level is found in the tail of the distribution. For example, in figure 10, approximately 78% of the wheel energy occurs at power levels less than 40 kW. Therefore, approximately 22% of the wheel energy occurs at power levels in excess of 40 kW. Of the vehicles in the RTS, a vehicle in the upper 10% percentile of the population of samples from the RTS will have more than 15% of its acceleration energy occurring at power level in excess of 40 kW. Conversely, the same 40kW wheel power limits on regeneration (regen) results in a loss of available regen energy for an upper 10% vehicle from the population of samples.

One interesting conclusion to be drawn from this data is that braking energy is distributed to higher power levels than evaluated in existing testing regimes. This type of customer behavior is not reflected in the testing procedures. With the power limits on current production hybrids, this may explain some of the differences between the testing fuel economy and customer's experiences [7].

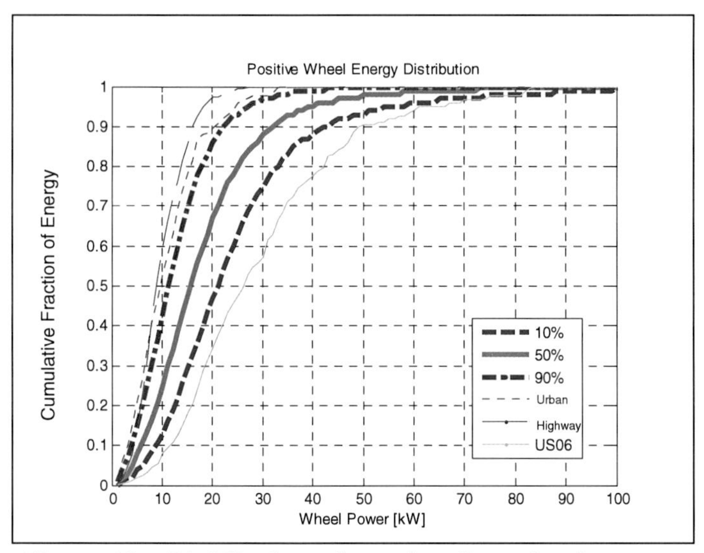

Figure 10 – Distribution of accelerating wheel energy versus wheel power.

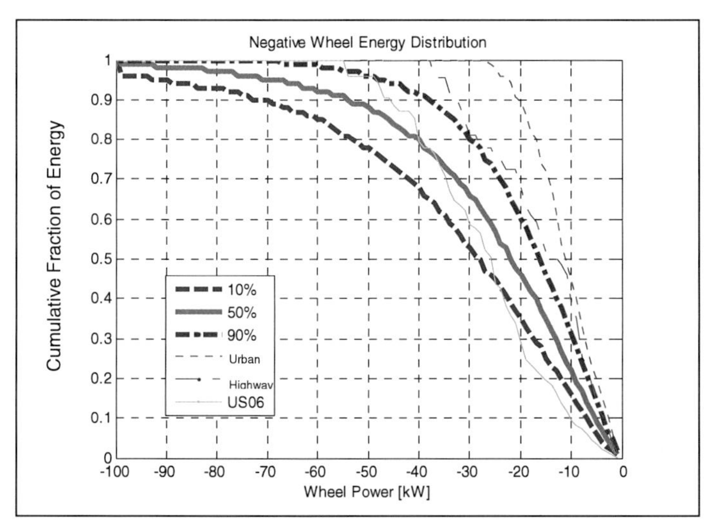

Figure 11 – Distribution of decelerating wheel energy versus wheel power.

Impact of Powertrain Constraints

In addition to considering the distribution of energy in the RTS data, another consideration is when the powertrain constraints will force the first engine start. This is important because it causes the vehicle to transition from EV operation to either blended or charge sustaining operation. While the transition from EV operation does not greatly affect the societal use of fuel, it does affect the regulated pollution emissions and provides a way to understand the energy security a vehicle offers an individual. Because of the scope and detail in the RTS, the impact of speed, power, and energy constraints can be considered. To visualize the impact of these constraints on a population of vehicles, the daily driving distances are normalized to a value between zero and one. Zero represents the start of driving for the day. One represents the end of driving for the day. The percentage of vehicles that can operate as an EV versus the percentage of completed daily driving is plotted in figure 12.

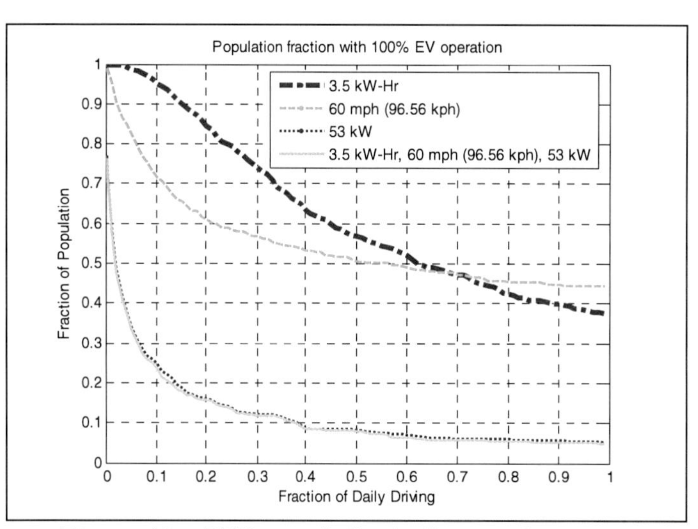

Figure 12 – RTS population's first engine start versus percent of daily distance for various powertrain constraints.

Consider some common constraint combinations for a mid-sized vehicle. Assuming 75% efficiency from battery to wheel for motoring and 60% efficiency from wheel to battery for braking regeneration, the first constraint considered is a battery energy constraint of 3.5 kW-hrs. This energy is sufficient to allow all vehicles to complete the first 5% of their daily driving while operating as an EV. After that point, an increasing percentage of the vehicles transition from EV operation to blended or charge sustaining operation. At the conclusion of driving for the day, less than 40% of the vehicles complete their entire day operating as an EV.

If the powertrain has only a 60 mph (96.56 kph) constraint, such that the engine is forced on when the vehicle speed exceeds 60 mph (96.56 kph), two significant things occur. Firstly, vehicles transition from EV operation earlier in the driving. Secondly, the number of vehicles which can complete daily driving as an EV changes to 43%.

In the absence of other constraints, if the battery and motor is limited to delivering only what is necessary to drive the urban driving schedule, in this case 53 kW, and the engine is used to make up for any differences in required power, the majority of vehicles start their engines before any appreciable driving is complete. Only 6% of the vehicles can complete their day as an EV. This single constraint dominates the transition from EV operation.

If all of the energy, speed, and power constraints are considered together, the resulting vehicle has slightly less EV operation than a vehicle which is constrained by power alone. For these constraints, similar to the limits of an urban-capable PHEV, the power constraint is dominant and limits the vehicle's ability to operate as an EV.

The combination of speed and power constraints has a considerable effect on limiting EV range driving over the

RTS data. It should be noted, that these constraints *would* permit a vehicle to complete the urban cycle as an EV. But the RTS data indicates that real world driving frequently exceeds the power and speed limits of the urban driving schedule. Additionally, the method used to calculate the RTS power levels in this study may underestimate real power variations since a level road grade is assumed. It is expected that were road grade data available, it would indicate higher power levels due to accelerations on grades. Therefore, this evaluation of vehicle operation against RTS data indicates that *only a small fraction of urban-capable PHEVs will actually operate as an EV.*

To compare the relative benefits of electrification, the following powertrains in a typical mid-sized vehicle will be considered:

- **Reference:** a conventional powertrain;
- **HEV:** a powertrain with a 40 kW electrical power constraint;
- **Conversion PHEV**: a PHEV powertrain with a 35 mph (56.32 kph) speed constraint, a 40kW electrical power constraint, and 3.5 kW-hrs of usable electrical energy;
- **Urban-Capable PHEV:** a PHEV powertrain with a 60 mph (96.56 kph) speed constraint, a 53kW electrical power constraint, and 3.5 kW-hrs of useable electrical energy;
- **E-REV**: a powertrain with 8 kW-hrs of useable electrical energy and EV capability not limited by electric power or driving speed.

These vehicles has coast down parameters with F0 equal to 10 N, F1 equal to 0.009 N-s/m, and F2 equal to 0.4392 N-s^2/m^2. The inertia is 1565 kg for the conventional vehicle and adjusted for additional hybrid powertrain and batteries mass.

Fuel Savings

To evaluate the benefits from the different powertrains, the RTS data was used in vehicle simulations to assess fuel savings. The study was performed on a subset of 175 vehicle-days of driving, each with a total distance less than 75 miles (120.7 km). This study required approximately 1500 hours of computer time to execute. The conventional, HEV, conversion PHEV, urban-capable PHEV, and E-REV vehicles were evaluated for each day's data.

Figure 13 shows the average fuel consumption for this population of vehicles. The HEV was able to improve the overall efficiency of operation and reduce the fuel consumption in this population of vehicles by approximately 25%. The PHEVs were able to displace approximately another 30% of HEV's fuel consumption. Significantly, for this test, the conversion PHEV and the urban-capable PHEV provided approximately equal benefits in fuel savings. Finally, the E-REV, with greater on-board energy storage and the ability to meet all power demands without starting the engine, was able to reduce the PHEV's fuel consumption by more than 50%.

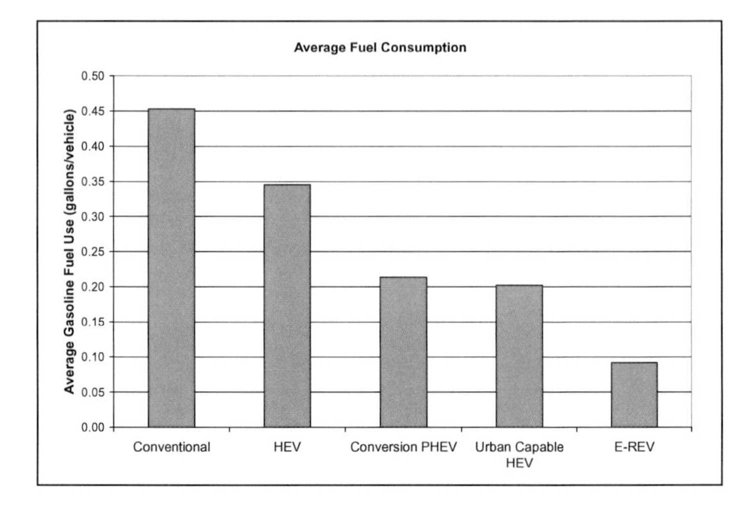

Figure 13a – Average fuel usage over the complete RTS dataset

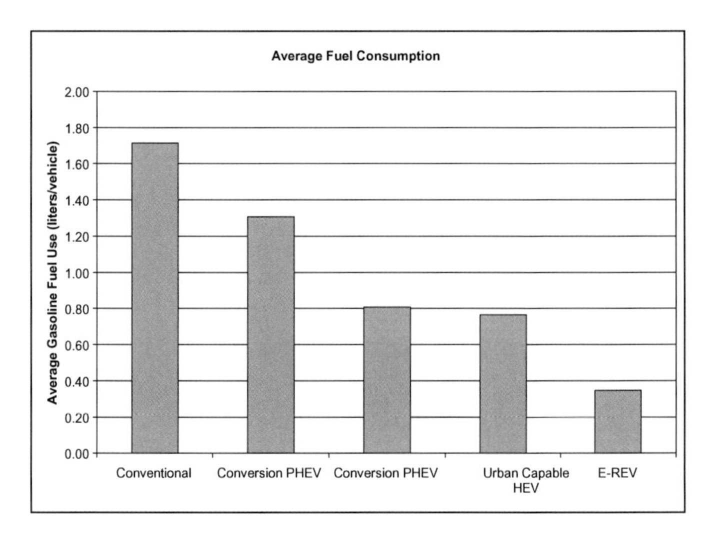

Figure 13b (metric) – Average fuel usage over the complete RTS dataset

Air Pollution Prevention

In addition to fuel usage, the emission of regulated pollutants is a significant concern. Modeling emissions is a complex issue. We propose the following method to simply evaluate the relative benefits of different powertrains.

Several authors have noted that the majority of tailpipe emissions occur within the first minute of engine operation [8,9,10]. After the powertrain thermally stabilizes, the additional emissions are often negligible. To assess the relative merits of different powertrain options, we propose counting whether there is an engine start on a trip. If one or more engine start occurs, then that trip contributes emissions. If no engine start occurs, no emissions are generated. These starts are referred to as initial trip starts (ITS). The relative benefits of different powertrains are determined by counting the number trips in a reference set and subtracting the number of initial trip starts. The difference between these two values provides a relative measure of pollution prevention.

Because of the previously mentioned issues with powertrain constraints, simply using the EV range to measure the effect of a plug-in vehicle on reducing emissions may be misleading. The EV range considers the energy constraint, but does not consider the impact of power and speed constraints. If those are considered, a data set like the RTS is necessary. Using the RTS, the reduction in initial trip starts (ITS) can be calculated over a population of vehicles.

A simple method to estimate the pollution prevention of a plug-in vehicle is to consider the number of initial trip starts from different vehicle classes. Figure 14 shows the number of initial trip starts for the reference, full hybrid and Plug–in vehicles when examined over the entire RTS dataset. The chart shows that PHEVs reduce the initial starts significantly when compared to a conventional full hybrid. An even more dramatic reduction in initial trip starts occurs when an E-REV is introduced.

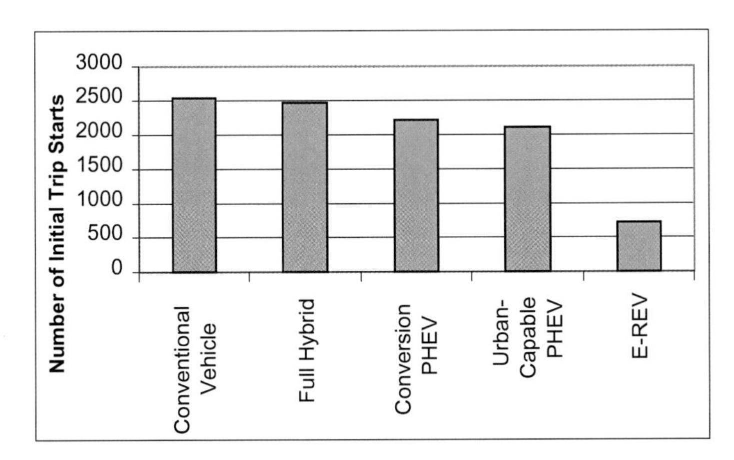

Figure 14 – Initial trip starts over the complete RTS dataset.

The higher energy content, higher power, and higher speed capability reduce and E-REV's initial trip starts by approximately two thirds compared to an urban-capable PHEV.

Another way to consider the effect on emissions of a particular plug-in vehicle is to examine the amount of engine-off driving. To illustrate the effect of differences in powertrain constrains (electric energy, power, and speed capability), a chart that shows the portion of the RTS that can be completed while operating as a pure EV without engine starts is used.

To generate this chart, each day's driving from the RTS data set was simulated to determine when the engine first starts based on the speed, power, and energy constraints. This distance is used to calculate the percent of daily driving while operating as an EV. The fraction of vehicles which operate as an EV along their daily driving is plotted in figure 15.

A further insight on the effect of the initial trip start on smog formation is the time of day of the initial start. While we note that E-REVs, to much a greater extent than PHEVs eliminate initial starts, it is also true that E-REV initial starts are much more likely to be much later in the driving cycle and contribute to far lower early morning emissions, particularly hydrocarbons (HCs). Further study is warranted into of the effects on peak ozone an the reduced and delayed HC emissions offered by PHEVs, and to a much greater extent E-REVs,

For example, when all PHEVs and E-REVs start their daily driving, 100% of the population is operating as an EV. When the sample of vehicles has driven 10% of the total distance for the day, only 10% of the conversion PHEV's still operate as an EV. In contrast, approximately 24% of the urban-capable PHEV's are still operating as an EV and 98% of the E-REV's are still operating as EVs. By the end of the day, about 3% of the conversion PHEV's completed operating exclusively as EV's, only 5% of the urban-capable PHEVs completed as EV's, and 64% of the E-REV's completed as EV's.

We believe that this is the basis of a significant finding. Even PHEVs modified to provide Urban Capability, 53 kW power and 60mph (96.56 kph) operation, will rarely operate as an EV for the full day's driving. On the other hand, the majority of E-REV drivers will experience a full day of EV driving.

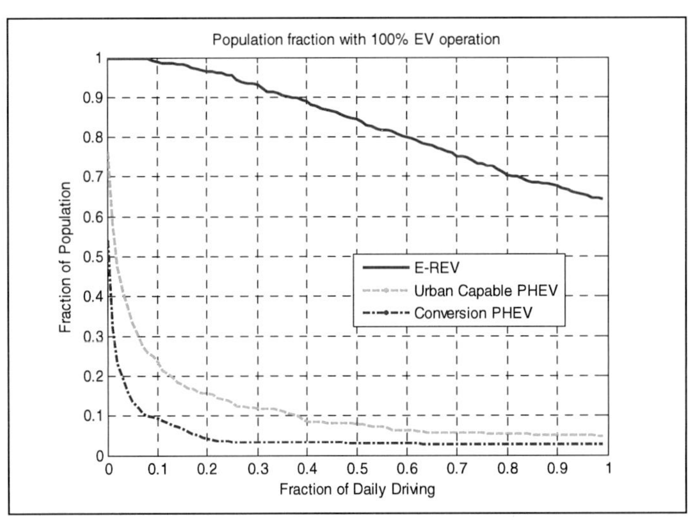

Figure 15 – Percent of population with EV operation through a day's driving.

CONCLUSIONS

As the electrification of the automobile progresses from HEVs, to PHEVs, to E-REVs, the environmental impact of the automobile significantly decreases and energy security increases. Just as the various PHEV's offer significant improvements over HEV's, the E-REV significantly improves over PHEV's.

The PHEV and E-REV are positive steps toward the goals of energy diversification, fuel savings, and emissions reduction. Furthermore and significantly, our findings indicate that:

1. The RTS data set of Southern California drivers contains widespread and significant driving at power levels and speeds beyond that represented by the urban driving schedule

2. An E-REV is more than ten times as likely to finish the day as an EV than an urban-capable PHEV derived from an HEV, when operated in the actual application, as represented by the RTS data set.

3. Similarly, an E-REV will consume, on average, less than half of the petroleum of a PHEV in the real world, if overnight charging is assumed.

4. An E-REV will reduce regulated emissions that are due to initial trip starts by more than 70% when compared to a PHEV in the actual application.

5. "Electric range" when operating on the urban schedule is not a direct measure of a plug-in vehicles' ability to run with the engine off, ability to displace petroleum or ability to reduce regulated emissions in the actual application. Rather, the ability to run with full performance on electric power alone leads to improvements which would be realized in actual application.

6. In the event of a petroleum disruption, an E-REV could support uncompromised vehicle operation for the majority of drivers.

We conclude that electrification that enables E-REVs may be well worth the effort. Specifically designed electric powertrains, incorporating higher power motors and thermal systems, higher energy batteries and integrating them into vehicle structures specifically designed for that purpose will be rewarded with societal benefits realized in real world use. While PHEVs can make improvements when compared to HEVs, An E-REV appears to realize a much greater portion of societal benefits.

Mild and Full HEVs continue to play a growing role in reducing overall consumption and setting commercial volumes to a level sufficient to develop electrification subsystems. But HEVs do not directly displace petroleum with other energy sources or lead to significant emissions reductions through all electric operation. Conversion PHEVs or even PHEVs with the ability to run the urban schedule are incremental toward these ends, but are of limited because is appears the real world application is in general, much more demanding.

By offering full-performance on electric power alone, the E-REV operates as an EV for the majority of real drivers. By retaining an ICE-powered charging capability, the E-REV overcomes the "range anxiety" limitations of earlier BEVs. We anticipate that the E-REV will be an important and practical step forward in the electrification of the automobile.

ACKNOWLEDGMENTS

We would like to thank the Keith L. Killough of the Southern California Association of Governments, and Tony Markel at NREL for their assistance in obtaining the data from the regional travel survey for these analyses. Thanks also to Ken Macklem of Softworks Corporation for his support in execution of the large scale studies in this project. Finally, we would like to thank Eric Rask of GM for his insights and efforts in vehicle simulation work.

REFERENCES

1. International Energy Outlook, 2007; Report #DOE/EIA-0383(2007); Energy Information Administration, US Department of Energy, May 2007
2. Advanced Energy Initiative, 2006,; White House National Economic Council, January 2006
3. D. Mc Cosh; We Drive the World's Best Electric Car; Popular Science, January 1994
4. SAE J1715 Information Report, Hybrid Vehicle (HEV) and Electric Vehicle Terminology, July 2007
5. F.R. Kalhammer, B.M. Koph, D.H. Swan, V.P. Roan, M.P. Walsh, Status and Prospects for Zero Emissions Vehicle Technology, Report of the ARB Independent Expert Panel 2007, April 13th,2007
6. Southern California Association of Governments; Year 2000 Post-Census Regional Travel Survey, Fall 2003
7. Duoba, M.J., et al., Issues in Emissions Testing of Hybrid Electric Vehicles, Global Powertrain Congress, Detroit, MI, May 6-8, 2000
8. K. Sano, T. Kawai, S. Yoshizaki, Y. Iwamoto; HC Adsorber System for SULEVs of Large Volume Displacement (SAE 2007-01-0929)
9. A.C. McNicol, et al.; Cold Start Emissions Optimisation Using an Expert Knowledge Based Calibration Methodology (SAE 2004-01-0139)

10. K. Nishizawa, et al.; Development of New Technologies Targeting Zero Emissions for Gasoline Engines (SAE 2000-01-0890)

CONTACT

E. D. Tate can be contacted at ed.d.tate@gm.com.

ADDITIONAL SOURCES

1. Federal Highway Administration; National Household Travel Survey, http://www.fhwa.dot.gov/policy/ohpi/nhts/index.htm
2. Southern California Association of Governments; Travel and Congestion Survey, http://www.scag.ca.gov/travelsurvey/

DEFINITIONS, ACRONYMS, ABBREVIATIONS

BEV: Battery Electric Vehicle

E-REV: Extend Range Electric Vehicle

ICE: Internal Combustion Engine

ITS: Initial Trip Starts

PHEV: Plug-in Hybrid Electric Vehicle

SCAG: Southern California Association of Governments

EV: Electric Vehicle

NPTS: National Personal Transportation Survey

RTS: SCAG Regional Travel Survey

The Engineering Meetings Board has approved this paper for publication. It has successfully completed SAE's peer review process under the supervision of the session organizer. This process requires a minimum of three (3) reviews by industry experts.

ISSN 0148-7191

Positions and opinions advanced in this paper are those of the author(s) and not necessarily those of SAE. The author is solely responsible for the content of the paper.

SAE Customer Service:
Tel: 877-606-7323 (inside USA and Canada)
Tel: 724-776-4970 (outside USA)
Fax: 724-776-0790
Email: CustomerService@sae.org
SAE Web Address: http://www.sae.org
Printed in USA

The CO_2 Benefits of Electrification E-REVs, PHEVs and Charging Scenarios

2009-01-1311
Published
04/20/2009

E. D. Tate and Peter J. Savagian
General Motors Corporation

doi:10.4271/2009-01-1311

ABSTRACT

Reducing Carbon Dioxide (CO2) emissions is one of the major challenges for automobile manufacturers. This is driven by environmental, consumer, and regulatory demands in all major regions worldwide. For conventional vehicles, a host of technologies have been applied that improve the overall efficiency of the vehicle. This reduces CO2 contributions by directly reducing the amount of energy consumed to power a vehicle. The hybrid electric vehicle (HEV) continues this trend.

However, there are limits to CO2 reduction due to improvements in efficiency alone. Other major improvements are realized when the CO2 content of the energy used to motivate vehicles is reduced.

With the introduction of Extended Range Electric Vehicles (E-REVs) and Plug-in HEVs (PHEVs), electric grid energy displaces petroleum. This enables the potential for significant CO2 reductions as the CO2 per unit of electrical energy is reduced over time with the improving mix of energy sources for the electrical grid. Vehicle electrification also introduces consumer choice in the CO2 content of the energy when electric power providers offer renewable power options for drivers.

To quantify the realizable reductions in transportation CO2 emissions, we consider several scenarios. We evaluate the CO2 reduction from the introduction of PHEVs and E-REVs onto the existing power grid. We consider the impact of consumer behavior and the availability of charging and the expected benefits based on vehicle operation. We evaluate the CO2 reductions possible due to a combination of changes in the power grid and in vehicle stock over time. Finally, we evaluate the impact of selective and voluntary consumer behaviors on CO2 contributions.

INTRODUCTION

GROWTH OF GLOBAL ENERGY USAGE AND AUTOMOBILE OWNERSHIP - Several studies have identified the growth of energy usage worldwide. The US Energy Information Administration (EIA) has recently adjusted the long term forecast, but the trend is clear, Energy use is projected to rise approximately 50% worldwide over the next 22 years [1].

A major portion of the forecasted increase in the use of energy is to support the growing mobility needs of a growing world population. As shown in Figure 1, estimates suggest that growth in the worldwide vehicle population will increase by approximately 60% over the next 22 years [2]. During this period, per capita worldwide vehicle ownership is projected to increase from approximately 12% to approximately 15%. Consequently, the majority of growth in vehicle ownership will occur in the developing regions of the world. This will happen as those regions become both more prosperous and more mobile. Indeed, private vehicle ownership and use is both a means to and a dividend of economic development. However, the means to power these vehicles and support projected growth remains a major challenge. Concerns about availability, affordability, energy security and environmental impact remain.

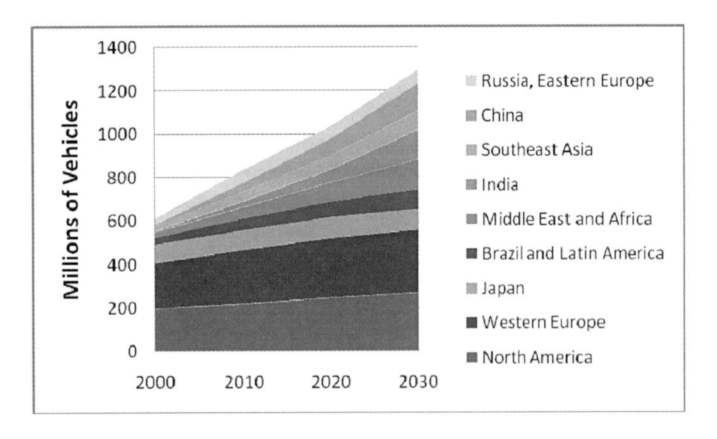

Figure 1 - Worldwide Light Duty Vehicle Ownership

ATMOSHPERIC CARBON DIOXIDE – Carbon-dioxide is a greenhouse gas (GHG) in which a great deal of attention has been paid in recent years. GHGs are emitted from a variety of natural and anthropogenic activities.

Automobile usage accounts for approximately 10% of those total anthropogenic CO_2 emissions. The remaining 90% of CO_2 comes from other forms of the world's industry (other forms of transportation (heavy duty trucks, rail air, and shipping), heating, and agriculture, etc. [3]

Automobile CO_2 emissions are regulated. Carbon-dioxide emissions are indirectly regulated in most vehicle markets today via fuel consumption or fuel economy standards such as the CAFE regulation in the US. More explicit regulations for CO_2 are in effect or proposed in other markets. The pending California ARB GHG regulation based upon Assembly Bill 1493 directly regulates CO_2, effectively regulating fuel economy.

Carbon-dioxide emissions are proportional to fuel consumption and inversely proportional to fuel economy in vehicles that use fossil fuel as its sole source of energy. Equation (1) is the chemical equation of hydrocarbon combustion.

$$C_xH_y + (x + y/4)\ O_2 \rightarrow x\ CO_2 + (y/2)\ H_2O + \text{energy} \quad (1)$$

If the fuel is gasoline, and x and y are on average 8 and 17 we get:

$$C_8H_{17} + 12.25\ O_2 \rightarrow 8\ CO_2 + 8.5\ H_2O + \text{energy} \quad (2)$$

For diesel fuel, if x and y are on average, 12 and 23, we get:

$$C_{12}H_{23} + 17.75O_2 \rightarrow 12\ CO_2 + 11.5\ H_2O + \text{energy} \quad (3)$$

The ratio of the reactant fuel to the product CO_2 is fixed by the chemical reaction. In the case of the gasoline example in equation (2), there are 8 units of CO_2 emitted for every one unit of gasoline consumed. For the diesel example in equation (3), there are 12 units of CO_2 emitted for every one unit of diesel fuel consumed.

One can readily see from equation (1) that the use of fuel and the production of CO_2 are simply different measures of the amount of combustion. Therefore CO_2 emissions requirements are generally equivalent to fuel economy regulations and technically, *one regulation is sufficient to control both measures*.

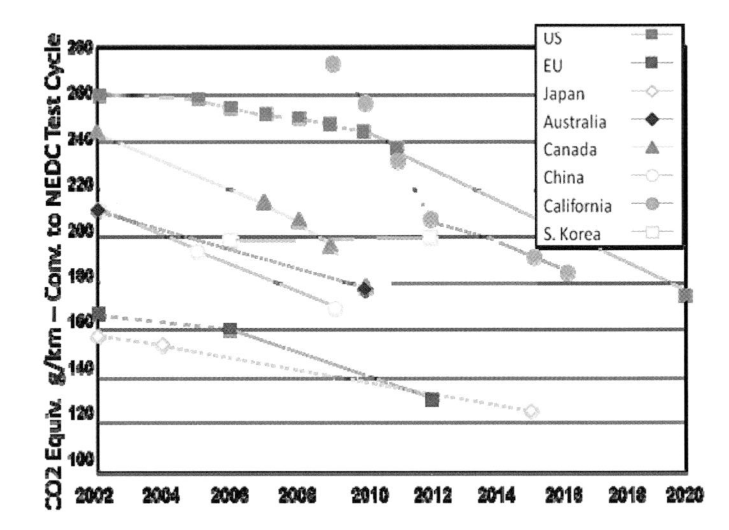

Figure 2 - Existing, Pending and Proposed Worldwide Vehicle Petroleum Consumption / CO2 Tailpipe Emissions Regulations

A comparison of major existing, proposed, and pending regulations were examined in [4]. The US CAFE regulation sets separate standards for cars and light trucks, and will include medium duty passenger vehicles beginning with the 2011 model year. Figure 2 shows the CO_2 equivalent average for a combined fleet of US light duty passenger vehicles and light and medium duty trucks through a GVR of 10,000 lbs. The California ARB GHG regulations would require two separate fleets by setting one standard for the fleet of passenger cars and small trucks (LDT1) and a second standard for the fleet of larger trucks (LDT2).

The trend is clear that all major markets have or are in the process of implementing new strict regulations. These regulations will drive major reductions in CO_2 emissions (or equivalently major gains in fuel economy) over a short time horizon. In some cases, greater than a 30% CO_2 reduction (more than a 45% increase in fuel economy) over a 10 year period is proposed. In the face of such challenges, incremental improvements in vehicle efficiency and vehicle size reductions may not be sufficient. Some automakers have begun to consider practical ways to implement vehicles with sources of energy other than fossil fuel to fundamentally change and the quantity of CO_2 produced by the vehicle operation. With these new technologies, CO_2 emissions would no longer be tied solely to equations (2) or (3) for energy.

ELECTRIFICATION AND AUTOMOTIVE CO_2 - GM's advanced propulsion strategy is to enable automobiles to shift significant portions of their required energy from petroleum to other sources. Figure 3 shows a network of the various energy sources, energy pathways, and possible on-vehicle energy storage media. Higher power electric machines, higher energy on-board electrical storage, and systems that allow for driving without a combustion engine enable vehicles that can use non-petroleum energy sources for transportation. We call the increase in electrical content and magnitude onto the vehicle "electrification".

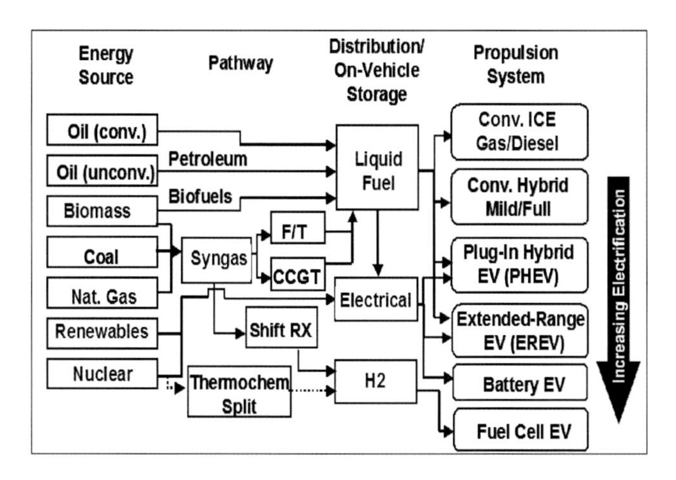

Figure 3 - Energy Sources, Paths, On-vehicle Storage and Propulsion Systems

In prior work, SAE 2008-01-0458 [5], the petroleum displacement and emission benefits of various types of electrification were evaluated. Special attention was paid to the differences between PHEVs and E-REVs.. In this study, that inquiry is continued by analyzing the CO_2 effects of various charging scenarios and consumer choices. The prior work considered only the scenario of single charge per day, where in this study we examine several scenarios that would involve more than one charge per driving day.

With electrification, different types of plug-in vehicles can use grid energy for partial substitution, preferential substitution, or wholesale substitution of fossil fuels.

Grid energy includes coal fired electrical power plants with various efficiencies, natural gas, nuclear, wind, hydropower, biomass, etc, each with varying CO_2 emissions associated with them. The mix of sources used is a function of region, season, and time of day.

A complete analysis that compares CO_2 emissions resulting from automotive use should consider not just tailpipe CO_2 emissions, but the emissions associated with the conversion of the source energy and its transmission pathways. This type of analysis is referred to as "well to wheel" analysis.

Automakers, regulators, and electrical infrastructure providers need to understand the various types of plug-in vehicles and their effect on the infrastructure. Insights on the well-to-wheels CO_2 emissions from the vehicles based on how they are driven and how they may be re-charged will be useful to:

1. Guide development toward the most effective plug-in vehicle types
2. Guide development of the most effective charging infrastructure
3. Guide the development of appropriate test methods and measures of merit for plug-in vehicles

This study uses real world measured US driving patterns of major populations of drivers to examine driving behavior and charging opportunities using large scale simulations.

STUDY APPROACH

The approach of this study was to examine PHEVs and E-REV vehicle types as model variants of a modern mid-sized sedan. This group of vehicles was scrimmaged against a variety of regional driving patterns, regional electrical power mixes and their CO_2 contributions. Their fuel displacement and CO2 reduction capabilities were compared to each other and to a reference conventional full hybrid. Various grid charging location and frequency scenarios were also explored. Regional driving patterns were statistically inferred from correlations between regional driving surveys and national driving surveys. Electric grid mix data comes from the current and projected status of the Emissions & Generation Resource Integrated Database (eGrid) [6].

VEHICLE TYPE DEFINITIONS - Various vehicle types were considered in the analyses done in this paper, the same as in [5]. Representative vehicles were modeled, configured as major sub-variants of a mid-sized sedan as typified by the 2009 Chevrolet Malibu. Charge sustaining fuel economy was approximately 36 mpg for each, and assumed so for the behavioral analysis.

Reference : Full Hybrid Electric Vehicle (HEV) – This vehicle is powered with a 2.4 liter four cylinder IC gasoline engine and uses the GM 2-Mode FWD hybrid system and a 1.8 kWhr, 30 kW battery pack. All vehicle energy ultimately comes from the gasoline, even though the engine was sometimes off when tractive effort is zero, negative or low. This vehicle was used as a control variable to define a vehicle with reduced CO_2 contributions, but coming entirely from liquid fuel.

Plug-in Hybrid Electric Vehicle (PHEV) – This vehicle is powered with a 2.4 liter four cylinder IC gasoline engine and uses the GM 2-Mode FWD hybrid system modified to 53 kW power and a 53 kW battery pack. Various battery energy capacities from 2 kWhr to 16kWhr were studied. This vehicle used energy preferentially from the battery pack when it is available and when tractive power and speed are below the FUD city schedule demands for this size vehicle (below 53 kW and speeds lower than 56 mph). This electric-only power and speed capability made the PHEV considered as "Urban Capable". After the battery was depleted of energy, this vehicle ran as a full hybrid.

Extended Range Electric Vehicle (E-REV) – This vehicle is powered with a 1.4 liter four cylinder IC gasoline engine and the GM E-REV propulsion system with a 120 kW electric power capacity. Various battery energy capacities from 2 kWhr to 16kWhr were studied. Energy was used exclusively from the battery when it is available under all driving conditions. After the battery is depleted of energy, this vehicle ran as a full hybrid.

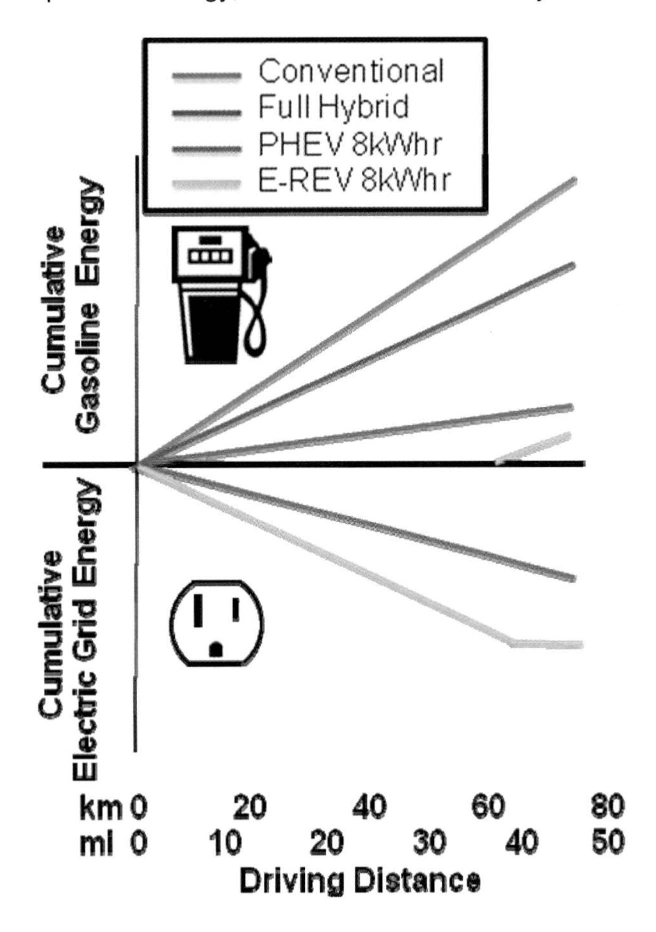

Figure 4 - A Comparison of Energy Usage by Vehicle Type

If the vehicles were run over a nominal controlled schedule, the cumulative energy consumption would appear as shown in Figure 4. Note that the full hybrid used approximately 35% less energy than the conventional non-hybrid vehicle, but both vehicles only used energy from gasoline. The PHEVs used energy preferentially from the electric storage on board the vehicle at first, but subject to the power and speed constraints of the PHEVs. The PHEVs eventually used only gasoline after the battery was depleted. The E-REV was similar to the PHEV, but there were no speed or power constraints to force the engine on. Gasoline was only used in the E-REV *after* the battery was depleted. The BEV only used energy from the electric storage, but could not necessarily complete the driving mission.

STUDY OBJECTIVES

The objectives of this study were to summarize the gasoline displacement, CO_2 displacement, and grid impact of various vehicle technologies.

STUDY METHODOLOGY

This study was performed by combing simple behavioral models of vehicles with trip information from the 2001 National Household Travel Survey (NHTS) [7]. The NHTS data includes trip distances, travel times, location of trip origination, and location of trip destination on almost fifty thousand vehicles and more than one hundred and thirty thousand trips. The NHTS includes vehicles which both move and remain parked during the travel day.

For each vehicle in the NHTS, a record of the vehicle's location and travel speed at each minute during the travel day was created. Both vehicles which moved and stayed parked during the travel day were included since average effects need to consider these parked vehicles. Because of significant differences between weekday and weekend travel, the travel days were organized to build a composite distribution of vehicle locations though out the week.

Figure 5 illustrates the distribution of vehicle locations throughout the week. These locations are used, in conjunction with charging restrictions, to determine when charging is possible, when the battery is discharged, and how fast the battery is discharged. Note that at any moment, the majority of cars are at home (green). The weekly commute and dwell at work (red) is also plainly visible. These are the obvious candidate locations for recharging the electric batteries on E-REVs and PHEVs.

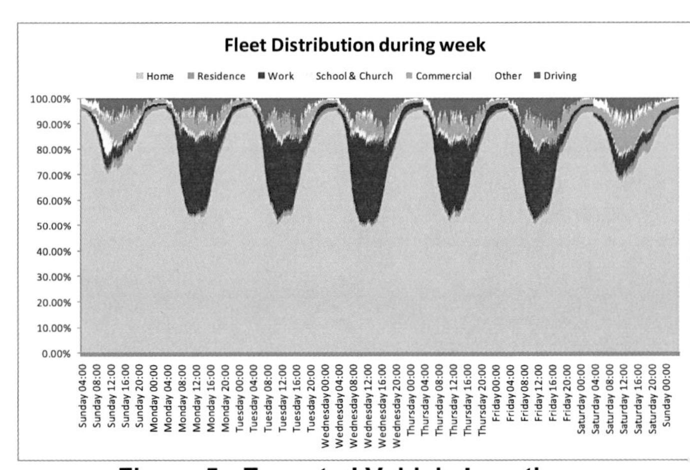

Figure 5 - Expected Vehicle Location

The vehicle models used in this study are behavioral models. They model the fuel, battery, and grid electrical energy consumption of the vehicle directly by defining the:

1. Electrical energy consumption per mile (km) when operating in a charge decreasing mode
2. Charge sustaining fuel economy
3. Fraction of fuel energy used during charge decreasing operation
4. Battery size
5. Charger power rate
6. Charger efficiency
7. Restrictions on charging time
8. Restrictions on where charging can occur.

For simulation purposes, it is assumed that the vehicle operator charges wherever and whenever charging is not restricted.

Since data on each vehicle only exists for a single day that runs from four AM to four AM on the next day, it is assumed that the vehicle is recharged at the start of the day. Since this may not always be achieved because of a combination of charger power, charging restrictions and battery discharge, the aggregate step change in charge from one travel day to the next is compensated.

With this simple behavioral model, conventional and Hybrid Electric Vehicles (HEVs) are modeled by their charge sustaining fuel economy only. PHEVs are modeled using a fixed power split ratio when charge decreasing. E-REVs are assumed to have pure electric operation when the charge is decreasing. Table 1 shows the values used for the vehicles considered in this study. It should be noted, while the behavior of these vehicles is very similar, they represent vehicles with potentially significant differences in design, architecture, cost, and features. However, for this study, only the behaviors of the vehicle are considered so the differences that emerge due to this behavior can be examined without the confounding impact of detailed vehicle design characteristics.

Table 1 - Vehicle Behavioral Parameters

	Conventional or HEV	PHEV	E-REV
Charge Sustaining Fuel Economy	36 [mpg] 6.53 [l/100 km]		
Charge Decreasing Electrical Energy Consumption [kWhr/mile]	n/a	0.350 [kWhr/mi] 0.2175 [kWhr/km]	
Charge Decreasing Electric Energy Fraction	n/a	0.7	1.0

The values chosen for Table 1 roughly correspond to the vehicles used in [5]. Of particular note is the electric energy fraction used for the PHEV behavioral model. This value is approximately equal to the vehicle-weighted average electric energy fraction observed, but unpublished, from [5]. The energy fraction is defined as the ratio of electrical energy to total energy used to move the vehicle while operating in a charge decreasing mode. A vehicle with a an electric energy fraction of one does not use the engine while charge decreasing. Similarly, a vehicle with an electric energy fraction of zero only uses the engine. This fraction is less than one for an Urban PHEV, indicating the engine is used along with the battery, because of either kinematic or battery power constraints that require engine operation to meet customer driving demands. Furthermore, since the objective of a PHEV is to minimize fuel consumption once the engine is on, it operates to minimize total losses. Therefore, once operational, the engine generally provides the majority of the tractive power. This energy fraction is higher if distance weighting is considered rather than vehicle weighting. This occurs because vehicles which travel a short distance typically have lower power and speed demands, which allow the electrical system to primarily move the vehicle a larger fraction of the time. Figure 6 illustrates the distribution of the electric energy fraction during charge decreasing operation for the Southern California Area Governments' Regional Travel Survey (SCAG RTS) with an urban capable PHEV [5].

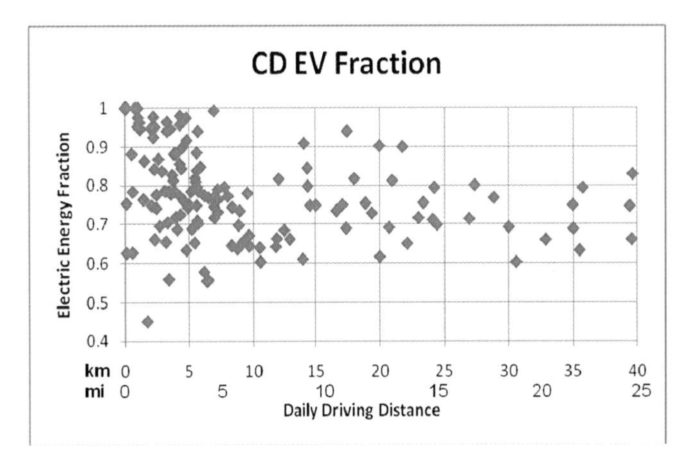

Figure 6- Charge Decreasing EV Fraction vs Daily Driving Distance

The electric energy fraction is not the same as the split between the average electric energy and the average fuel energy delivered to the wheel while the battery is discharged. This split is calculated using Equation 4.

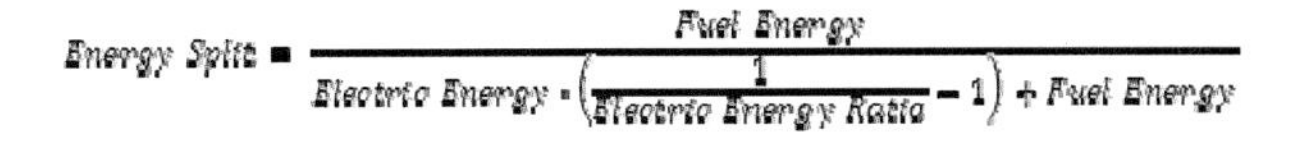

$$\text{Energy Split} = \frac{\text{Fuel Energy}}{\text{Electric Energy}\cdot\left(\frac{1}{\text{Electric Energy Ratio}} - 1\right) + \text{Fuel Energy}}$$

Equation 4 - Energy Split & Ratio

This equation has the property that for all ratios, a PHEV and an E-REV with identical electrical power and fuel consumption per mile will consume the same quantity of electrical energy and fuel once the battery is depleted.

If a vehicle has a battery, it is initialized at four AM on each day to have a full charge. When the vehicle moves, the battery discharge is calculated until the battery state of charge (SOC) goes to zero. This mode of operation is referred to as charge decreasing operation. When charge decreasing, the charge-decreasing energy split determines what fraction of travel is powered by electrical and by fuel energy. Once the battery is depleted, the vehicle is powered by fuel alone and the charge-decreasing fuel economy determines the fuel consumption. Useable battery storage was modeled ranging from zero kWhrs to 16 kWhrs.

When the vehicle is parked, three charging scenarios were considered. The most restrictive case only allows charging at home between the hours of nine PM and nine AM. This is equivalent to one charge per day for the vast majority of drivers. The most liberal charging scenario permits charging anytime the vehicle is parked, at work, and at all other stationary locations. A middle case of unrestricted charging at home and while at work was also considered. A fourth charging scenario of unrestricted charging at home was also considered, but the benefits were similar to restricted home charging, therefore we have omitted those results. For each scenario, a charger that draws 1.1, 3.3 and 6.6 kW from the wall to charge with 90% efficiency was considered. Furthermore, to simplify modeling, the charger was assumed to operate at full power until the battery is either charged or the charger is disconnected. Where the start times are restricted, a random delay is introduced for each charger to prevent high aggregate peaks in the grid.

Table 2 - Charging Scenarios

	Scenario 1	Scenario 2	Scenario 3
Charging Locations	Home	Home & Work	Everywhere
Charging Time	9:00 PM thru 9:00 AM w/ random 2 hour start delay	Anytime	Anytime

One problem with modeling charging using the NHTS is that the data is only for a 24 hour period that starts and ends at 4:00 AM. Therefore in the modeling, some vehicles started with too high of an initial SOC (e.g. vehicles away from home at 4:00 AM) and some vehicles did not complete charging before 4:00 AM (e.g. large battery vehicles with small chargers). To compensate for these sources of 'energy leakage' in the modeling, once the population of vehicles was simulated, the differences in average SOC that occurred on daily boundaries were compensated by adjusting the average grid energy calculations.

For each vehicle, the three charging scenarios, at three charger power levels, and a battery that ranged from 0 to 16 kWhrs were considered. All vehicles in the NHTS were evaluated for fuel and electrical use. The expected (or average) value from this fleet of vehicles was evaluated to determine the expected (average) per vehicle effects.

STUDY RESULTS & DISCUSSION

The expected reduction in fuel use is influenced by battery size, powertrain architecture, and the charging scenario. Figure 7 illustrates the expected daily fuel use for the various cases considered. When the battery size is zero, all vehicles behave like an HEV which has a 36 mpg (6.53 l/100 km) fuel economy.

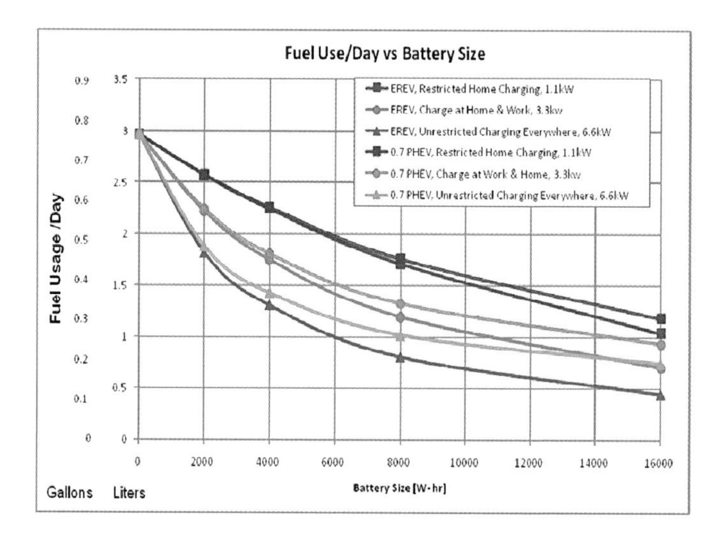

Figure 7 - Fuel Usage vs. Battery SIze

From these plots, a distinct difference between the PHEV and the E-REV appears. Because the PHEV uses both fuel energy and electrical energy to move the vehicle when charge decreasing for larger batteries, the PHEV is not able to effectively use the energy stored in those batteries. However, for small batteries and restrictive charging schemes, the E-REV and PHEV obtain approximately the same reductions in fuel consumption. For very large batteries (e.g. >= 16 kWhr), the PHEV's marginal improvements in fuel consumption for marginal increases in battery size are almost zero.

The electrical energy used from the grid has two characteristics of interest. The first is peak expected electrical power draw from a population of vehicles. The second is the expected energy use by the fleet of vehicles. To calculate the expected peak electrical power draw, the minute-by-minute charging for each vehicle in the NHTS was determined. To calculate the expected power at any minute, the average power draw over the population of vehicles was calculated. An example of this calculation is illustrated in Figure 8. In this example, the expected power for a fleet of 8 kWhr E-REVs charged at 3.3 kW at both work and home is shown. The peak expected power is 813 watts. When the peak power is evaluated over the battery sizes, charging scenarios, and vehicles considered in this study, the graph in Figure 9 is obtained.

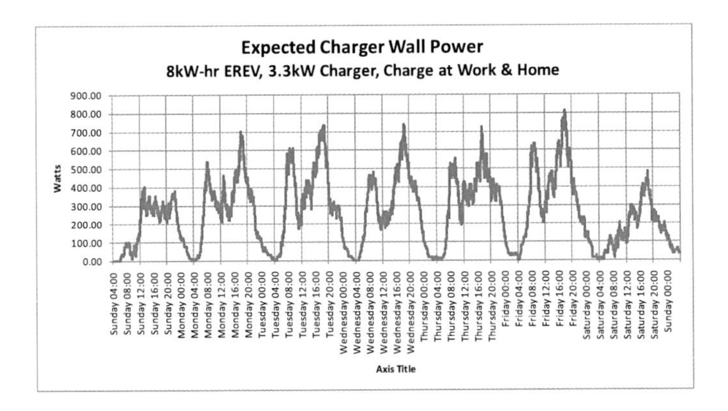

Figure 8 - Expected Charger Wall Power Example

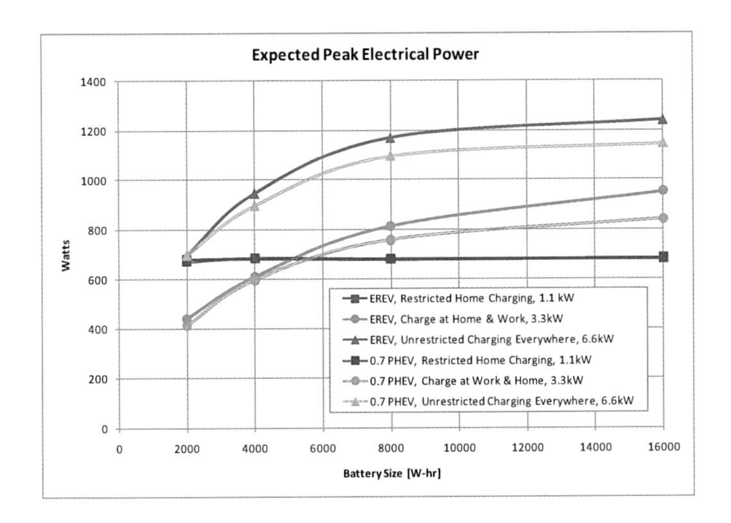

Figure 9 - Peak Grid Loading

One interesting behavior shown in Figure 9 is that the peak grid loading is strongly affected by both charging location and charging time restrictions. For example, allowing 4 kWhr vehicles to charge at work and home at 3.3 kW results in less peak grid loading than restricting charging to 1.1kW between the hours of 9 pm and 9 am. To examine what is occurring, consider two 4kWhr E-REV's. Let one vehicle be charged with a 1.1 kWhr charger only at home between the hours of 9 pm and 9 am subject to a random 2 hr start delay. Let the other vehicle be charged with a 3.3kW charger at both work and home. Figure 10 illustrates the differences in charging time and charging power between these two vehicles. The differences illustrate that if vehicles are allowed to charge at multiple opportunities and locations without restriction, the expected loading on the electrical grid is much lower than the charger's power rating.

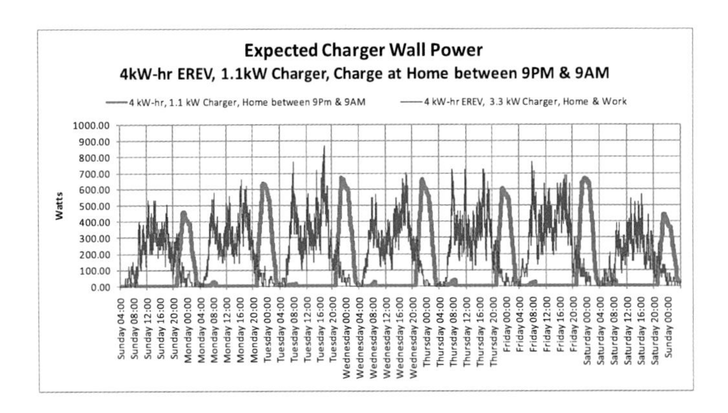

Figure 10 – Effect of charging restrictions on peak grid power.

While the peak grid loading affects infrastructure, the electric grid's CO_2 emissions are primarily a function of the amount of electricity used. Figure 11 illustrates the expected grid energy use per day based on the battery size, the charging scenarios, and the vehicle architecture.

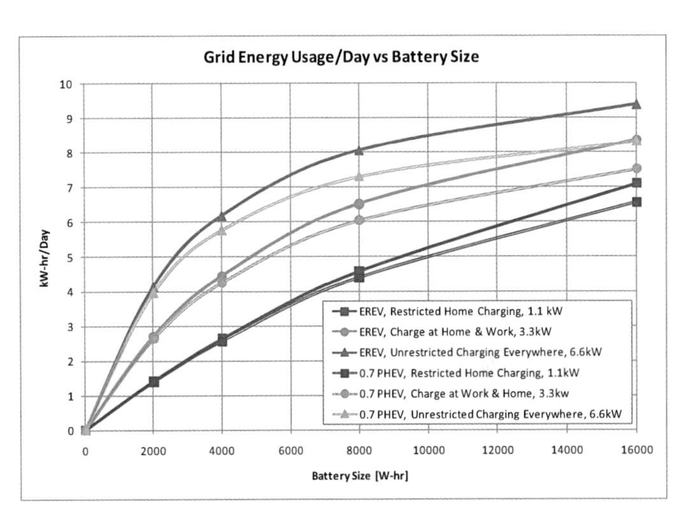

Figure 11 - Grid Energy Usage

The CO_2 emissions attributable to the vehicle depend on the fuel usage and the grid energy usage. The EPA estimates that the regions in the US vary in carbon intensity from 0.23 to 0.89 kg CO_2/kWhr [9]. For reference, the 2005 EPA estimate of California grid emissions is used. Using a well-to-wheel CO_2 factor for gasoline of 11.34 kg/gallon (2.995 kg/l) [8] and a CO_2 factor of 0.32 kg/kWhr for power plant emissions, the emissions estimate in Figure 12 is obtained. By normalizing on the conventional vehicle, the relative reduction in CO_2 is obtained and plotted in Figure 13.

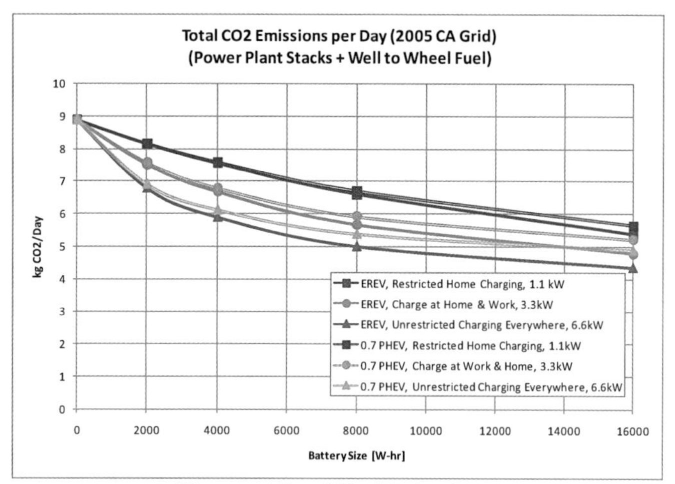

Figure 12 – CO_2 Emissions

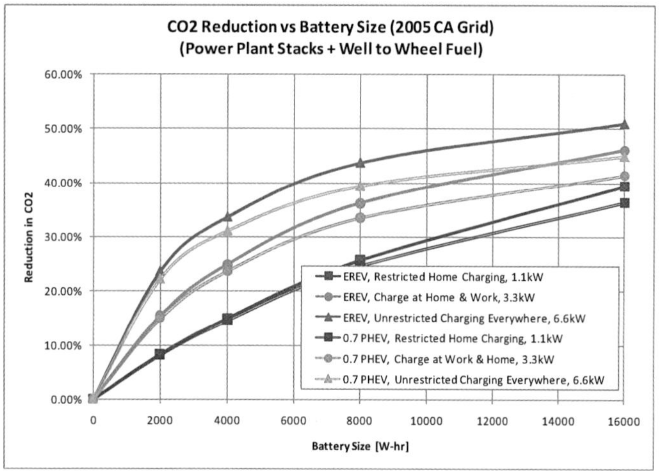

Figure 13 - Percentage Reduction in CO_2

Since the total CO_2 emissions are a function of the fuel usage and electric grid usage, the differences seen in fuel use and grid usage are reflected in the CO_2. For small batteries, the PHEV and the E-REV deliver effectively similar results. However, as the battery size increases, the E-REV advantages over the PHEV performance become clear.

One unique aspect of the E-REV is the ability to complete a driving day using only electrical energy. Of the vehicles that are driven, Figure 14 illustrates the fraction that can complete the day using only electricity. One interesting characteristic of this chart is the charging scenarios influence the fraction of full EV days almost as strongly as the battery size.

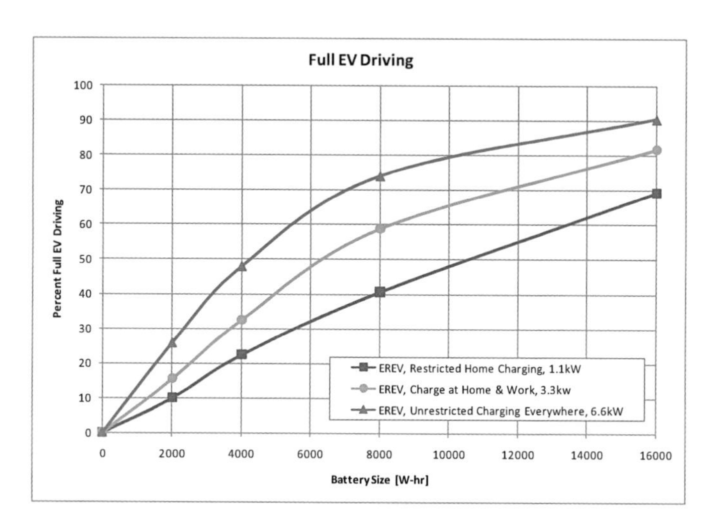

Figure 14 - EV Driving

To more closely consider the differences between the PHEV and E-REV, consider their behavior when charged at 3.3 kW at both work and home. As seen previously for smaller batteries, the PHEV and the E-REV offer similar benefits. However, once larger batteries are considered, the E-REV offers appreciable difference in fuel and CO_2 reduction.

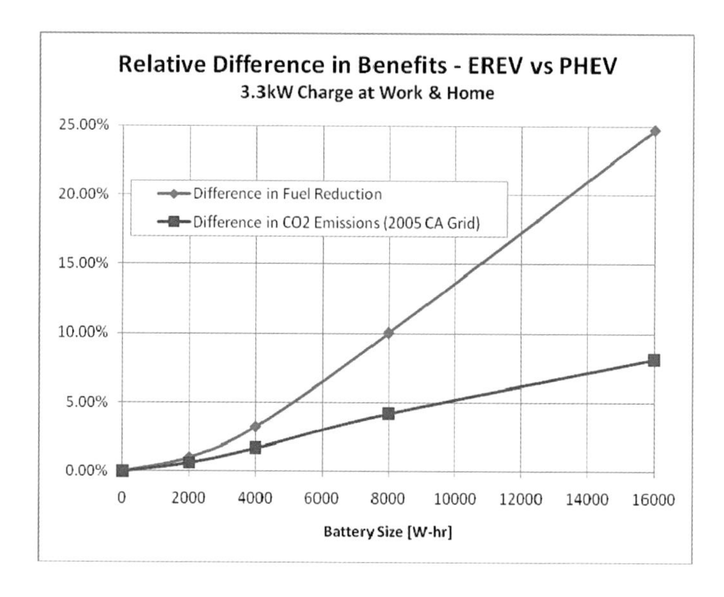

Figure 15 - Relative Differences in Fuel & CO_2 benefits

While this study was quite exhaustive in the vehicle samples used, there are several enhancements that could improve this work. If publicly available data sets that provided summary data similar to the NHTS along with detailed second-by-second velocity information were available, detailed simulations could be performed instead of the behavioral models used here. This would provide more confidence in the results and allow a deeper understanding of the details in operation. Additionally, in the NHTS, a lack of multi-day data on individual vehicles made it difficult to evaluate the impact of different charging schemes that might be used to reduce peak electrical grid loading or take advantage of grid variations in CO_2 intensity throughout the day. Multi-day data would also eliminate the need to correct for SOC differences between days, as noted earlier.

Additionally, this study considered the use of PHEV's and E-REV's as a substitution for any vehicle in use today. However, the benefits of these vehicles are best realized when they are used by a customer whose driving patterns match the benefits of the vehicle. If the deployment of these vehicles is concentrated on customers who get the maximum benefit, the benefits outlined in this study can be significantly higher.

THE EFFECT OF CUSTOMER CHOICE IN ELECTRICITY

In addition to the choice to drive a PHEV or E-REV, a customer can also make a choice in the electricity they use. The EPA has programs for renewable and green electric energy that have much lower CO_2 emissions when compared the conventional electric grid [9,10]. This energy can be purchased by retail electric customers through various programs [11]. These programs sell electric energy which is produced from renewable sources and is nominally carbon neutral [12]. Assuming carbon neutral electricity, these vehicles can offer large CO_2 reductions. The possible CO_2 reduction by purchasing this electricity is illustrated in Figure 16. Surprisingly, the market cost to use this electricity adds a minimal expense to the operation of the vehicle. Carbon neutral electricity [13] can be purchased for a differential of about \$0.015 to \$0.120 per kWhr. Therefore, for an 8kWhr E-REV, charged at work and home, a year's worth of carbon neutral electricity can be purchased for between \$40 and \$250 per year. For most parts of the country, the differential is about \$60 per year.

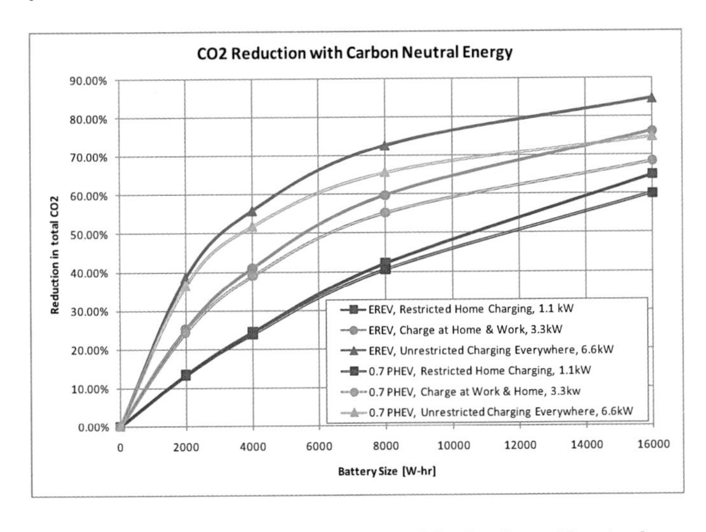

Figure 16 - CO_2 Reduction with Carbon Neutral Electricity

THE EFFECT OF CUSTOMER DRIVING SELECTION

The distribution of daily driving distances in the USA has a small fraction of vehicles used to cover very large distances. For any calculation of average benefits, a very long distance driver sees very little benefit from the electrification of the vehicle. In other words for a driver who travels 1000 miles (1609 km) in a day, only 4% of the fuel would be displaced by an *kWhr PHEV. The benefit of the batteries would be minimal. These drivers benefit most from fundamental improvements in base fuel economy, and cannot take much advantage of vehicle electrification.

The benefits of electrification are greatest for shorter distance drivers. With the rational, self-selection of vehicles based on daily driving distance, the electrification benefits of PHEV s and E-REVs improve dramatically. When the top 8% of daily driving, distances greater than 100 miles (161 km), are omitted from this analysis, there are dramatic changes in the behavior of the fleet. As can be seen in Figure 18, removal of the top 8% of long distance travel days eliminates about 35% of total miles traveled. This reduction in travel is reflected in a reduction in both fuel and CO_2.

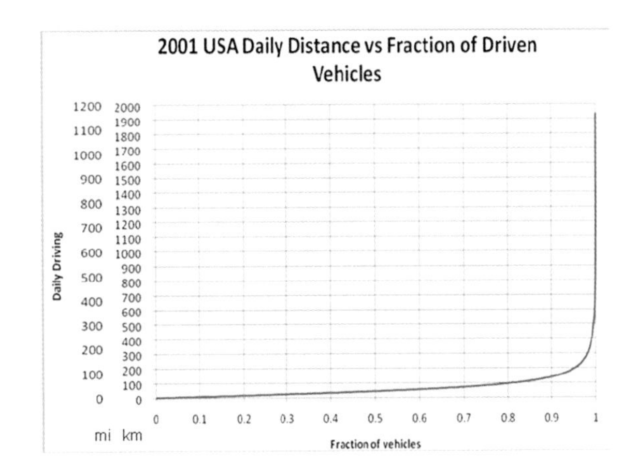

Figure 17 - Driving Distances

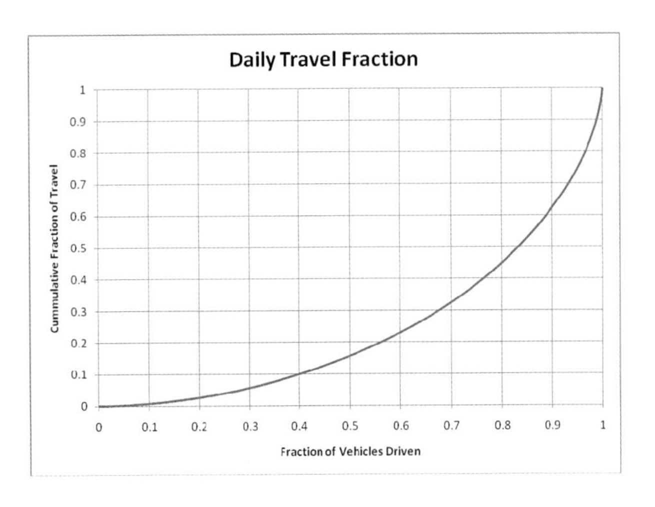

Figure 18 - Cumulative Travel

For this restricted group of vehicles, an 8kWhr E-REV with charging at home and work can eliminate almost 80% of fuel usage. This is illustrated in Figure 19. Furthermore, because of the increased fraction of electrical use, the same vehicle and charging scenario results in almost a 50% reduction in CO_2 emissions. With this restricted set of drivers, the relative difference between a PHEV and E-REV is increased. In fact, for these drivers, the baseline E-REV reduces almost 25% more fuel than the equivalent PHEV. Additionally, it reduces CO_2 by about 8% more than the PHEV.

If drivers, who drive fewer daily miles than the national averages, choose E-REVs, 'green power', and make arrangements for charging at work and home, an 8kWhr E-REV can potentially reduce CO_2 by approximately 80% compared to a baseline vehicle and reduce CO_2 emissions by 25% more than an equivalent PHEV.

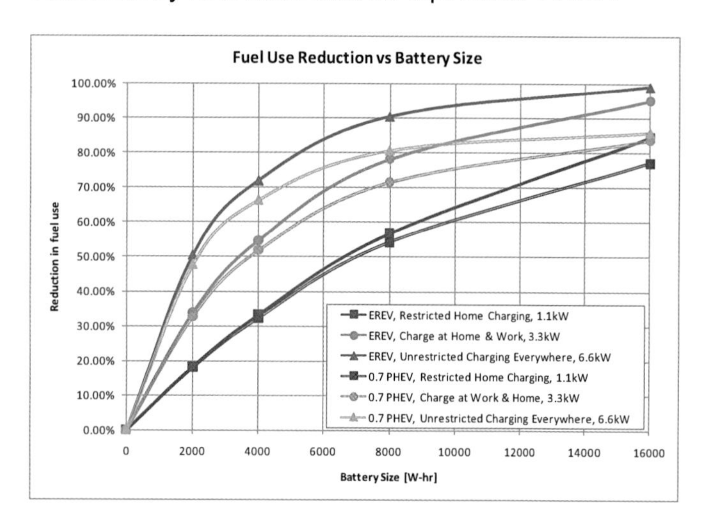

Figure 19 - Fuel Use Reduction, Restricted Driving Set

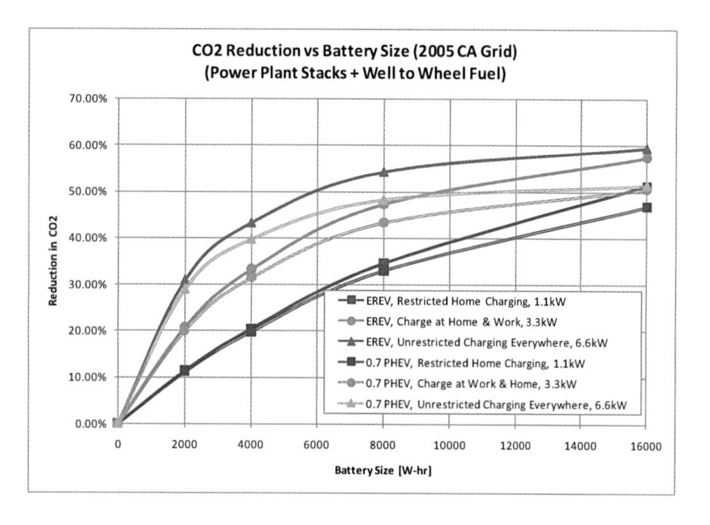

Figure 20 - CO_2 Reduction, Restricted Driving Set

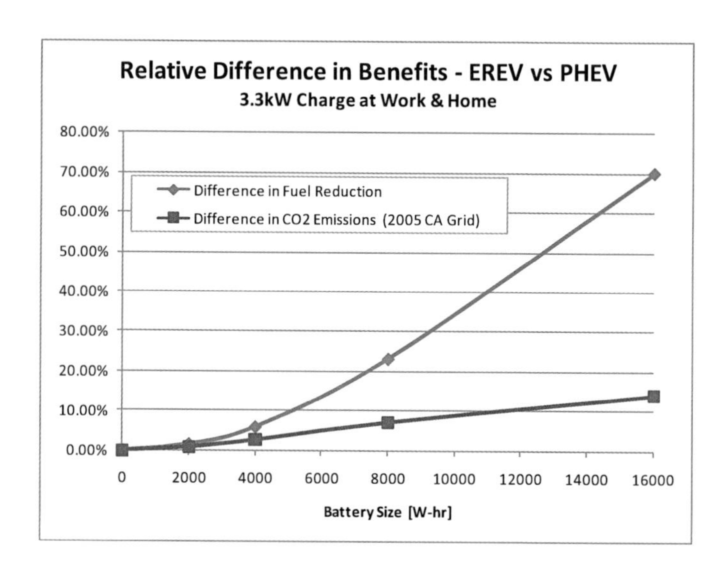

Figure 21 - Relative OHEV & E-REV Difference, Restricted Driving Set

CONCLUSIONS

For the vehicles considered in this study, the following conclusions are drawn.

1. Overall, E-REVs and PHEV electrification have real fuel savings and CO_2 benefits when compared to conventional full hybrid vehicles. For greater than 90% of drivers, those who drive less than 100 miles per day, the expected fuel savings can be greater than 70% and near term CO_2 reductions can be greater than 40% (CA grid).

2. E-REVs hold an advantage over PHEVs for fuel displaced with electrical energy and CO_2 reductions, based on the vehicle type alone. This advantage becomes more pronounced with larger battery sizes. For instance, on the restricted data set, an 8 kWhr E-REV will give a 22% reduction in liquid fuel consumed and an 8% reduction in CO_2 relative to an 8kWhr PHEV. (3.3 kW charging at home and at work, CA grid)

3. Considering the complete vehicle proposition (since E-REVs are more likely to be produced with larger batteries than PHEVs), E-REV hold a large advantage when compared to PHEVs. For instance, on the restricted driving set, an 8kWhr E-REV will use less than half the liquid fuel and emit less than one third of the CO_2 of 4kWhr PHEV. (3.3 kW charging at home and at work, CA Grid).

4. The aggregate grid capacity impact of PHEVs and E-REVs is not, in general, large. An 8kWhr E-REV using a 3.3kW charger has an expected peak grid load of roughly 800 Watts. This is equivalent to a 50 inch plasma TV [15] or a few high wattage light

bulbs. The introduction of these vehicles does not appear to significantly strain the existing US electrical grid.

5. Making charging available at work significantly improves the ability of an E-REV or PHEV to displace fuel and lower CO_2 emissions. For instance, on the restricted driving set, an 8 kWHr E-REV charged at 3.3 kW at home and at work will lower liquid fuel consumption and CO_2 produced by 50% and 30% respectively when compared to 1.1kW charging at home during restricted to evening and early morning charging (CA grid).

6. Since the benefit of charging at work and elsewhere is large, test methods that measure the merits of PHEVs and E-REVs should consider the probability of multiple charges per day.

7. The electric power system will increasingly use renewable energy due to legislative mandates. By 2015, the carbon intensity of the grid is expected to be reduced by meeting renewable portfolio standards [17]. Long term, grid CO_2 emissions are expected to continue to fall. This decrease in carbon intensity will decrease the CO_2 impact of E-REVs and PHEVs vehicles over time. In the limit of carbon neutral energy, an electrified vehicle's CO_2 emission benefits will match that vehicle's liquid fuel displacement benefits.

ACKNOWLEDGEMENTS

The authors would like to thank the EPA's Green Power Partnership personnel and Alex Pennock from the Center for Resource Solutions for their time in explaining the various renewable and green energy programs.

REFERENCES

1. International Energy Outlook, 2008; Report #DOE/EIA-0484(2008); Energy Information Administration, US Department of Energy, September 2008
2. World Oil Outlook, 2008; Organization of Petroleum Exporting Countries, January 2008
3. H. Lenz, C Cozzarini; Emissions and Air Quality; Society of Automotive Engineers, Inc., 1999.
4. F. An, D. Gordon, H. He, D. Kodjak, D. Rutherford; Passenger Vehicle Greenhouse Gas and Fuel Economy Standards: A Global Update; The International Council on Clean Transportation, July 2007
5. E.D Tate, M.O. Harpster, P.J. Savagian; The Electrification of the Automobile: From Conventional Hybrid to Plug-in Hybrids, to Extended Range Electric Vehicles, SAE Congress, Detroit, MI, April 2008.
6. Emissions & Generation Resource Integrated Database (eGrid), http://www.epa.gov/cleanenergy/energy-resources/egrid/, Dec 10, 2008.
7. 2001 National Household Travel Survey, http://nhts.ornl.gov/, Dec 10, 2008.
8. Argonne GREET Model, http://www.transportation.anl.gov/modeling_simulation/GREET/index.html, Sept 5, 2008, Dec 15, 2008.
9. EPA Green Power Partnership, Green Power Market, http://www.epa.gov/greenpower/gpmarket/index.htm, Dec 28, 2007, Dec 10, 2008.
10. Green-e, http://www.green-e.org/, Dec 10, 2008.
11. EPA Green Energy Locator, http://www.epa.gov/greenpower/pubs/gplocator.htm, Oct 27, 2008, Dec 10, 2008.
12. Climate Protection Partnership Division of Office of Atmospheric Programs, US EPA, Climate Leaders Greenhouse Gas Inventory Protocol Optional Modules Methodology for Project Type: Green Power and Renewable Energy Certificates (RECs) (http://www.epa.gov/climateleaders/documents/greenpower_guidance.pdf), Nov 2008.
13. Plasma TV Energy Consumption Testing, http://www.plasmadisplaycoalition.org/results/power.php, Dec 10, 2005.
14. EPA, Renewable Portfolio Standards Fact Sheet, (http://www.epa.gov/CHP/state-policy/renewable_fs.html), Oct 15, 2008. Dec 10, 2008.

CONTACT

E. D. Tate can be contacted at ed.d.tate@gm.com.

ADDITIONAL SOURCES

1. Federal Highway Administration; National Household Travel Survey, http://www.fhwa.dot.gov/policy/ohpi/nhts/index.htm
2. Southern California Association of Governments; Travel and Congestion Survey, http://www.scag.ca.gov/travelsurvey/

DEFINITIONS, ACRONYMS, ABBREVIATIONS

BEV: Battery Electric Vehicle
E-REV: Extend Range Electric Vehicle
ICE: Internal Combustion Engine
ITS: Initial Trip Starts
PHEV: Plug-in Hybrid Electric Vehicle
SCAG: Southern California Association of Governments
EV: Electric Vehicle
NPTS: National Personal Transportation Survey
RTS: SCAG Regional Travel Survey

2011 Chevrolet Volt Specifications

GM Vehicle Program Code: D1JCI

Vehicle Description

5-door hatchback, 4-passenger, front-wheel-drive, extended-range electric vehicle (E-REV)

Exterior Dimensions

Overall length (in/mm):	177/4498
Wheelbase (in/mm):	105.7/2685
Width (in/mm):	70.4/1788
Height (in/mm):	56.3/1430
Front overhang (in/mm):	39/993
Rear overhang (in/mm):	32.2/820
Front track (in/mm):	61.2/1556
Rear track (in/mm):	62.1/1578

Interior Dimensions

Headroom (in/mm)	
Front:	37.8/960
Rear:	36.0/915
Shoulder room (in/mm)	
Front:	56.5/1436
Rear:	53.9/1369
Hip room (in/mm)	
Front:	53.7/1365
Rear:	51.2/1301
Leg room (in/mm)	
Front:	42/1068
Rear:	31/787
Cargo volume (ft^3/L):	10.6/301

Capacities

Curb weight (lb/kg):	3781/1715
GVWR (lb/kg):	4583/2079
Fuel tank (gal/L):	9.3/35.2
Generator cooling (qt/L):	7.7/7.3
Battery pack cooling (qt/L):	7.4/7.0
Power electronics cooling (qt/L):	3.1/2.9
Engine oil w/filter (qt/L):	3.7/3.5
Drive unit fluid (qt/L):	8.9/8.45

Powertrain

GM Voltec rechargeable electric propulsion system with integral ICE powered generator

Combustion Engine

Architecture:	GM "Family Zero" inline 4-cylinder; cast-iron block, aluminum head
Bore x stroke (mm):	73.4 x 82.6
Displacement (in^3/cm^3):	85.3/1398
Compression ratio:	10.5:1
Valvetrain:	DOHC, 4 valves per cylinder; variable intake and exhaust valve timing
Throttle control:	Electronic
Rated power (hp/kW, SAE net):	84/63 (est.)
Max. rpm:	4800
Required fuel:	Premium gasoline; E85-capable from MY2012
Fuel delivery:	Electronic sequential-port, returnless fuel rail

Battery Pack

Configuration:	T-shaped module with integrated thermal management
Case construction:	Glass-filled polyester composite
Total rated energy (kW·h/MJ):	16/58
Total usable energy (kW·h/MJ):	9.4/34
Total pack volume (L):	100
Pack length (ft/m):	5.5/1.67
Total pack mass (lb/kg):	375/170
Total number of cells:	288
Cooling medium:	Electrolyte-based liquid
Minimum operating temperature:	0-10°C (32-50°F)

Battery Pack cont'd

Maximum/minimum state of charge (%):	85/30
Max. state-of-charge utilization (%):	65
Pack warranty:	8 yrs/100,000 mi (160,000 km)

Battery Cells

Manufacturer:	LG Chem/Compact Power Inc.
Chemistry:	Lithium-ion manganese-spinel ($LiMn2O4$)
Architecture:	Prismatic with proprietary ceramic-coated Safety Reinforced Separator (SRS)
Cooling medium:	Liquid

Electric Transaxle

Traction motor:	3-phase ac induction type, air cooled
Rated peak output (kW/hp):	111/149
Rated peak torque (lb·ft/N·m):	273/370
dc generator:	High-flux permanent-magnet type, liquid cooled
Rated peak output (kW/hp):	55/74
Final drive ratio:	2.16
Hydraulic clutches:	3
Driver-controlled operating modes:	Normal, Sport, and Mountain

Off-car Charging System

Standard voltage:	110/120 V, 15 A, ac (U.S.); 220/240 V ac (Europe)
	Optional "quick-charge" voltage, SAE Level 2: 240 V ac
Charge coupler:	SAE J1772-compliant standard connector with cord

Emissions Rating

U.S. EPA:	NA at time of publication
CARB:	Super-Low Emissions Vehicle (SULEV)

Body & Chassis

GM body architecture:	Delta II
Vehicle classification:	SAE midsize J1100 EPA class
Body structure:	Steel; 80% high-strength alloys
Weight distribution unladen, front/rear (%):	51/49
Coefficient of drag (Cd):	0.281
Chassis control:	All-speed traction control, StabiliTrak; drag control

Suspension

Front: Independent MacPherson struts, coil springs, anti-roll bar; hydraulic ride bushings

Rear: Compound-crank torsion-beam with semi-trailing arms, coil springs, tubular shocks, hydraulic ride bushings

Power Brakes/ABS

Front rotors (in/mm):	11.8/300, vented
Rear rotors (in/mm):	11.5/292, solid
Front calipers:	Sliding type with two 60-mm opposed pistons
Rear calipers:	Sliding type with single 38-mm piston
Assist type:	Electrohydraulic, 4-channel ABS
Friction/regen blending:	TRW Slip Control Boost

Steering

Type:	Rack-mounted ZF electric power steering (EPS), variable assist
Turning circle, curb-to-curb (ft/m):	36/11
Ratio:	15.36
Tires:	Goodyear Assurance Fuel Max, all-season, low-rolling-resistance
Front:	P215/55R17
Rear:	P215/55R17
Spare tire:	Standard tire inflation kit

Production

Assembly plant: GM Detroit-Hamtramck, Hamtramck, MI

Planned volume (units): CY2011 = 10,000; CY2012 = approx. 45,000

Fuel Efficiency & Performance

EPA city/highway:	NA at time of publication
***AEI* tested fuel economy,** mpg**, 150-mi loop**:	51
Claimed range, EV mode (mi/km):	25-50/40-80
***AEI* tested range, EV mode** (mi/km)**, 150-mi loop**:	44.5/71.6
Claimed range with range-extender (mi/km):	+350/483
Est. average annual electric energy useage:	2520 kW·h
Top speed (mph/km·h):	101/161

Safety Engineering (std.)

Dual-stage frontal airbags; side-impact and knee bags for driver and front passenger; roof-rail (side curtain) side-impact bags for front and rear outboard seating positions. Front airbags include Passenger Sensing System. Front seatbelt dual pretensioners and force limiters. Pedestrian alert (horn chirp) for use in EV mode, driver activated by turn signal lever. StabiliTrak stability control system with brake assist and traction control. Optional: Rearview camera and front/rear park-assist package.

Navigation/Information System

GM OnStar: Standard – 5 years of Directions and Connections plan includes automatic crash response; emergency services, crisis assist, stolen vehicle assistance including stolen vehicle slowdown and remote ignition block; remote door unlock; turn-by-turn navigation with destination download and OnStar eNav (where available), OnStar vehicle diagnostics, roadside assistance, remote horn and lights, hands-free calling.

Pricing (U.S.)

2011 MSRP: $40,280 (base); Volt qualifies for the maximum $7500 federal tax credit ($33,500 with the credit)

Standard features: 17-in alloy wheels; automatic headlights; heated mirrors; keyless ignition; remote ignition; automatic climate control; cruise control; auto-dimming rearview mirror; six-way manual front seats; tilt-and-telescoping steering wheel; cloth upholstery; Bluetooth; OnStar; touchscreen navigation system with voice controls and real-time traffic; 6-speaker Bose audio with CD/DVD player, auxiliary audio jack, iPod/USB interface and 30 GB of storage.

Optional features: Premium Trim package adds leather upholstery, leather-wrapped steering wheel, and heated front seats. Rear Camera and Park Assist package adds a rearview camera and front/rear parking sensors.

CHAPTER TWO:

Creating the heart of Volt

GM's battery requirements meant creating new state-of-the-art in-vehicle energy storage—and doing it in less than four years. Top GM and supplier engineers reveal how they did it.

Volt's Li-ion battery pack is a monument to systems engineering. Note the electric-power and coolant inlet tubes in the front cover.

Creating the heart of Volt

The lack of a suitable battery made electric vehicles niche players for most of the last century. For General Motors' Volt team, finding a battery with the energy capacity, performance, durability, reliability, and package density needed for use in a high-volume passenger car with the range-extender powertrain presented a significant challenge.

"In April 2007 when GM put out its solicitation and requirements for Volt, nobody had a battery that would meet the requirements—including us," said Prabhakar Patil, CEO of Compact Power Inc. (CPI), which supplies Volt's battery cells.

Patil recalled that CPI's parent company LG Chem, the South Korean lithium-battery giant, had a cell that could meet GM's power requirements or energy requirements—but not both.

"Typically lithium-ion's performance characteristics are kind of a trade-off between power and energy. GM's battery requirement for Volt meant raising the state of the art of the entire power/energy envelope," he said.

Volt's technology linchpin, and its biggest chunk of supplier involvement based on cost, is its lithium-ion battery pack. CPI's development bogey was to achieve long life, super-robust thermal stability, and high energy density.

"The challenge for advanced automotive batteries is the superb energy content of liquid-petroleum motor fuels. The standards for energy and power density set by the ICE are very tough to beat," observed Dr. Ann Marie Sastry, a University of Michigan engineering professor and expert on advanced battery technologies.

Regarding reliability, GM needed to warranty the pack for the life of the vehicle. The aim was to validate Volt's technology with consumers and also to meet the California Air Resources Board's Advanced Technology Partial Zero Emissions Vehicle (AT-PZEV) standards—an early strategic goal that did not pan out.

The car carries a Super-Low Emissions (SULEV) certification at launch, and GM plans to add emissions equipment to meet AT-PZEV. Its 8-year/100,000-mi (161,000-km) powertrain warranty is equal to Toyota's Prius warranty and is three years longer than GM's standard powertrain warranty.

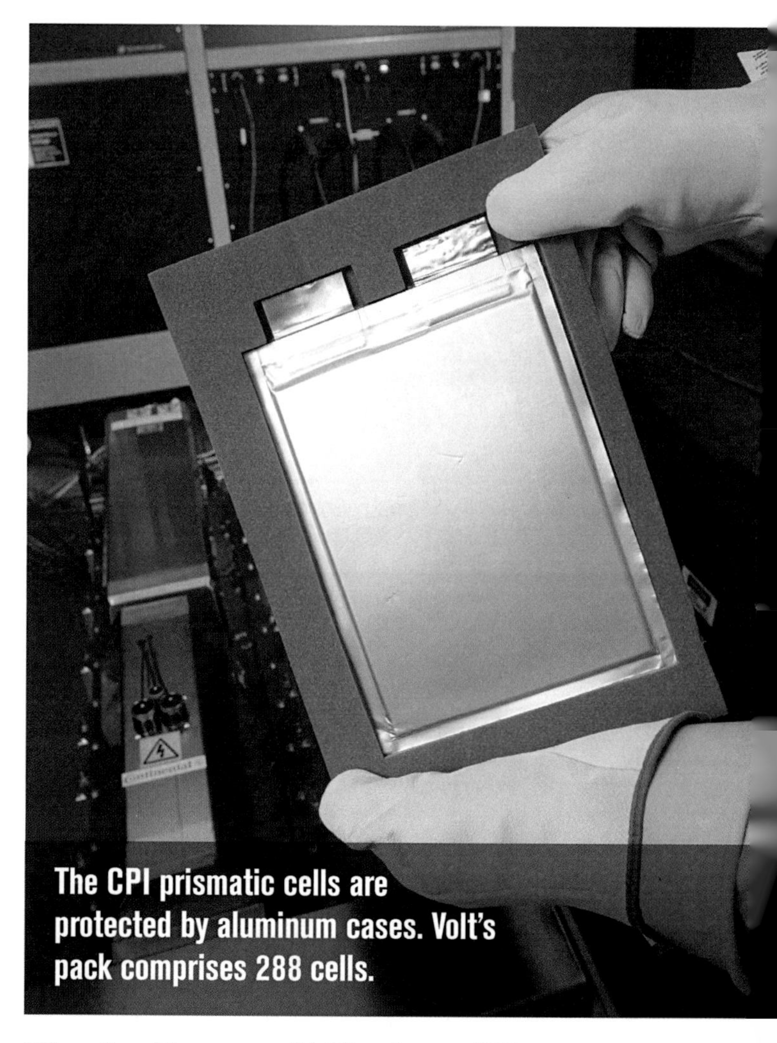

The CPI prismatic cells are protected by aluminum cases. Volt's pack comprises 288 cells.

The battery cell "bake off"

To establish Volt's battery-cell chemistry, and to select the program's cell supplier, GM began with deep-dive investigations into more than 25 different lithium-based battery chemistries, according to Andy Leutheuser, Lead Battery Systems Engineer on Volt.

He described the lengthy evaluation as "a rigorous, multi-phase process." Phase Zero involved prescreening more than 155 chemistries submitted in reports from makers and technologists across the advanced-battery spectrum. In Phase 2, that universe of candidates was reduced to 60 potential cell chemistries GM's experts felt were promising for automotive use. Cell qualification followed in Phase 3, which left CPI and A123 Systems as finalists.

Compared with nickel-metal hydride (NiMH), the incumbent chemistry long proven in hybrid-vehicle use, lithium-ion cells pack up to three times as much power (110-130 W·h/kg) in a much smaller package. Lithium cells are also more configurable, according to experts, and are prone to less discharge when not in use. They're also 50% lighter and potentially less expensive.

Electric-powertrain engineers appreciate lithium's ability to be "tuned" to increase cell power or energy densities, depending on the vehicle application. Leutheuser said the Volt's 288 cells have been tuned to deliver a balance of power and energy.

The LG Chem prismatic (flat package) cells use a carbon-graphite anode and manganese cathode, separated for protection against thermal "runaways" by the company's proprietary separator within the cell. GM battery engineers said this combination of chemistry and architecture provides a very stable, robust cell. The Volt cells are encased in polymer-coated aluminum housings and are stacked vertically within the pack.

"Lithium is unquestionably where the industry's going," said Mark Verbrugge, Director of the Chemical Sciences and Materials Systems Lab at GM R&D. He explained that finding ways to reduce cell and pack cost through engineering and manufacturing is an ongoing priority.

The final evaluation phase for Volt's cells was Pack Qualification, where LG Chem's manganese-spinel cells were pitted against those from Massachusetts-based A123 working in partnership with Continental Automotive (serving as pack integrator).

"In selecting our battery partner, we had to consider many factors," Verbrugge told *AEI*. They included having significant R&D depth in materials, design, and process; understanding battery life, failure modes, and abuse tolerance; endurance for long development periods; relationship to raw materials suppliers; and high-volume cell manufacturing experience.

He noted that a cell supplier's overall quality and process control must be near-perfect because failure of a single cell in use can cause failure of the entire pack.

When GM announced the results of its "battery bake-off" in January 2009, *AEI* talked with Denise Gray, Director of Hybrid Energy Storage Systems, about her company's decision.

"It was pretty close between the two candidates, and there wasn't a single 'standout' element for CPI," she said. "Their cell design works really well with our pack design, and LG Chem has a lot of experience in the battery business. We also were a bit more familiar with their cells because we'd used them in most of our early testing."

GM's pack-assembly secrets

GM originally planned to have CPI manufacture the Volt's 5.5-ft-long (1.7-m) battery pack as well as supply the cells. That strategy changed in order to give the automaker more design latitude in terms of

Lithium-ion manganese spinel prismatic cells make the Volt's pack (shown at right) 75% lighter and 40% smaller than the 2nd-gen NiMH pack used in EV-1 (shown at left), with the same energy capacity.

integrating the pack with the vehicle, as well as greater control over intellectual property.

"We were developing some thermal management and power control solutions that we thought were pretty slick," explained Volt's Vehicle Line Executive, Tony Posawatz. "Pack management and integration are critical for us, as we look to apply the Voltec powertrain to other programs."

Posawatz noted that keeping pack engineering in-house enabled GM to develop some automated pack-assembly processes at its Brownstown Township plant near Detroit that he believes will help save manufacturing costs and help improve the design of future battery packs.

Within the T-shaped pack are nine interlinked, liquid-cooled battery modules, each containing 32 prismatic cells. The cells themselves are less than 0.25 in (6.35 mm) thick, measure approximately 5 x 7 in (127 x 178 mm), and weigh about 1 lb (0.45 kg).

The Volt's battery pack is a very sophisticated piece of systems engineering. It includes integrated thermal management and power control, and its design enables it to fit longitudinally along the Volt's high-strength-steel central tunnel and under the rear seats. Besides storing electrical energy, the battery pack serves as a semi-structural element of the vehicle structure, contributing to the body's claimed 25-Hz bending stiffness.

Volt's energy management system is designed so the battery does not have to endure the deep-cycling that shortens life. Control algorithms allow the pack to operate within a conservative 65% state-of-charge (SOC) window. In more extreme vehicle duty cycles, such as driving in Mountain mode, the SOC will raise the lower limit to ensure there is adequate power when needed. The battery's upper and lower "buffer zones" help ensure long life.

To keep its cool, and warmth, GM engineers opted for a liquid-cooled pack rather than air cooling for durability, thermal stability, and packaging benefits. Li-ion automotive batteries must endure very high dynamics. Momentary peak loads, such as created during brake-energy regeneration and acceleration, generate powerful electrical currents. The currents create internal resistance, which causes significant warming of the cells.

GM battery systems engineer Bill Wallace has seen the size of GM's battery-development group grow from less than 30 to more than 200.

Temperature extremes can diminish a battery's efficiency and rapidly accelerate battery aging, noted Frank Weber, Volt's enthusiastic and laser-focused Global Chief Engineer who departed the program last year for Opel.

"For example, the delta between 70°F (21°C) and 90°F (32°C) can be critical to battery life," he asserted. The battery is designed to work while plugged in, at temperatures from -13°F (-25°C) to +122°F (+50°C). The permitted temperature gradient within a battery cell, and from cell to cell, is 5 to 10 K.

A 50:50 glycol mixture is actively circulated through 144 metal "fins" between each of the Volt's 288 cells. The fins are 1-mm-thick (0.04-in) stamped aluminum plates that conduct heat. The Volt's pack has five thermal management circuits to handle the multiple subsystems. The system uses multiple electric coolant pumps (12- and 50-W) supplied by Buehler Motor of Germany. The pumps feature brushless dc motors and integrated electronics, and are designed to run extremely quietly, explained Robert Riedford, President of Buehler Motor Inc.

In cold weather, the battery is preheated during charging. Inputs from 16 temperature sensors are sent to a heating coil that warms the coolant, regulating individual cells' temperature, said Bill Wallace, GM's Director of Global Battery Systems.

GM engineers worked with heat-transfer-systems expert Behr America to develop Volt's front-end cooling module (FECM) as well as to introduce the industry's first use of "chiller" technology in a production electrified vehicle. (See sidebar, page 33.)

Volt's battery management system runs more than 500 diagnostics at 10 times per second, continuously monitoring the battery in real time. GM engineers said 85% of the diagnostics ensure the pack is operating safely, while the remaining 15% track battery performance and life.

GM engineers have performed more than 1 million mi (1.6 million km) and 4 million hours of validation testing on Volt's battery packs since 2007.

For the engineers, scientists, and chemists who developed Volt's battery pack, the project "was a giant leap of faith for everybody involved," said CPI's Patil. He said the key to the program meeting its November 1, 2010, start-of-production date—effectively launching a battery that didn't exist in 2007—was the teamwork among his company and GM.

"A window of 3½ years is tough enough to meet on a conventional vehicle program," he noted. "But trying to develop new technology and put it in a production vehicle...there were a lot of skeptics who said it would never work."

Battery development: "not much room for error"

Prabhakar Patil

A robust, reliable, high-performance Li-ion battery was critical for the Volt program's success. General Motors engineers say they could not have done it without their partnership with Compact Power Inc. (CPI), part of South Korean battery giant LG Chem. CPI's CEO, Prabhakar Patil, is no stranger to electrified vehicles, having led the Escape Hybrid program while an engineer at Ford. Here he recalls challenges of developing the "heart" of Volt.

Q: What was the Volt program's key technology takeaway for CPI?

Patil: Systems engineering. Often you cannot predict the level of interaction of subsystems until you put them together—putting the cells into modules and the modules into a pack, for example. So we had to rely quite a bit on analytical models, many of which we developed for this program. They were also critical for developing the liquid-cooled thermal management system. There simply wasn't enough time to do a build-and-test kind of approach.

The battery's calendar life had to be modeled, because we couldn't wait for 10 years of actual testing, over the required miles. We had to have 'acceleration factors' that gave us confidence that we could warranty the packs for 150,000 miles.

Analytical modeling of vehicle batteries is still evolving. The USABC [U.S. Automotive Battery Consortium, an industry group] played an important role with us because they have the benefit of looking across different manufacturers and technologies, and identifying common factors.

Q: What are the CPI cells' competitive advantages that clinched the Volt supply contract?

Patil: Ours was the only technology GM evaluated that has two layers of protection inside the cell. One layer is the manganese-spinel-based chemistry, which A123 also has. The other is the internal separator [between the anode and cathode], which is our innovation. Our other strong point is the flat-format prismatic cell architecture, whose failure mode is naturally benign. Its robustness from a manufacturability standpoint is inherently good.

Another asset was LG Chem's experience in making these cells in high volume. That's a track record you need when you have over 200 cells per pack. A pack is only as good as its weakest cell, so the OEM needs proof that you can deliver cells with extremely high consistency. Being part of LG Chem, our intimate knowledge of the cell characteristics allowed us to do the module and pack development quickly.

Q: What has the Volt program meant to you as an engineer?

Patil: It's been very challenging and gratifying at the same time. The program really pushed technology from many sides—at the cell level, getting power and energy together. It pushed packaging technologies. It pushed development of liquid cooling in an EV pack of this complexity. It pushed our use of analytical methods.

With so much riding on the Volt program, there was not much room for error. We delivered the first pack to GM on Halloween, 2007. It was developed in six months, and that same pack is still running in GM's battery lab.

In the early tests that battery's performance came in line-on-line to what GM predicted. Such longevity and performance are very gratifying for me as an engineer.

Voltec Battery Design and Manufacturing

2011-01-1360
Published
04/12/2011

Robert Parrish, Kanthasamy Elankumaran, Milind Gandhi, Bryan Nance, Patrick Meehan, Dave Milburn, Saif Siddiqui and Andrew Brenz
General Motors Company

doi:10.4271/2011-01-1360

ABSTRACT

In July 2007, GM announced that it would produce the Chevy Volt, the first high-production volume electric vehicle with extended range capability, by 2010. In January 2009, General Motors announced that the Chevrolet Volt's lithium ion Battery Pack, capable of propelling the Chevy Volt on battery-supplied electric power for up to 40 miles, would be designed and assembled in-house. The T-shaped battery, a subset of the Voltec propulsion system, comprises 288 cells, weighs 190 kg, and is capable of supplying over 16 kWh of energy. Many technical challenges presented themselves to the team, including the liquid thermal management of the battery, the fast battery pack development timeline, and validation of an unproven high-speed assembly process.

This paper will first present a general overview of the approach General Motors utilized to bring the various engineering organizations together to design, develop, and manufacture the Volt battery. Furthermore, aspects of battery design features will be provided with some greater depth. On a more technical discovery the battery system will be explored including an overview of the electrical and mechanical systems that comprise the assembly.

INTRODUCTION

The Rechargeable Energy Storage System or RESS is the fundamental battery device utilized by the Chevrolet Volt. The RESS evolved from proof of concept development, integration, manufacturing concept and finally a production ready design in just over two years. The lofty task was facilitated by a cross-functional team formed by tapping key expertise in systems engineering, controls, heat transfer, electronics, high voltage wiring, manufacturing, quality, purchasing and project management. Members of this team were pulled from GM Organizations around the world, including Mainz-Kastel, Germany, Honeoye Falls, New York and various other areas in Michigan and California. Much of the team was co-located to provide access to test / validation labs, personnel and hardware, facilitating significant consolidation of the project timeline.

The fundamental design of the RESS is broken into three key disciplines: Mechanical, Electrical and Chemical. This work is intended to provide the technical audience a broad view and understanding of these systems.

ORGANIZATIONAL STRUCTURE AND PRODUCT DEVELOPMENT PROCESS

To execute this ambitious plan, the RESS Assembly was strategically broken down into defined Product Development Teams (PDTs) which would take on the tasks required to design, develop, manufacture, test and validate the subsystems and components that comprise the RESS Assembly. PDTs were established for the following components and subsystems: cells, modules/sections, electrical hardware, electronics and structure. These PDTs maintained coordination and focus via an RESS System Integration Team (SIT) and RESS Integration Team (RIT). The SIT also monitored and managed issues with pack-level performance to requirements and imperatives such as mass, cost, quality and build ability. The RIT managed the physical (solid UniGraphics virtual model) integration of the RESS.

To realize a production ready RESS Assembly, a non-standard product development process, more akin to an advanced technology work study, was utilized early on. As the project progressed, this product development cycle took on the appearance of a highly time-compressed, multiple-iteration development program. Through the course of the program, five high-level RESS design progressions were utilized, from early supplier concepts defining the limits of the suppliers' capabilities, to the first production-intent designs, to the test-improved final production intent test properties, and finally culminating in the production approved parts manufactured in the GM production facility. Using an out-of-the-box, "can do" mentality, the Product Development Teams, Manufacturing and RESS Program Management were able to mow down obstacles and take extraordinary steps to bring the RESS Assembly to fruition.

MECHANICAL CONSTRUCTION

RESS mechanical components consist of the tray, cover, bulkhead [part of the Bulkhead and Battery Disconnect unit (BBDU)], Manual Service Disconnect (MSD) Plug (part of the MSD Assembly), hold-down brackets and isolators [Fig. 1]. Each component plays an integral part in satisfying the requirements of the complete housing package.

The battery tray was designed as the backbone of the RESS battery. The design needed to be structurally sound to be able to carry the 140kg weight of the battery sections, and to tie in as an integral part of the vehicle structure. After researching various materials for the tray, steel was chosen as the material due to its stiffness and strength advantages. Structural reinforcements on the tray were designed to increase the strength and stiffness of the tray [Fig. 2]. The BBDU, current sensing bracket, and hold-down brackets are fastened to the tray using nuts and weld studs on the tray. Attached to the bottom of the steel tray is an aero shield made out of a lightweight, reinforced thermoplastic that reduces aerodynamic drag created under the vehicle.

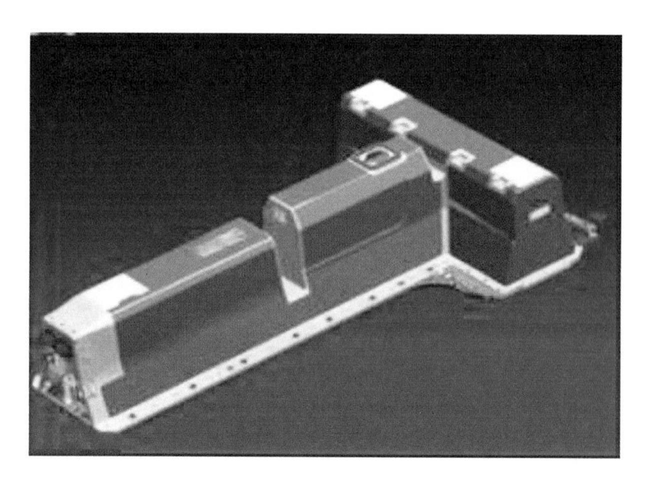

Figure 1. RESS battery assembly.

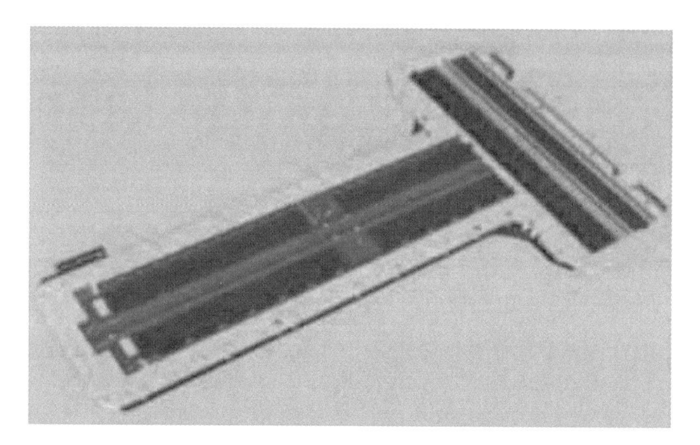

Figure 2. RESS tray assembly.

Rubber isolators are placed between the battery sections and the surface of the tray to increase the friction force between the sections and the tray. The additional friction force is needed to secure the battery and eliminate any movement of the battery during shock, vibration and impact. The isolators also absorb the dimensional variation of the sections.

The battery sections are secured on the tray by steel hold-down brackets. There are three short brackets that secure Sections Two and Three and a long bracket which secures Section One of the battery. The design of the bracket serves a dual purpose in locating the sections on the tray and securing them to the tray. Each hold-down bracket assembly [Fig. 3] consists of wedge blocks that are attached to the bracket. The nylon wedge blocks are installed into slots on the brackets. The oversized slots allow the wedge blocks to be positioned based on the location of the battery sections. The wedge block pushes the section into the stationary c-channel on the opposite side of the section on the tray. The lip of the bracket also captures the feet of the repeating frame and eliminates any z-direction (vertical) movement.

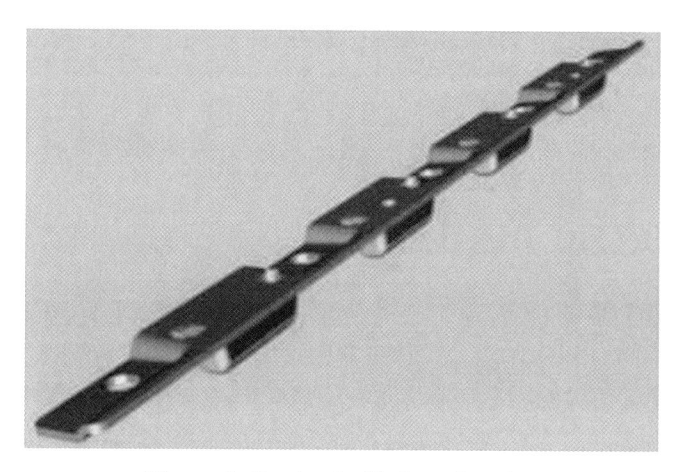

Figure 3. Section hold-down bracket.

The battery sections and the pack's internal components are enclosed by a cover [Fig. 4] made of a sheet molded compound. The cover's primary duty is to resist any foreign body or liquid from entering the pack. A silicone seal on the cover also seals the cover to the battery tray. The pack's exterior is covered by a heat reflecting aluminum foil (modified polyethylene copolymer film with scrim reinforcement). The need for insulation was driven by specific performance requirements for the exposure temperature for the lithium ion battery cells and was further defined by limited packaging space.

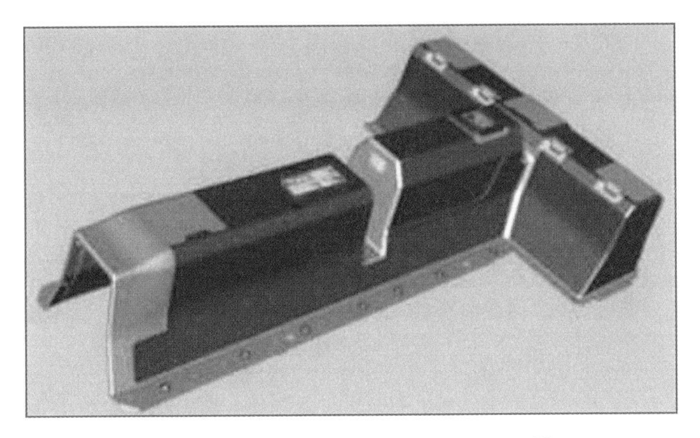

Figure 4. RESS battery cover assembly.

Similar to other complex systems, there are points of interface between expertise and teams. Specifically in the mechanical enclosure this is observed in the Bulkhead Battery Disconnect Unit (BBDU) and the Manual Service Disconnect (MSD). These parts, in addition to the cover and tray, combine to provide a sealed RESS unit defined by specific sealing requirement. Also from a mechanical perspective, the BBDU assembly provides a novel interface to vehicle high and low voltage in addition to fluid connections to the vehicle's heating and cooling system. Unlike the cover and tray, the BBDU and MSD also have very special electrical purposes, are discussed later in the electrical section of this work.

BATTERY SECTIONS AND SYSTEM COOLING

The battery pack has three battery cell sections connected by bus bars and coolant hoses. Each section is comprised of multiple modules [Fig.5]. There are two unique module sizes -- one module has 18 cells and the other 36 cells. Each module consists of two end plates with a thermistor, cooling fins (nine or 18), foam separators (eight or 17), and repeating frame elements (eight or 17) to hold the battery cells. Each module is connected to the adjacent module within a section with bus bars and structural bars.

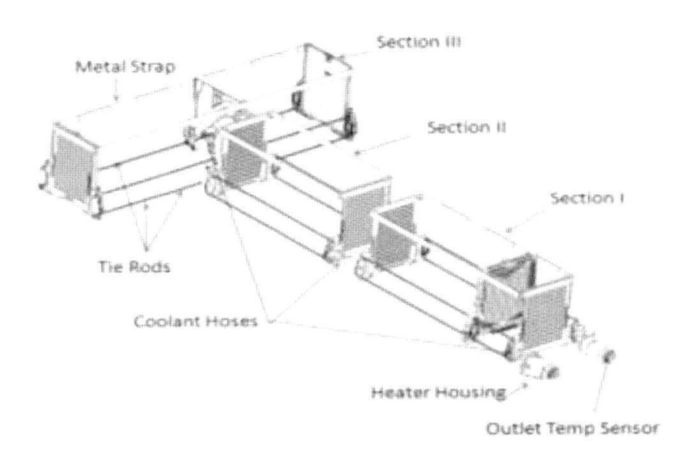

Figure 5. Fluid handling and section structure.

Structural end caps with hose bibs mark the ends of section. On the top of each module there is a molded plastic interconnect board which is plastic-welded to the module. The interconnect board contains sense leads and connections for the cells. The section is held together by an external metal strap on the top and four tie rods at the bottom. An injection molded cover is placed at the top of sections to cover the electronics and electrical connections.

The battery pack is conditioned (cooled and heated) with 50/50 de-ionized water and Glycol mix. The battery pack has a 1.8 kW heater to heat the battery in cold climates. The cooling loop outside the pack has a chiller to cool the battery in hot climates. The coolant is supplied to the battery pack via coolant hoses which pass through a heater module with an integral temperature sensor (housed in the BBDU) to measure coolant inlet temperature. Silicone hoses connect the heater to each of the sections. The hoses are clamped to the hose fittings with worm gear clamps.

A cooling fin with inlet and outlet ports is placed between each battery cell such that it is adjacent to one side of each cell. The cooling fin has a Mylar layer backing for electrical isolation. The cooling fin has nine equal length cooling channels that take the fluid from the inlet manifold to the outlet manifold, ensuring consistent heat transfer capacity across the face of each cell. The manifold is made up from many sets of elastomeric seals contained in each repeating frame element [Fig.6]. The narrow passages in the cooling fin and the contact surface area of the elastomeric seals demand a very tight cleanliness requirement to be maintained throughout the coolant system, both during manufacturing and in service. A serviceable filter is part of the cooling circuit which ensures the seamless operation of the cooling circuit in the face of contaminants.

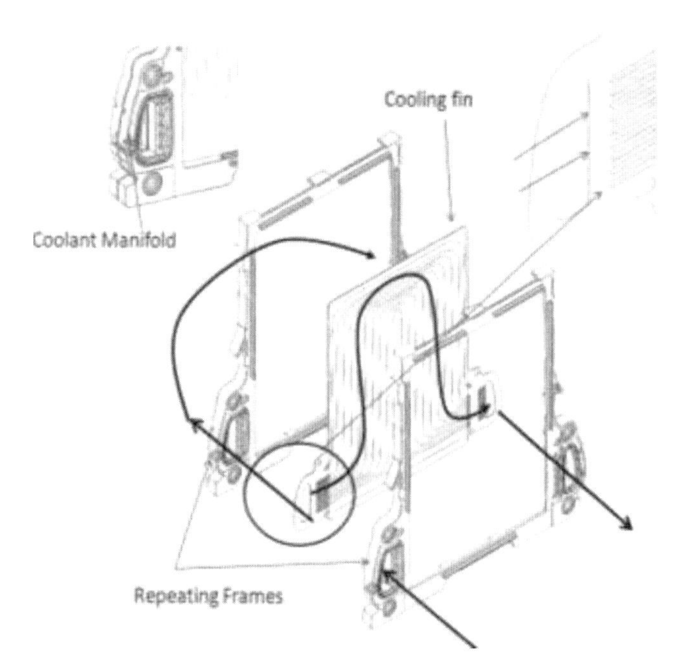

Figure 6. Repeating frame and cooling fin (typical).

The pressure loss across the cooling fin is an order of magnitude higher than the manifold pressure. This enables a uniform distribution of flow to each cooling fin. All the cooling fins are in parallel and the flow goes from the inlet to outlet manifold. This feature ensures a uniform temperature distribution of cells throughout the battery pack. The coolant in the outlet manifold then exits the pack through the BBDU. The coolant's exit temperature is measured via a temperature sensor at its point of exit.

TEMPERATURE SENSING

The RESS pack has one coolant inlet temperature sensor in the heater module and one outlet coolant temperature sensor, both part of the BBDU, that measure coolant inlet and outlet temperature. There is a thermistor at both ends of each module, of which 16 are monitored. All 18 of these thermistors are actively monitored during charging or use of the pack. The thermistors in the modules are in direct contact with cells. These provide the temperature of cells to the thermal control system, where the sensor in the heater and the outlet provide the coolant temperature to the thermal control system.

POWER DISTRIBUTION

The primary HV requirement for the RESS is delivery of a nominal 360VDC leg to leg with peak current of 400A for 30 seconds. Getting that power out of the individual cells requires a robust joining strategy of the battery cell tabs. This is achieved by ultrasonically joining the positive and negative tabs of three cells together in parallel to form a cell group, which is joined in the same joining process to an interconnect board (ICB) [Fig. 7]. The ICB takes the six or 12 three-cell groups per module and joins them in series to achieve the power requirement of the RESS. Three ICB layouts exist to accommodate two different module sizes and two connector orientations on the larger-size modules.

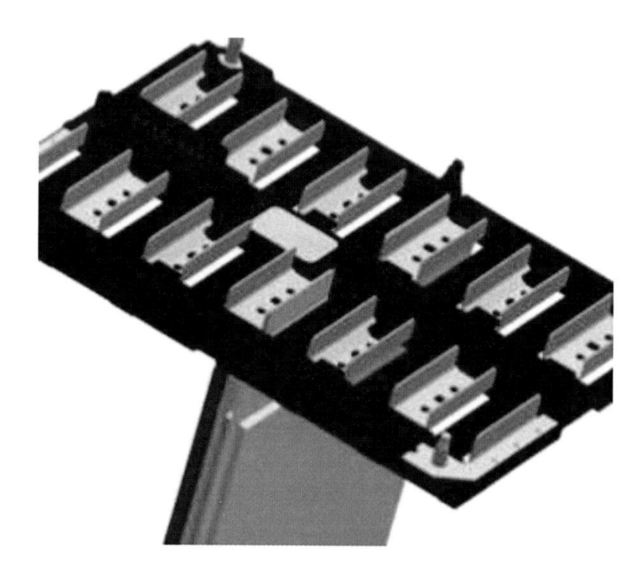

Figure 7. Inter-connect board (shown with one three-cell group attached)

In a given module, which will have either six or 12 groups of cells, the position of the anodes and cathodes of each group is reversed from group to group. The interconnect board ties the anode of one group of pouch cells to the cathode of the adjacent group and so on throughout the module. Every group has a separate connection to its anode for voltage sensing. These are routed to a connector on the ICB.

Two, three or four modules are tied together electrically to form a section using a nickel-plated, solid copper bus bar. The three sections in the Volt battery are configured as follows. Section One (forward in pack) uses one six-group (of three cells each) and two 12-group modules. Section Two (mid-pack) uses two 12-group modules. Section Three (rear of pack, cross-car) uses a six-group module and three 12-group modules. Each interconnect board contains threaded studs at the corners of one long side to accomplish intra-section and section-to-section connections.

Section-to-section electrical connections are accomplished using formed bus bars [Fig. 8]. Separate from the module-to-module bus bars, these bus bars consist of multiple copper strips that are encased in protective, orange insulation. The rigid, flat construction allows for tighter packaging when necessary. Compared to a $35mm^2$ round cable, a bus bar is 10.3mm wider but 4.7mm thinner and contains similar current-carrying ability.

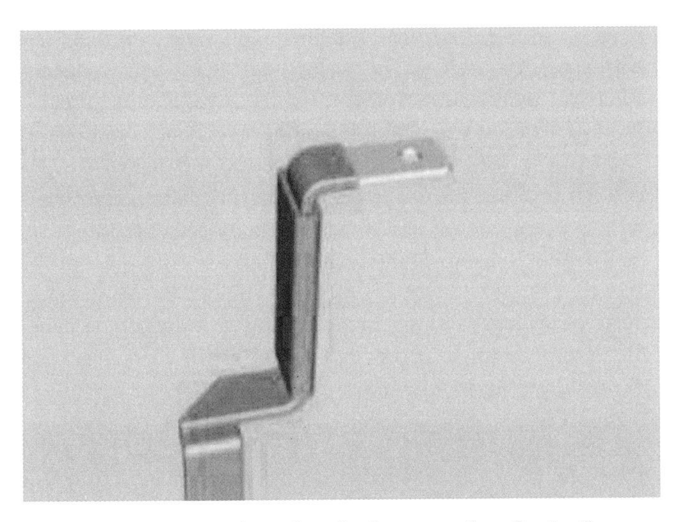

Figure 8. Bus bar electrical connection (typical).

Figure 9. Manual service disconnect and receptacle.

At the ends of each bus bar, the individual laminations are welded together and pierced with a 6mm hole for termination with a nut. The orange color of the insulation highlights that these are high-voltage components and that proper care should be taken when near them. Each bus bar is hi-pot tested to guarantee proper insulation coverage and isolation. There are four flexible bus bars in the pack that serve to connect the following components: BBDU positive terminal to Section One, Section One to Section Two, Section Two to Section Three, and Section Three to BBDU negative. Physically located on top of Section Two and electrically located between Section Two and Section Three is a circuit protection device called a Manual Service Disconnect (MSD). The MSD [Fig. 9] is accessible from the passenger compartment beneath the center storage console. It houses a 350A fuse and is designed to withstand peak current (400A for 30 seconds). If the MSD ever needs to be disconnected a user must disengage primary and secondary latches. When the primary latch is disengaged, two interlock circuits are broken which signals the vehicle's energy management subsystem to separate the battery from the vehicle electrical system by opening the main contactors. These interlocks prevent high-voltage terminals from ever being exposed and accessible while the battery is live (i.e. contactors closed).

The BBDU as described in the Mechanical section sits in front of the Volt battery pack. This is the device through which the battery's high voltage is routed [Fig. 10]. High-voltage connections on the BBDU include a large, shielded connection to the propulsion system, a fused, smaller gage connection to the vehicle's accessory power module (for conversion to 12V loads), and a similar connection to the on-board charging module.

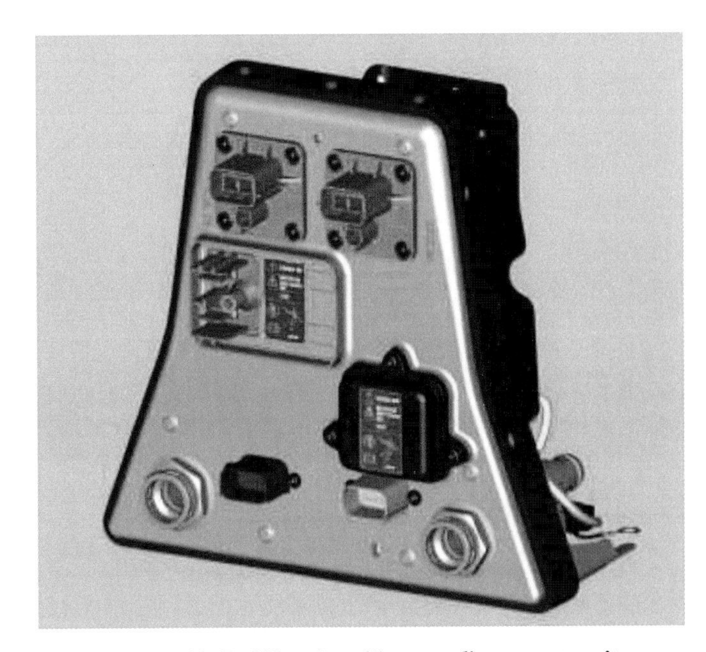

Figure 10. Bulkhead and battery disconnect unit.

Inside the BBDU are components that allow safe switching and distribution of pack energy. Upon starting the vehicle and preparing for propulsion mode, a pre-charge of the power train is necessary. This pre-charge process brings the capacitive power train load up to pack voltage without passing a large spike of current through the main contactors. If large spikes are passed through the main contactors, the contacts can spot weld and become permanently closed. A pre-charge resistor [Fig. 11] and FET (Field Effect Transistor) are used in conjunction with the main negative contactor and a specific multi-use contactor. With the main negative contactor engaged, the battery and power train are connected on their negative legs. The multi-use contactor is closed providing positive battery to the pre-charge resistor which is in series with the pre-charge FET. Once the FET is closed, the battery and power train are a complete circuit and in a fraction of a second, the power train voltage is brought up to pack voltage. The main positive contactor can then be engaged without fear of a large spike being passed through it. This pre-charge process requires delivering a lot of energy in a short time so the pre-charge resistor is sized appropriately. Although only a 35 Watt device, it is large enough to dissipate the heat that can be generated during repeated pre-charge attempts if they are ever necessary.

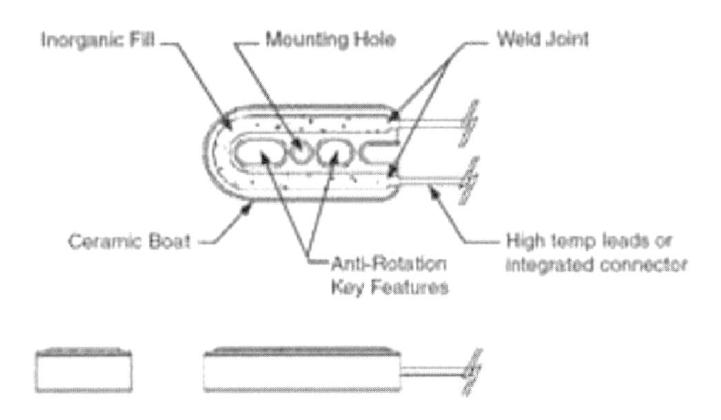

Figure 11. Pre-charge resistor.

Another function of the BBDU is to facilitate charging of the battery. When the battery is being charged, two other contactors are utilized. These are smaller and require lower current consumption than the main contactors. Again, the multi-use contactor is closed and the charger connection on the bulkhead feeds the battery while bypassing the main contactors. A fuse in series with the multi-use contactor protects the pre-charge elements as well as the heater function - the final high-voltage function of the BBDU.

To maintain proper pack temperature, a controller external to the pack sends a signal to the pack to exercise a heater contained in the coolant inlet port, also contained in the BBDU. This pulse-width modulated signal drives a specific heater FET which is in series with an 1800 Watt heating element. The FET is situated across the battery positive and negative legs and also requires that the multi-use relay be engaged.

LOW VOLTAGE ARCHITECTURE

Operation of the battery's high-voltage subsystem in a safe and reliable manner is dependent on a separate, low-voltage architecture. Measurements, control signals, interlocks and communications are the four main functions the low voltage subsystem provides.

As described in the Temperature Sensing section, there are thermistors inside sections that provide temperatures. These devices are electrically attached to traces molded into corresponding ICB's. Between the ICB and respective electrical module referred to as a Voltage Temperature Sub Module (VTSM) is a standalone harness that completes the circuit. The other measurement performed using the low-voltage architecture is overall current sensing. The sensor, which provides high and low resolution current readings, is a Hall-Effect sensor situated on the bus bar between Section One and Section Two [Fig. 12]. It supports measurement of positive or negative current depending on whether the pack is delivering power under propulsion mode or receiving power during charging or regenerative braking.

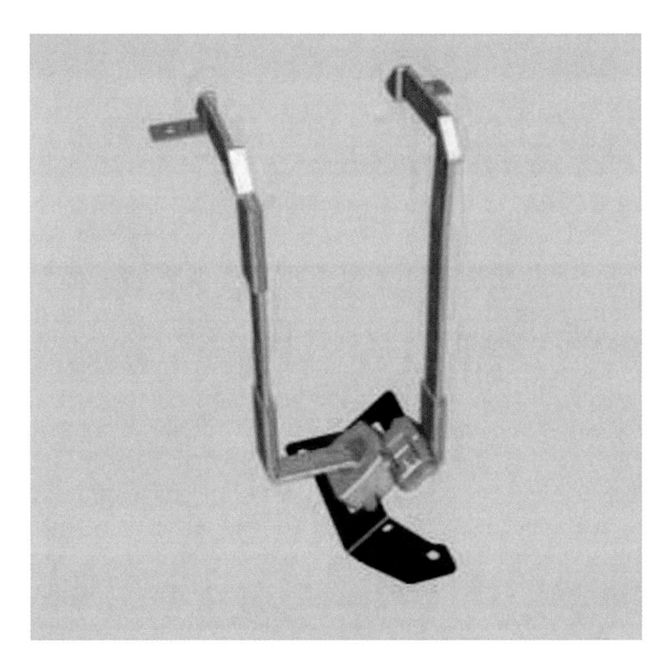

Figure 12. Hall-effect sensor and integral bus-bar assembly.

As mentioned before, interlock circuits run throughout the pack to signal to the contactor control subsystem that the contactors be opened. Two interlock circuits exist in the vehicle and are supported with appropriate connections within the battery. One of the interlock loops enters the auxiliary power module connection (at the top of the bulkhead) and the other enters at the charger connection next to it. Both loops route through the fuse cover on the front of the bulkhead, then to the MSD, and back to their respective connector. Therefore, the interlock loops are broken when 1) a high-voltage connector is removed, 2) the fuse cover on the front of the bulkhead is removed, or 3) the MSD is removed. These are all operations that could be performed during a normal service of a vehicle.

The contactors and FETs require external signals for operation. All five contactors and both FET controls are manipulated by an external controller using a high-side or low-side drive. Additionally, in the case of the heater control a PWM (Pulse Width Modulation) signal is used.

Communication of vital cell and health data from the pack is accomplished on two CAN-based (Controller Area Network) systems. Both networks utilize twisted pair wiring and follow General Motors' requirements for overall network length including stubs. The energy management bus provides data specific to battery functions and exists on a smaller, limited network. The other network communicates to the broader vehicle and provides diagnostic and information to the driver for example charge level of the battery based on cell open-circuit voltages. A redundant circuit provides an error signal to an external controller indicating a non-specific over voltage condition.

LITHIUM ION CELL

The battery cell used in the Chevy Volt has been designed for three main purposes. The first purpose is to store electrical energy from the vehicle high-voltage system and from the charger. The battery cells are capable of doing this by converting recharge and regeneration electrical energy into chemical energy. The second purpose is to provide electrical energy to the vehicle high-voltage system. The third purpose is to provide electrical power for propulsion of the vehicle, cranking of the internal combustion engine and operation of other vehicle components. The battery does these last two functions by converting stored chemical energy into electrical energy.

There are four different types of battery cell designs: cylindrical, button, prismatic, and pouch. General Motors chose to use a pouch cell [Fig. 13] for the Chevy Volt. The pouch cell has several advantages to the other battery designs, such as being more efficient in terms of mass and volume compared to the other designs.

The pouch cell design used by General Motors which is currently manufactured by LG in Ochang, Korea, and will also be manufactured in Holland, Michigan, can be described by its external and internal design. The external design of the battery cell includes a positive terminal made of aluminum and a negative terminal made of nickel plated copper. Both terminal tab leads were positioned on the same end of the battery. The internal portion of the cell is surrounded by a laminate aluminum packaging sealed on all four sides. Isolation tape was used on areas of the packaging that are in proximity to metallic module parts. Both terminals were designed to employ the use of insulating tape.

The inside of the battery has a positive and negative collector foil made of aluminum and copper, respectively. There is also positive and negative electrode film, electrolyte solution, and separator film inside this battery.

Figure 13. Cell packaging (typical).

There are 288 cells used in the battery pack. The pack is setup to use three cells in parallel, for a total of 96 sets of three parallel cells. The design of the pack has been configured to use the least number of cells and yet still achieve all safety and performance requirements of the pack, while at the same time minimize mass, packaging space and cost. Further advantage is gained through the optimized of cell count, thereby reducing complexity and improving cell balancing - a key battery performance metric.

Prior to the sale of the first Volt, thousands of cells have cumulatively run through millions of hours of testing so that the unique performance characteristics could be measured and validated versus the requirements. There are also industry standard safety requirements such as UN3480, for transporting lithium ion cells which were met. Due to the design of the battery pack which uses liquid coolant, some testing took place in the presence of this coolant to ensure that any presence of coolant would not have a negative effect on the cells. The long term durability testing used for these cells took into account, among other variables, a wide range of temperature, battery states of charge, and rates of charge and discharge at these different temperatures.

BATTERY ELECTRONICS

The monitoring electronics subsystem in the RESS is referred to as the Voltage Current Temperature Module (VITM) system. The RESS monitoring electronics primarily monitor battery internal parameters and report the data via serial data messages to an off-battery controller. The external module functions as the interface between the RESS and the vehicle, and houses the battery control algorithms, such as battery state estimator, contactor control, cell balancing algorithms, etc.

A semi-distributed architecture was selected for packaging the electronics in the RESS. This approach uses a central VITM for pack-level functions, and four distributed VTSMs for the cell voltage and temperature sensing functions distributed within the RESS [Fig. 14].

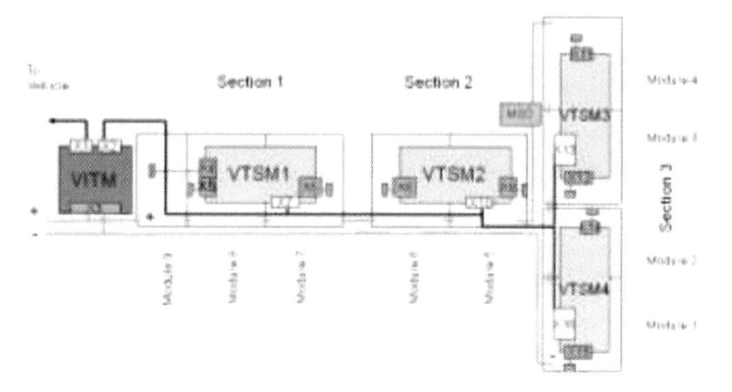

Figure 14. VITM system architecture.

The VITM is mounted near the front of the RESS for close proximity to the main contactors for pack voltage measurements, and to the RESS front bulkhead with the vehicle interface connectors.

The VITM measures the overall pack voltage with sense lines routed from the main contactors to the VITM. These high-voltage inputs to the VITM are electrically isolated from the rest of the 12 volt powered circuits in the VITM. The VITM measures the high-voltage bus current by reading analog outputs from an inductive current sensor, with one signal for full pack current range and a separate signal for more accurate readings of low currents. It also reads coolant temp sensors mounted at the RESS liquid coolant inlet and outlets. The VITM contains an active isolation detection system to monitor the isolation resistance between the high-voltage bus and chassis ground.

The VITM manages power moding by detecting three different hardware wakeup lines from the vehicle. It also supports the OBD-II diagnostic requirements for the measurement circuits, communications circuits, etc.

The VITM reports data to the vehicle over CAN. The VITM communicates on three separate CAN busses. One is internal to the RESS for communications between the VITM and the distributed VTSMs. Two are used for VITM communications to the vehicle; the High-voltage Energy Management Expansion Bus is the primary data path for normal functional messaging with the vehicle. This includes the sensor data read directly by the VITM, as well as data it has gathered from the VTSMs. The High Speed GMLAN bus is used for redundant messaging, for service modes, and module reprogramming.

The VITM has three electrical connectors; one for low voltage signals from the VITM to vehicle, which route through the BDU to connectors on the bulkhead at the front of the battery. One is for the RESS internal low voltage harness, which connects the VITM to the VTSMs, current sensor and coolant temperature sensors. One is for the high-voltage sense lines for pack level voltage measurements.

The four distributed VTSMs are mounted on modular sections of cells, with short jumper harnesses for signal lines to each module of cells monitored by that VTSM. This minimizes the sense lead wiring lengths and proximity to chassis ground. Also, the sense leads from modules at higher or lower voltages in the in the pack string are not in close proximity to each other. The VTSM measurement circuits are powered by the groups of cells they are monitoring. The serial data is then communicated across an isolation barrier to the pack internal RESS CAN for communications to the VITM.

The VTSMs each measure the cell voltages for the cells in the sections they are mounted too. Cells are grouped with three in parallel, each of the 96 three-cell groups are monitored. Due to the different numbers of cell groups in the sections, there are three configurations of VTSMs. One has 30 input channels, one has 24 and is used in two locations, and one has 18 channels. The VTSMs also measure the cell module temperatures by reading thermistors mounted in the module end frames.

Besides the primary measurement path for cell voltage monitoring, there is a redundant system that monitors for cell overvoltage, and drives a dedicated hardware signal to the external battery controller.

The VTSMs provide dissipative cell balancing for each of the 96 cell groups. The cell balancing algorithm runs in the external control module, which can command individual cell balance gates in the VTSMs to turn on via serial data messages sent to the VITM.

The modules include diagnostic systems that continuously monitor for faults. VTSM diagnostics include monitoring for open cell sense lines, inputs that are out of range (low or high), failed cell balance gates or other internal faults.

MANUFACTURING SYSTEMS

The battery assembly process is a combination of conventional automotive assembly techniques and processes more commonly found in the medical and electrical industries. The reason for this is the varying degrees of precision and speed in order to complete the assembly of the battery. By typical throughput standards, these numbers are analogous to what would be found in a low-volume assembly facility, but due to the quantities of repeating elements such as cells, frames, and cooling fins that are found in a pack, the upstream cycle times must be shorter.

The battery assembly is divided into three main areas based on the types of operations and required cycle times. The first area is the section subassembly line, which assembles all of the repeating elements into the three sections of the pack. The second area is the section final assembly area in which the electrical connections of the section are performed as well as certain performance checks. The third area is the pack main line. In this area the sections are assembled on the tray, and the final thermal and electrical connections are made. The cover is installed to the tray completing the assembly of the RESS Assembly. Final functional checks are performed before the pack is loaded to a carrier for delivery to Detroit-Hamtramck.

The section subassembly line runs at a relatively high rate due to the overall number of parts required to build each section. Prior to assembling the cells into the sections, several quality checks are performed. Additional cell preparation tasks are performed to enable electrical integration. Subsequent to these tasks, the cells are assembled to the frames and thermal management components in sequential order. At the end of the line, non-repeating components are assembled which maintain the dimensional integrity of the section and integrate the thermal management system of the pack.

In the section final assembly area, the inter-connect boards are installed, and the section-level high-voltage connections are made including cell tab joining. A quality check of the section is performed and the cover is installed prior to the fully-assembled section being delivered to the pack main line.

The pack main line is the final area of the plant. In this area, the sections are assembled to the tray and integrated mechanically, thermally, and electrically. The first step is taking the sections from the section assembly conveyor and installing and securing them to the tray. Hoses are installed and clamped to the sections and high-voltage bus bars and low-voltage harnesses are connected. Upon the completion of these assembly steps, the battery undergoes a series of electrical and functional tests. The pack cover is installed, and the fully-assembled battery is set in a dimensional check fixture as the final inspection.

CONCLUSIONS

Requirements for the Chevy Volt's RESS were established very early on in the development cycle, while the Volt was initially being announced. Given the revolutionary nature of the concept and the vast learning garnered through the multiple preplanned iterative test and development cycles, these initial requirements matured considerably. In spite of the ultra-short development cycle time, the product design and manufacturing strategy made quantum leaps forward to match these changes in product requirements, all-the-while ensuring these changes are fully validated to ensure not only that the product meets all requirements, but that it is of extremely high quality, reliability and durability. The RESS that has been described herein is the culmination of efforts by the consolidated brainpower of the multi-faceted GM Team and its suppliers. Time will prove how prescient the concept of the Chevy Volt truly is, aided in no short order by how well this vision, and specifically the RESS has been executed.

CONTACT INFORMATION

Robert (Rob) Parrish, P.E.
robert.1.parrish@gm.com

Patrick Meehan
patrick.meehan(@gm.com

Saif Siddiqui
saif.siddiqui(@gm.com

ACKNOWLEDGMENTS

The authors of this paper would like to thank the leadership of General Motors Battery Systems Engineering and Manufacturing for the opportunity to have a place to technically participate in the re-invention of automotive propulsion during a very special time in history for the United States automakers. Moreover, we appreciate and thank our families for their understanding and sacrifice at times to provide dedicated long hours and weekends that made this system a success.

The Manufacturing Systems portion of this material is based upon work supported by the Department of Energy under Award Number DE-EE0002217.

DISCLAIMER

The Manufacturing Systems portion of this report was prepared as an account of work sponsored by an agency of the United States Government. Neither the United States Government nor any agency thereof, nor any of their employees, makes any warranty, express or implied, or assumes any legal liability or responsibility for the accuracy, completeness, or usefulness of any information, apparatus, product, or process disclosed, or represents that its use would not infringe privately owned rights. Reference herein to any specific commercial product, process, or service by trade name, trademark, manufacturer, or otherwise does not necessarily constitute or imply its endorsement, recommendation, or favoring by the United States

Government or any agency thereof. The views and opinions of authors expressed herein do not necessarily state or reflect those of the United States Government or any agency thereof.

DEFINITIONS/ABBREVIATIONS

RESS
Rechargeable Energy Storage System

PDT
Product Development Team

RIT
RESS Integration Team

SIT
System Integration Team

BBDU
Battery Bulkhead Disconnect Unit

MSD
Manual Service Disconnect

HV
High Voltage

LV
Low Voltage

ICB
Inter-Connect Board

FET
Field Effect Transistor

CAN
Controller Area Network

PWM
Pulse Width Modulation

VITM
Voltage / Current / Temperature Module

VTSM
Voltage / Temperature Sub Module

AGC
Automated Guided Carriers

BAP
Battery Assembly Plant

The Engineering Meetings Board has approved this paper for publication. It has successfully completed SAE's peer review process under the supervision of the session organizer. This process requires a minimum of three (3) reviews by industry experts.

ISSN 0148-7191

Positions and opinions advanced in this paper are those of the author(s) and not necessarily those of SAE. The author is solely responsible for the content of the paper.

SAE Customer Service:
Tel: 877-606-7323 (inside USA and Canada)
Tel: 724-776-4970 (outside USA)
Fax: 724-776-0790
Email: CustomerService@sae.org
SAE Web Address: http://www.sae.org
Printed in USA

VOLTEC Battery System for Electric Vehicle with Extended Range

2011-01-1373
Published
04/12/2011

Roland Matthe and Lance Turner
General Motors Company

Horst Mettlach
Adam Opel AG - General Motors Company

doi:10.4271/2011-01-1373

ABSTRACT

Mid 2006 a study group at General Motors developed the concept for the electric vehicle with extended range (EREV),. The electric propulsion system should receive the electrical energy from a rechargeable energy storage system (RESS) and/or an auxiliary power unit (APU) which could either be a hydrogen fuel cell or an internal combustion engine (ICE) driven generator. The study result was the Chevrolet VOLT concept car in the North American Auto Show in Detroit in 2007.

The paper describes the requirements, concepts, development and the performance of the battery used as RESS for the ICE type VOLTEC propulsion system version of the Chevrolet Volt.

The key requirement for the RESS is to provide energy to drive an electric vehicle with "no compromised performance" for 40 miles. Extended Range Mode allows for this experience to continue beyond 40 miles. Multiple factors helped refine a requirement of at least 8 kWh usable energy, and 115 kW discharge power over the applied battery state of charge range. The Chevrolet Volt vehicle is based on GM's global compact vehicle platform. Aggressive targets for mass, volume, and timing have been considered for impact beyond start of production (4Q2010).

A battery cell providing both, very high energy density and high power density at the same time had to be developed and validated applying the latest Li-Ion technology.

Integration into the car should allow for good aerodynamics, provide the best crash protection and have low impact on customer useable space. The battery must also be able to perform in all typical automotive atmospheric conditions. An inter-cell thermal system was sized and balanced, to efficiently manage temperatures within the battery and help lengthen battery life.

New tests and methods had been developed for battery systems development in the lab, in the vehicle and in models. Data and examples will be shared. Specific vehicle/battery test activities will be introduced. Finally performance results demonstrating the characteristic(s) of the system will be shown.

INTRODUCTION

Electric Vehicle with extended range a new vehicle class

A vehicle that functions as a full-capability electric vehicle when energy is available from a rechargeable energy storage system (RESS), e.g. a high voltage battery and having an auxiliary energy supply that is only engaged when the battery energy is not available is called Electric Vehicle with extended range (EREV).

Figure 1. Chevrolet Volt Concept car 2007

An EREV can travel the initial range electrically with energy from the "plug-in" charged battery and additional range based on a second fuel (e.g. Gasoline, E 85, Diesel, Hydrogen, CNG) converted by an engine generator set or a fuel cell system. The first operation phase is called electric vehicle mode and represents charge depletion for the electrical energy storage, the second phase is called extended range mode and the energy storage system is in charge sustaining mode. In both modes the vehicle performance shall be equal so the vehicle dynamics are not influenced by the mode transition. In the EREV concept the RESS needs to provide the full power for transient operation in charge sustaining and charge depletion mode. The vehicle has to contain two energy storage systems a rechargeable energy storage system and a system for the "second fuel" which could be for example gasoline, Bio-ethanol, Diesel, Hydrogen or natural gas. The RESS provides energy in the charge depletion phase and to support acceleration and regenerative braking in the charge sustaining mode. The vehicle requires a capability to be recharged from an electricity source outside of the vehicle through plug-in. An on-board charger allows the use of the existing electrical infrastructure providing 100V to 240 V outlets in households of most countries.

The intention of the new propulsion system is to generate the maximum miles driven electrically to replace petroleum based energy (gasoline, Diesel) through electric energy, which should be generated from different sources and from renewable energy eventually. Most vehicles are operated less than 64 km (40 miles) per day in the USA and other parts of the world (Europe, Asia), but in many cases longer distances have to be travelled requiring driving range for a single drive of more than 200 km (124 miles) before the driver and the passengers need to stop for human needs. The idea had been, to come up with a concept which covers more than 75% of all driving overall with electric energy and maximize the replacement of petroleum based fuel for cars.

MAIN SECTION

Development of Performance Requirements

When the Chevrolet Volt concept had been developed in the year 2006 a number of key vehicle parameters had been established. Those drove the requirements of the rechargeable energy storage system. The vehicle was intended to set a new standard for electric vehicles in regard to acceleration, durability and customer value. A vehicle that shall meet the needs most drivers, is fun to drive, is energy efficient and very attractive. A vehicle which provides access to alternative energy sources and options to deal with increasing uncertainty of affordable supply of gasoline. A vehicle making a step toward sustainable individual transportation.

Vehicle Range

Studies of General Motors based on the real customer usage information provided by the Federal Highway Association and the Southern California Association of Governments (SCAG) -Regional Travel Survey (RTS)[1] showed that 80% of the petroleum can be replace by vehicles capable of 40 miles electric range. In addition sufficient energy (second source) for longer trips should be stored in the vehicle to make the vehicle usable for all trips. 300 miles for an extended range trip followed by the option to refuel the car in 3 minutes or recharge it with electric power make the vehicle capable to be the single car for every driver.

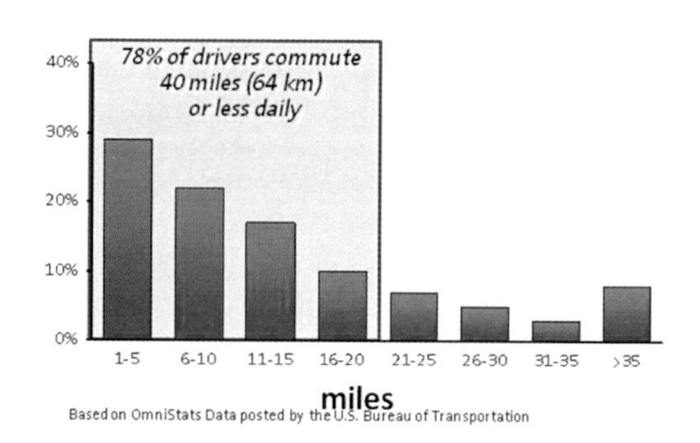

Graph 1. Typical one-way miles from home to work

Energy (Electric)

The amount of energy stored depends on the range required (km or miles) and the energy consumption (Wh/km). The energy consumption of a vehicle is dependent on vehicle characteristics (Drag resistance, rolling resistance, mass, propulsion efficiency), the driving style (acceleration, speed, profile), the ambient conditions (temperature, sunshine, rain) and the topographic profile. Driving Style can be represented by drive cycles, for example the EPA (Environmental Protection Agency) urban-, highway- or US06 drive cycle. The urban drive cycle with an average speed of 32 km/h and a top speed of 90 km/h represents the low end of performance and energy use for an electric vehicle, the US06 drive cycle with an average speed and 77 km/h and a top speed of 129.2 km/h represents a faster, more dynamic driving style requiring more power and more energy use per mile. The Urban drive cycle represents only a few percent of all driving recorded in the regional travel survey of the SCAG. The power required for the US 06 cycle would cover more than 90% of all drives in the study.

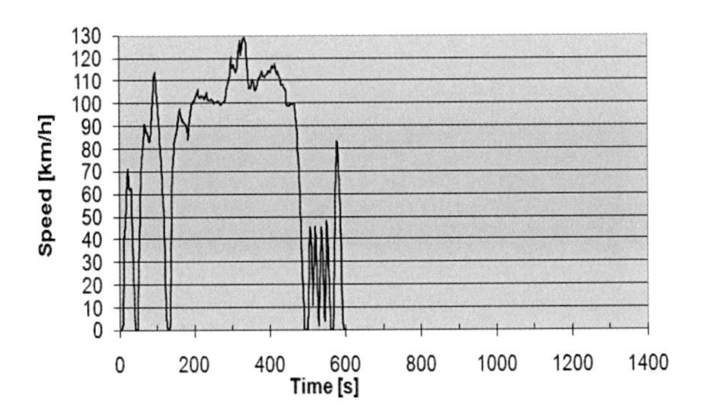

Graph 2. US06 drive cycle: Speed versus time

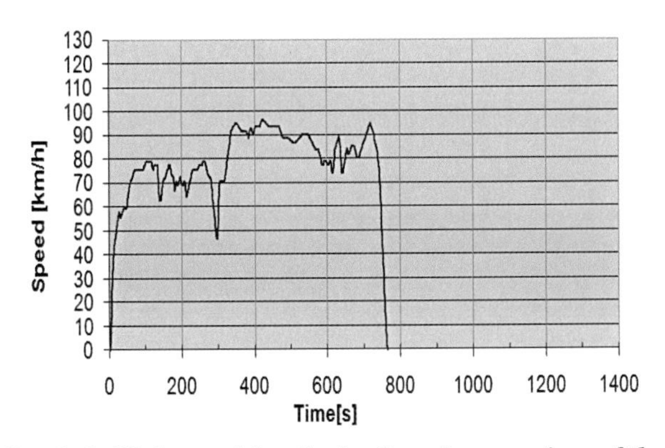

Graph 3. Highway drive Cycle: Speed versus time of the

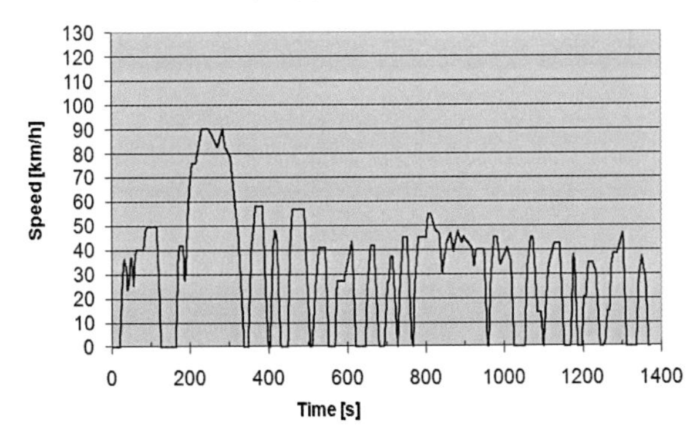

Graph 4. Urban drive cycle: Speed versus time

The vehicle requirement had been defined as 40 miles (64 km) electric range in urban driving style. To develop the specifications toward energy for the rechargeable energy storage systems, simulations of the EPA drive cycles and the preliminary Volt vehicle parameters had been performed. The urban drive cycles net 120 Wh/km energy consumption. And the US06 drive cycle required 180 Wh/km. To enable 64 km of driving in the urban drive cycle 8 kWh of energy was needed. Advance Lithium Ion battery systems can achieve energy densities of up to 100 Wh/kg, which is higher than most other storage systems such as electrochemical capacitors. The capacity and energy of a battery will decrease over time as well as through charge and discharge cycles.

Sizing for the total energy of the battery system would depend on 3 factors:

- Useable energy at end of life
- Power requirements over the operating state of charge (SoC) range
- Battery temperature during driving, charging and "off-time"

Considering these effects, and balancing it versus the cost, mass and volume demands, the total begin of life energy requirement of 16 kWh was resolved. This nominal energy was defined for normal conditions (23°C, 1 hour discharge rate C/1)

Range ("second fuel" gasoline)

The vehicle has to be capable to drive more than 480 km (300 miles) in extended range mode until refueling the storage system, e.g. gasoline tank. The battery is affected as it still has to provide power for acceleration as well as recapture electric energy during deceleration ("regenerative braking"). Energy is still required to supplement the generator power, but demand is lower than the full propulsion power. In the extended mode, as multiple accelerations are addressed, equivalent charge by the generator or regenerative braking is induced to sustain driving performance. This energy requirement is dependent on the operation strategy of the engine-generator drive unit, which over longer periods would define an energy reserve of 300 to 600 Wh.

Vehicle Performance

Appealing performance had been an early attribute of the Chevrolet Volt, expressed by a targeted acceleration time of less than 9 seconds from 0 km/h to 97 km/h (0 to 60 miles per hour). Passing performance measured as the acceleration time from 80 km/h to 120 km/h (50 mph to 75 mph), this is referenced as elasticity, and is an important metric in Europe. Other vehicle performance attributes that were dependent upon battery power were top speed and speed at grade. The vehicle has to be able to accelerate to 160 km/h (99 mph) and sustain this velocity continuously. For a grade of 6%, the speed of 112 km/h (7 mph) should be maintained over several km's or miles. This emulates ascent of interstate highways in mountainous areas.

The Chevrolet Volt has to be able to perform the US06 drive schedule in charge depletion and charge sustaining modes. It must be capable of Autobahn environments, where it is typical to drive up to 160 km/h. A recorded speed trace from typical German Autobahn test drive is shown in figure 5.

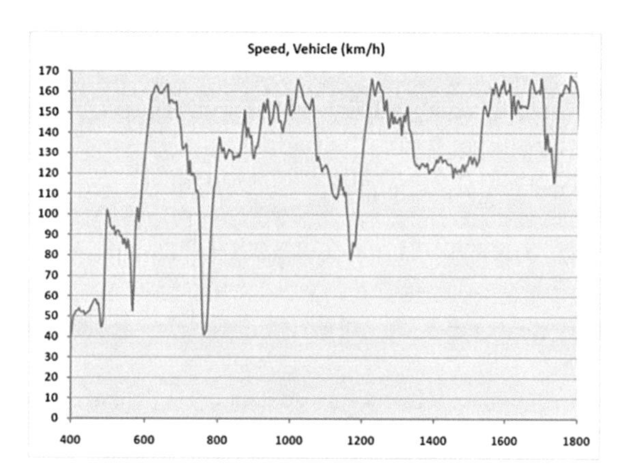

Graph 5. Electric vehicle on the German Autobahn: Speed(km/h) over time (s)

It may not be obvious that the speed trace is influenced by traffic, speed limit and road layout. In addition the topographic profile, not shown, will impact the power consumption. Comparison of the 3 emission based drive cycles and the recorded Autobahn data, show that a system has to meet real world power requirements, in addition to the emission based drive cycles. The root mean square value of the power (P rms) is a measure for the usage of the battery and its relation to the defined schedule.

Table 1. US emission cycles in comparison, in addition real drive data.

	Duration	Distance	Speed	Distance	Speed	CD P rms	CS P rms
	s	km	km/h	Miles	mph	kW	kW
US06	600 s	13 km	77 km/h	8.02	48.1	27.4 kW	24.9 kW
Hwy	765 s	16 km	76 km/h	10.1	47.5	13.0 kW	8.6 kW
Urban	1372 s	12 km	32 km/h	7.5	19.7	9.0 kW	9.8 kW
Autobahn example	600 s	19 km	114 km/h	11.8	70.7	37.7 kW	24.5 kW

Battery Discharge Power

The urban drive cycle based power profile does only require up to 40 kW peak discharge power.

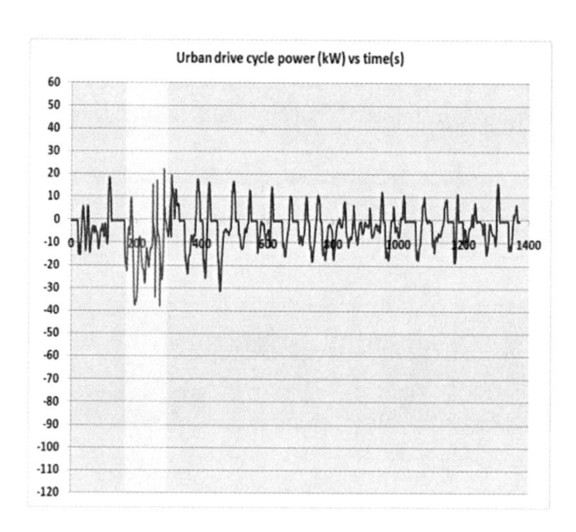

Graph 6. Urban Drive Cycle: Power (kW) over time (s)

To perform the US06 drive cycle, in charge depletion and sustaining mode, and achieve acceptable vehicle acceleration, the battery peak discharge power was established as 115 kW.

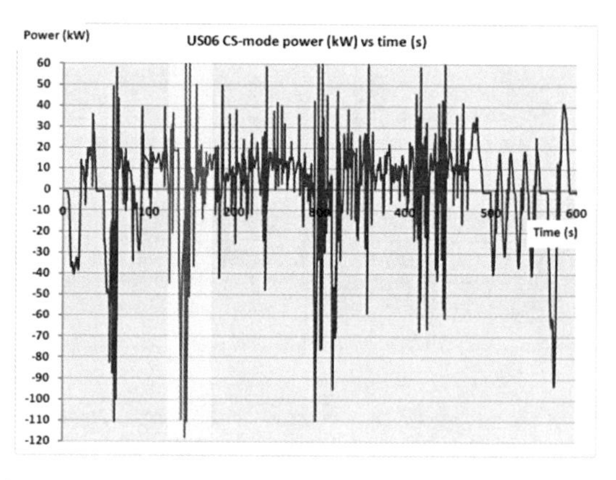

Graph 7. US06 drive cycle: battery power over time

The Autobahn power trace shows a higher power demand, and longer durations, but still satisfied by previously defined requirements.

As a reference point, performance shall be met when the battery is at temperatures above 20°C. Based on anticipated behavior in other ambient conditions, for instance, 10°C, 0°C and −25°C, the team was able to constrain the selection of the battery chemistry and system design.

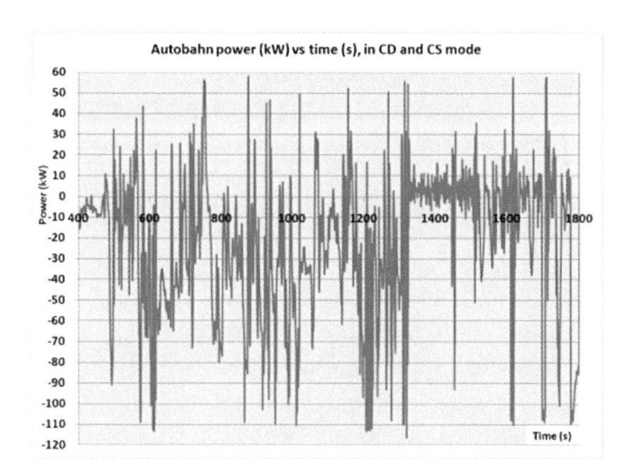

Graph 8. Electric vehicle Autobahn in Germany, power profile in CD and CS mode.

Battery Charge Power (plug-in)

Energy from the electric grid shall be used as much as possible. The vehicle integrated on-board charge module (OBCM) is designed for level 1 (120V AC single phase 12 A) and level 2 (240V AC 15 A) charging. The resulting wall consumption could ramp up to either 1.2 kW or 3.2 kW as DC charge power demands are directed by on board electronics. The DC power, however, will depend on the actual input AC voltage. Charge duration can vary from 3.5 hours up to 10 hours, depending on AC supply voltage, auxiliary power use, temperature, and beginning SOC. The battery charge rate can vary from 0.2 C down to 0.05 C rate (C-rate = nominal capacity /1 hour).

Battery charge power (Regenerative)

During regenerative braking charge power of up to 60 kW can be directed to the battery. This represents a charge rate of almost 4 C. Longer downhill braking and may see transient operation of up to 60 kW shall be captured for up to 10 seconds when the battery temperature is above 20°C. During Charge Sustaining operation the battery has to be able to be charged with 60 kWh and provide up to 115 kW discharge power.

Electrical Interface

The Battery system would have separated high voltage (HV) and low voltage (LV) interfaces. All high voltage connections include a high voltage interlock loop (HVIL), which will induce contactors to open when pack is being serviced.

High voltage interfaces are:

- Connector for traction inverter
    - Maximum current 400 A for 30 sec
    - Maximum continuous current I rms 135 A for 60 minutes
    - Maximum Voltage compatibility 420 V
- Connector for on-board charger module
- Connector for auxiliary power module (APM) Manual service disconnect, which allows manual intervention of the series string of cells, within the battery.

Low voltage connectors include:

- 12 V power for battery sensors and controllers
- 12 V power for contactors, wakeup and HVIL
- CAN-bus for battery controls

Life

Due to the high value of the battery system, the design intent was for a 10 year life and 240 000 km (150 000 miles) in vehicle performance is required. Batteries have the characteristic of degradation over discharge/charge cycles and time. The capacity degradation may have an effect on range. Over time, the charge sustaining range could decline, but the generator can compensate this effect for the driver. The internal resistance could also increase over time. Exposure to this effect can be avoided by over sizing the battery, so that at beginning of life, battery provides more than adequate power that is governed. As pack approaches its end of life, constraints are relaxed to provide sufficient power. Degradation is also accelerated at elevated temperature. Over the vehicle life, one might expect, more than 5000 full charge depletion cycles and additional charge sustaining. The conservative assumption for battery usage is 80 % of distance driven in charge depletion mode and 20 % in charge sustaining. The actual cycle number can vary as electric range can vary between 40 km and 80 km (25 and 50 miles) as a result of driving style and ambient temperature conditions.

Automotive Requirements

Operating Temperature

Automotive systems have to operate over a wide temperature range. The battery system would need to meet the US06 driving cycle performance from 0°C to 32°C battery temperature and still be operational from −25°C to 52°C ambient temperature. Exposure or storage shall be possible in the range of −40°C to 55°C. This impacts the requirements of all sub-components like the electronic control systems, electrical/mechanical actuators such as main contactors, and the housing of all RESS components. The system would have to perform over many years in areas similar to Phoenix or in Detroit. Phoenix represents a challenge as the high temperatures and sun radiation can drive battery temperatures into degradation ranges. In Detroit, over a long period, the temperatures could fall below 0°C. At low temperatures the system power can be reduced and battery life preserved to avoid lithium plating.

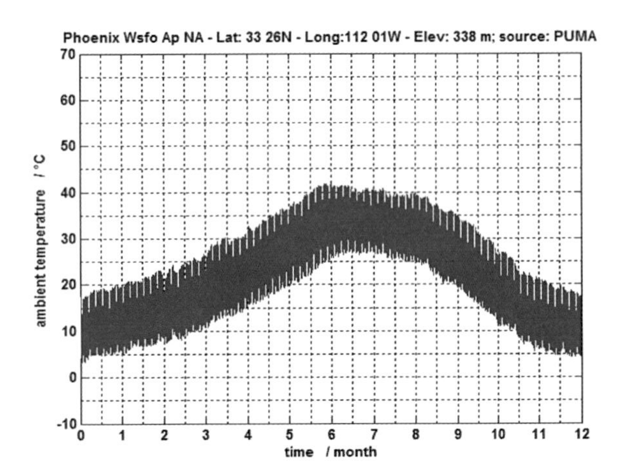

Graph 9. Average Temperatures for Phoenix, Arizona

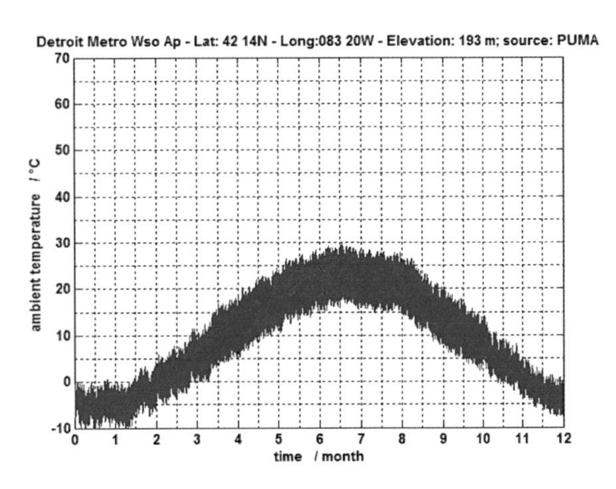

Graph 10. Average Temperatures for Detroit, Michigan

Altitude

Driving in high altitude requires the battery system to be able to operate down to 59 kPa ambient pressure. Transition back to normal ambient (102 kPa), within 15 minutes, must also be achievable, without effect on the battery sealing integrity.

Water (flooded road, Carwash)

The vehicle has to be capable of driving in all weather and road conditions. This includes the flooded road and car wash with underbody spray. To meet this vehicle requirement the system has perform under standards close to the IEC class of IP6K9K for water protection.

Vibration & Shock

Vibration profiles and mechanical shock exposure caused by road conditions and installation conditions in vehicle have been modeled. The battery system has to perform worst case vibration profiles without degradation.

Vehicle Integration (Physical)

Mass & Volume

As in most electric cars, battery system is a major contributor to vehicle mass. To enable the Chevrolet VOLT to evaluate its efficiencies and support vehicle dynamics, the initial mass target for battery system had been set to 160 kg. As the battery structure became more of a vehicle structural element, the equivalent part weights were integrated into the design target numbers and revised to 180 kg later to 190 kg. The vehicle concept of the Chevrolet VOLT required ample space for four persons and a large luggage space. The battery pack should be designed as compact as possible. To meet these requirements the battery pack system a volume target given was 140 dm^3.

Crash

The vehicle would also perform all FMVSS and ECE crash tests. Before, during and after these barrier events, the battery system should not expose high voltage or gaseous emissions into the occupant cabin. The team had decided the most optimal battery location was to be outside of crushable zones. Centered on vehicle, fore/aft and side to side, pack was also to be loaded from under vehicle below the cabin floor. Due to this packaging, an additional layer of separation between battery system, passenger compartment, and underbody components was developed in the form of a cover.

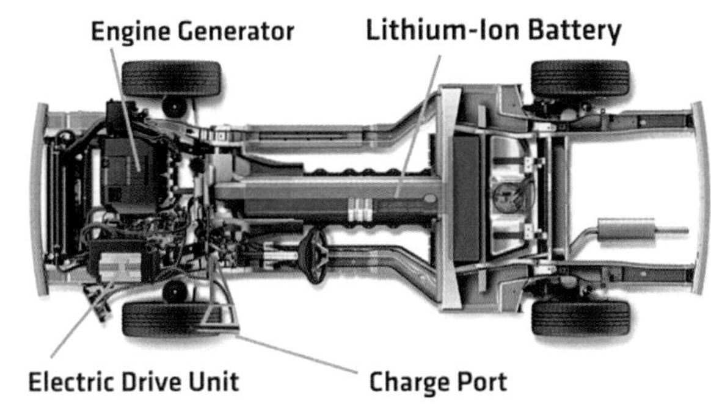

Figure 2. Position of VOLTEC Battery System in Chassis

Vehicle Assembly

The battery system shall be installed at one station in a normal GM vehicle assembly line. It is required to fulfill every assembly step within 60 seconds.

Figure 3. Chevrolet Volt on assembly line

Defining the solution

Cell technology down select base

As the requirements for the Volt battery system are quite demanding, it was quite clear that from the cell chemistries available today, only the lithium-ion technology would have a chance to satisfy the demand. Nearly all available lithium-ion combinations were evaluated during multiple workshops, resulting in a field of 16 different suppliers. The majority of cells had been based on a graphite anode. Only a few were applying lithium titanate as active anode material. On the cathode side, Li-NiMn, LiFeP, Li-Mn based, Li-Mn-spinel, Li-NCA (Nickel Cobalt-Aluminum) and Li-NCM (Nickel-Cobalt-Manganese) were considered. Table 1 shows the different anode and cathode active materials that were subject to investigation.

Table 2. Active materials evaluated

Anode material	Cathode material
Graphite	Li-NiMn
Hard carbon	Li-FeP
MCMB	Li-Mn-based
$Li_4Ti_5O_{12}$	Li-Mn-spinel
	Li-NCA
	Li-NMC
	Li-CO_2

The combination of the different anode and cathode material provided a large choice of chemistries making it quite difficult to select the best suited technology, to go forward with, for the Chevrolet Volts battery system. To support the down select and follow up evaluations, the different chemistries proposals were subjected to a Pugh matrix.

Criteria established for existing technologies

A Pugh matrix with the following categories was applied for the selection of the most suitable battery proposals:

Safety is priority

For the Volt battery system, safety is and was a priority. Therefore, the abuse tolerance rankings were a very important criterion for the battery selection. The cathode material plays a significant roll toward influencing cell behavior when subjected to abuse conditions. The properties associated to different cathode materials are shown in figure 1 which depicts the results of a differential scanning calorimetry (DSC).

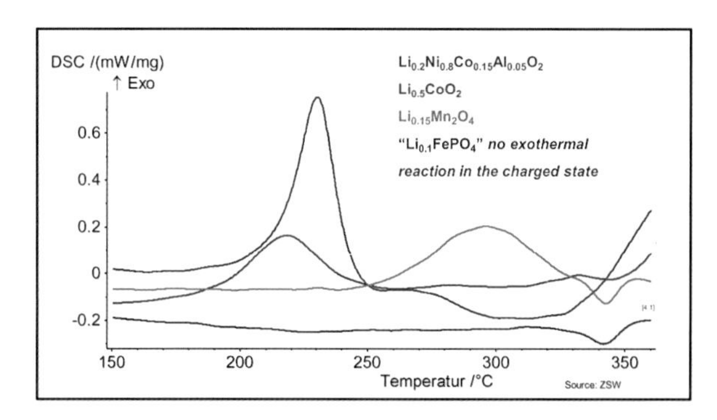

Graph 11. DSC results of different cathodes

The Li-Co_2 and the Li-NCA cathode show a strong exothermal reaction starting at around 180°C or 200°C, while Li-Mn-spinel is stable until around 250°C. Note Li-FeP shows a lack exothermal reaction when exposed to maximum tested temperature of 350°C. LFP and Li-Mn-spinel cathodes provided the best abuse tolerance in comparison with NCA and LCO. This thermal stability of LFP and Li-Mn-spinel is a clear advantage with regard to other abuse related areas. Remaining cell hardware, such as, separator, anode, electrolyte and cell design have significant contributions toward abuse tolerance of a lithium-ion cell. The team was therefore investigating additional safety reinforcing features, such as, separators with improved thermal stability.

Energy density

The major challenge, as typical for EV's, is to package enough energy for the specified electric range for the vehicle. For the Volt, a battery roughly 16kWh of energy, needed to be packaged into the vehicle thus demanding for battery cells with quite high energy density.

The first evaluation showed that chemistries using lithium-titanate anodes would have a difficult job to meet the energy

Table 3. Pugh Matrix to down select battery system proposals

Category	Proposal 1	.....	Proposal 16
Abuse Tolerance			
Power			
Energy			
Package Layout			
Thermal Capability			
Software and Controls			
Life Estimation			
Technology Maturity			
Mass			
Part Count / Complexity			

density requirements, due to the low cell voltage in the range of 2/3 of cells with graphite based anodes.

From the initial supplier workshop, a mass and volume target of 160 kg and 115 dm3 was provided as a starting point. This results in a specific energy of 100Wh/kg and an energy density of 140Wh/l (scaled to address system level needs). Assumptions were made for the cells to contribute to approximately 70% of the battery pack mass, and around 55% of the total pack volume. This relates to cell specific values of roughly 145Wh/kg or 250Wh/l. These values are quite challenging, which may also be obvious when comparing with the USABC requirements for a 40 mile PHEV.

Table 4. Comparison Chevrolet Volt vs. USABC PHEV energy requirements

	VOLT system level requirement	VOLT cell level requirement	USABC 40mile PHEV requirement
Specific energy [Wh/kg]	100	145	142
Energy density [Wh/l]	140	250	230

Power

One of the development goals for the Volt was to create a "no compromise vehicle" which implies outstanding performance and no reduction of vehicle performance in the EV mode. As a consequence, a battery system with 115 kW peak power was needed. In addition, this power has to be provided over a wide SoC range to allow full power in EV mode and sufficient power to support the internal combustion engine (ICE) in Extended-Range Mode. A specific power of roughly 700 W/kg and power density of 1000 W/l on system level was needed. Introducing the same scaling factors as already used to determine the energy density on cell level, a cell specific power of 1000 W/kg and 1750 W/l is required. To make this requirements even worse, this power requirements need to be fulfilled at a voltage level matching the needs of the electric motor and power electronics.

The following table shows a comparison of the Volt power requirements with the USABC power requirements for a 40 mile PHEV.

Table 5. Comparison Chevrolet Volt vs. USABC power requirement

	Volt system level requirement	Volt cell level requirement	USABC 40mile PHEV requirement
Specific power [W/kg]	700	1000	317
Power density [W/l]	1000	1750	–

Life

Cell calendar life is following the Arrhenius law which means that high temperatures have a severe influence on cell life. At higher temperatures not only the Li-Ion transport inside the battery cells is improved - which leads to higher power and reduced losses - but also aging is accelerated. typical calendar life vs. temperature behavior is shown in graph #1.

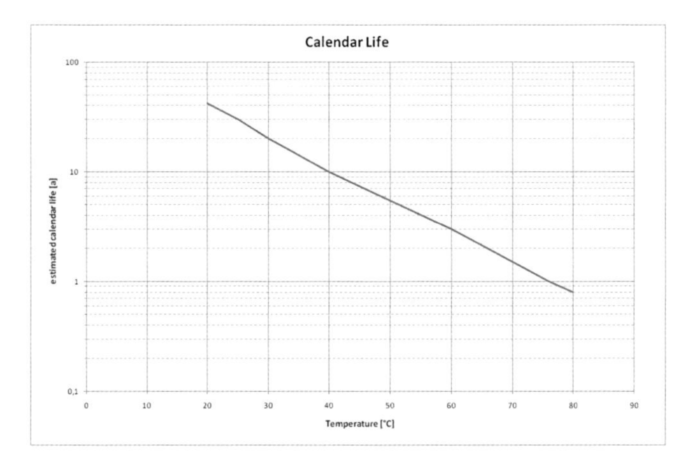

Graph 12. Calendar life vs. temperature (Example)

The exact achievable calendar life depends also on the chemistry and the applied SoC and may differ from the years shown in graph #1. However, it is obvious that temperatures above 35°C to 40°C should be avoided or at least reduced to a short time only. This operating strategy will enable a projected calendar life of at least 10 years. The consequence for this maintenance is invoke multiple resources toward a dynamic internal thermal management system for the VOLTEC battery.

Performance (power, energy, Li plating) and Temperature sensitivity

As typical for batteries also the Li-Ion battery cells used for the Chevrolet Volt have a power and energy performance that depends on the temperature. At low temperatures the internal resistance of the cells rises and therefore the power is reduced. This is true for both charge and discharge power. Graph #2 shows an example for 25°C and 0°C.

The low temperature charge capability is very sensitive. If the charge power or better the charge current is too high, there is the risk that not all Li-Ions can be intercalated to the carbon anode. As consequence Li-plating can occur and damage the cell. Therefore the charge power at low temperatures is strictly controlled and kept below the limit of a risk of Li-plating.

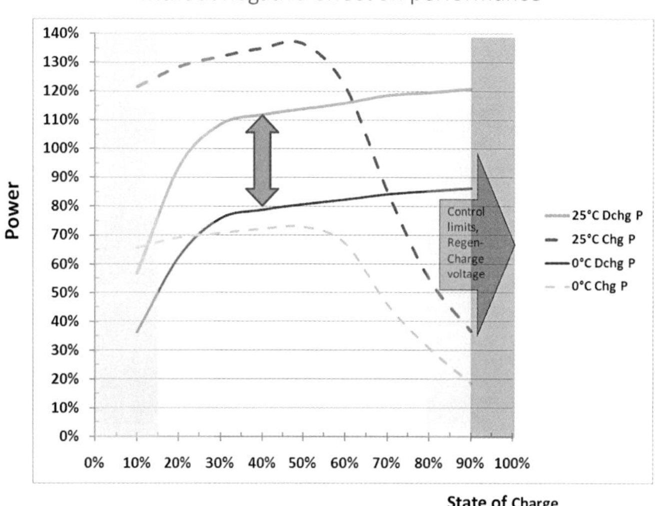

Graph 13. Battery system power as function of State of Charge (SoC) and Temperature (°C)

Beside the temperature influence on the performance, there is the influence of the battery state of charge (SoC) on discharge and charge power capability. Typically, the discharge power is the highest at high SoC and decreases towards low SoC. The charge power is more or less vice versa. So, at low and medium SoC we see a good charge power capability which is reduced towards high SoC. Details can be seen in Graph #2. Similar to the charge power at low temperature, the charge power at high SoC needs to be controlled quite precisely.

As consequence, the battery cannot be operated over the complete SoC range in the vehicle. The low SoC is excluded due to insufficient discharge power and the high SoC can at least not be used for regenerative braking. Therefore the usable energy is only a part of the total energy. One definition

Table 6. Comparison of key cathode material characteristics

Cathode Criteria	NCA	Mn-based	LFP
Energy density	Good	Good	Medium
Power density	Good	Excellent	Good
Battery life	Good	Medium	Good
Abuse tolerance	Poor	Excellent	Excellent

is the SoC range where both, the discharge power and the charge power requirement is met.

Available technologies

With a main focus on safety, the Mn-based and iron phosphate technologies looked most promising. Table #4 shows the performance of these two technologies in comparison with the nickel based chemistries in the key categories.

Li Mn2O4- based

For the manganese based chemistry a decoupled development project was started with the supplier LG Chem. LG Chem developed a cell with a pouch form factor. For enhanced abuse tolerance, manganese based cathode chemistry is applied and the cell features a safety reinforced separator (SRS™). Picture #2 shows Volt battery cell

Figure 4. LG-Chem pouch cell

The cell was developed according to the GM requirements and footprint to allow packaging into the Chevrolet Volt. During the development project performance and life of the battery cell was optimized in several iterations.

LiFePO4

Due to the challenging requirements of the Volt battery, it seemed to be advisable to rely not on one development path only. A second cell development path was started with A123 System following the lithium-iron-phosphate technology that provides an excellent abuse tolerance from a cathode point of view. The cell development was very similar to the advanced phase of the cell development carried out with LG Chem and several iterations were performed. Since the iron-phosphate technology has a lower nominal voltage than the manganese based chemistry, roughly 10% more cells were connected in series for the battery pack using the iron-phosphate technology.

A balanced decision was made for the manganese based cell from LG Chem based on performance, production readiness, vehicle integration, efficiency and durability. Importantly, LG Chem demonstrated superior manufacturing readiness for prismatic cells to meet the Volt program timing.

Electrical Architecture

A battery system for automotive applications needs to ensure the safe operation of the battery cells and take care of connecting and pre-charging the vehicles high-voltage system.

The electrical architecture of the Volt battery system in addition connects the on-board battery charger and the auxiliary power modules (APM) to the battery. In addition to the fuse and the manual disconnect that disconnects the battery for maintenance, the electrical architecture comprises in total 5 contactors. Two main contactors connect and disconnect the traction inverter of the electric motor and the APM to the battery. One contactor together with a pre-charge resistor manages the pre-charge of the vehicle's high voltage system. Two additional contactors ensure that the on-board battery charger is only connected during recharge of the battery.

The batteries controls architecture on the other hand measures redundantly each single voltage of the 96 cell groups (3 cells connected in parallel) in the Volt battery. In addition cell temperatures are measured all over the battery pack. These values are communicated together with pack voltages and current are the battery controller. An overview of the electrical architecture is shown in Figure x.

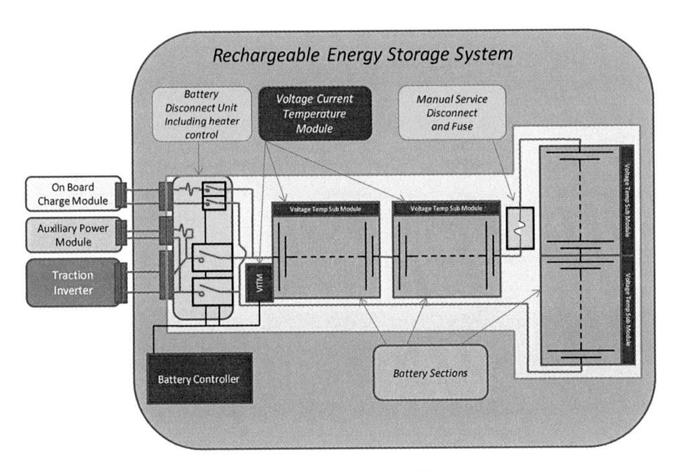

Figure 5. Electrical architecture Volt battery system

Operation strategy of the Chevrolet Volt

The Chevrolet Volt has three different operation modes to allow operation as an electric vehicle and providing a range comparable with an internal combustion engine (ICE) vehicle.

Figure X shows the different operation modes. Normally, the operation starts with the charged battery in electric vehicle (EV) mode. All propulsion power is provided by the battery and braking energy can be recovered by charging the battery. In this mode, up to 40 miles (64 km) range is provided and the battery is depleted to a lower SoC threshold.

When reaching this lower threshold, the engine / generator unit will automatically kick in and provide the average power for the vehicle propulsion. However, the power for dynamic acceleration will be provided by the battery. Also, the battery will recover the braking energy like in the EV mode. This mode is called Extended-Range Mode and the battery is operated in a small SoC window. In this mode, a vehicle range of several hundred miles is possible.

When the trip is finished, the battery will be recharged by the on-board battery charger from a standard 120V or 240V wall outlet. This mode is called Charge Mode and will recharge the battery in approximately 10 or 4 hours depending on the wall outlet. The Operating strategy of the Chevrolet Volt requires that the battery system can provide the peak power still at low SoC. On the other hand in extreme weather conditions like temperatures of −20°C, the engine-generator will provide sufficient power and the Chevrolet Volt will perform much better than a pure electric vehicle.

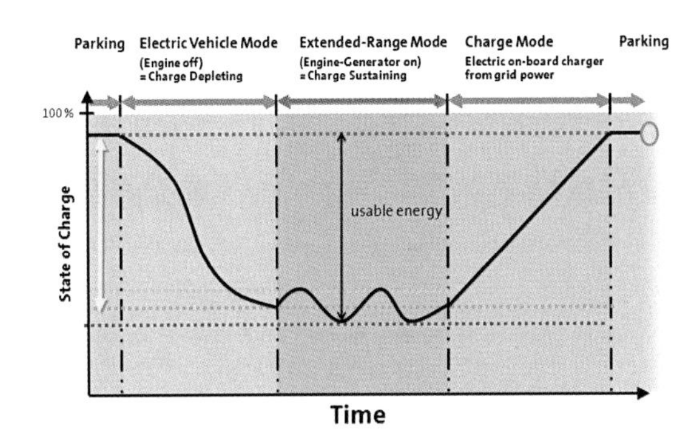

Figure 6. Operation Modes of the Chevrolet Volt

Dealing with "the temperature requirements and the battery temperature behavior"

As already discussed, the battery performance strongly depends on the temperature. At low temperatures the performance is reduced for both energy and power. However, the power depends much stronger on the temperature than the energy.

Best performance of Li-Ion cells is reached in a very narrow temperature band. Figure x shows a typical example for available power versus temperature.

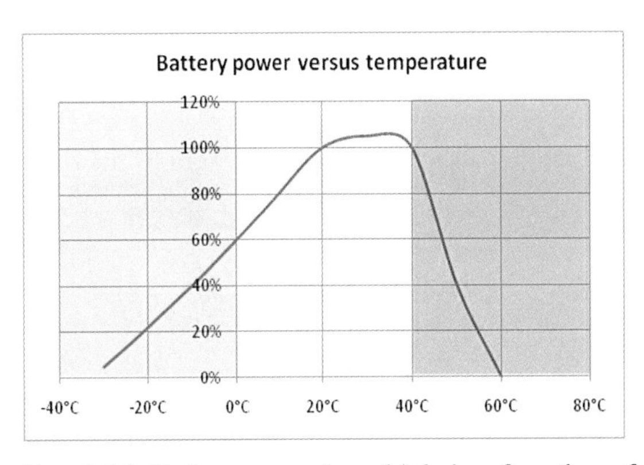

Graph 14. Battery power in vehicle is a function of resistance and control limits

At high temperature (40°C and above) the battery shows a quite low internal resistance and ample power, so it should perform very well. Unfortunately, operation at high temperatures is not recommended to avoid overheating of the battery and to reduce the aging of the battery that is highly accelerated at high temperatures. Therefore it is necessary to introduce a thermal management to the battery pack.

Thermal system architecture

For applications with a high power density, the thermal management has an important influence on the overall performance of the battery system and its reliability and durability. A poor thermal management system can lead not only to thermal problems like overheating of the cells but also to "electrochemical problems" of the battery like diverging SoC or even diverging capacity and internal resistance.

The VOLTEC battery system has both, a high power density and (from thermal point of view) a difficult package situation. Due to the T-shape battery pack, the battery cells are distributed over a total length of roughly 2 meters. Therefore, the form factor of the cell should have the best starting conditions. Pouch cells have a higher surface to volume ratio than prismatic can or cylindrical cells thus they transfer the heat, produced during operation, very well to the ambient. So, consequently the Volt battery system makes use of the superior thermal properties of pouch cells and combines this with a sophisticated liquid thermal system. The system is designed such that all cells are cooled "in parallel" and much effort was spend to bring the liquid coolant as close as possible to the source of heat generation, the cell. A liquid thermal management system allows much better than an air thermal management system to equally cool al cells with minimum temperature imbalance in a difficult and volume constraint package situation. Since Li-Ion cells operate only in a small temperature window with full performance and full durability, it is necessary to condition the cells. At high ambient temperatures and during heavy load cycles the battery cells are cooled by the thermal management system. At cold temperatures, the thermal management system can be used to heat up the cells in order to bring the cells faster to their optimum temperature and full performance.

Figure x shows the principal layout of the liquid thermal management of the Volt battery system.

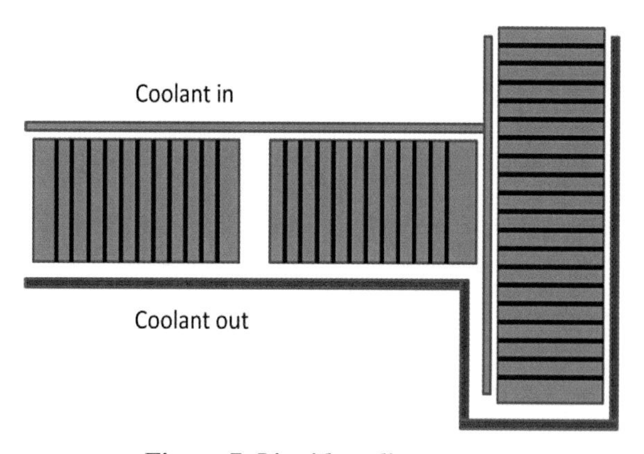

Figure 7. Liquid cooling system

The thermal system of the Volt battery is designed to condition the battery cells during EV mode but also to cool the cells during extended range (E-REV) mode. Since the extended range mode is not timely limited by the battery capacity and can continue for several hours it was no option to rely on the thermal mass of the battery cells. To rely on the thermal mass only could be a risk for battery life even in pure EV mode, especially in areas with high ambient temperatures. The Volt battery thermal management however will handle all these situations.

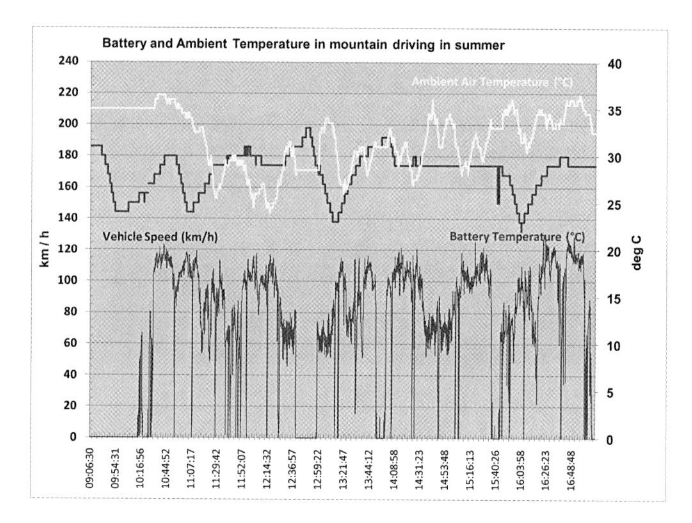

Graph 15. Liquid cooling system

Test data recorded with a Volt integration vehicle during a summer drive in mountain area demonstrate the thermal system capability. Even at a long fast uphill and downhill drive the battery temperature is controlled between 20°C and 30 deg C, even when ambient temperatures are above.

Thermal interface

The liquid thermal management system of the Volt battery system is fully integrated into the vehicle's heating ventilation air conditioning (HVAC) system via its thermal interface, a coolant in and coolant out hose. The coolant is normally cooled by a low temperature radiator specially dedicated to the battery. This ensures a good efficiency and minimum energy use for the thermal management which is important for a good range in pure electric mode. At high ambient temperatures, the coolant heated up in the battery system is cooled down in the radiator to a first temperature level and further cooled down to a second temperature level via a chiller that is connected to the vehicle's AC system. Thus, the Volt battery system can be cooled at nearly all ambient temperatures and vehicle overall efficiency is optimized. In addition, at low temperatures it is possible to make use of the vehicle's HVAC system for slightly heating up the battery system. This increases the battery performance at the beginning of the trip and allows the battery system to faster reach its optimum temperature.

Developing the Solution

Battery packs for electric vehicles are considered to be "energy batteries" or ~60Ah. Applications for start/stop/ launch may be better suited as "power batteries" or ~6Ah. A little know fact that early development of the EV1 (Impact 4.0) allowed for engineers to utilize trailers equipped with gas powered generator generators. Only a few vehicles operated with this modification, although integration was simple. All that was required was a breakout box between the battery and inverter, and High Voltage cables equipped with interlock safety loop. Fundamental operation was governed by crude controls via remote (driver operated) dongle. The energy flow was uni-directional (generator to battery), this allowed for "all day experiments" with EV1 as a "series hybrid electric car".

Constraints were still governed by ambient temperatures, as max pack temperatures, delta T, and module voltages influenced various diagnostic fault over-rides. Lead Acid was the prevailing technology, and as a energy battery rated for 40 Ah made for short discharge and long charge times. Standard capacity test were ran at C/3, and based on an air-cooled system, required ~1/2 day to settle back to a uniform thermal gradient. Note, a single fan was used to address all 26 batteries, which were configured into a 2 levels to form a "T". A liquid managed battery system is more compact and results in high specific energy density and allows to control consistent temperature throughout the pack.

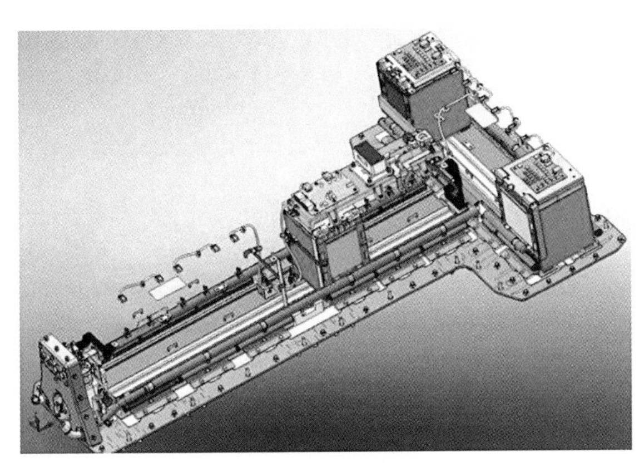

Figure 8. Proof of concept battery system with liquid thermal system

Cells Testing

A pack is only as good as its weakest "link". Considerable focus is directed toward the cell hardware, anodes, cathodes, separators, electrolyte, tabs, bags or packaging. Abuse tolerance studies, fluid compatibility, and performance over multiple parametric experiments were discussed and disclosed during the decoupled development activates between GM and selected vendors. The processes established to combining these parts are certainly of significance, and based on the early interactions, both parties were able to identify areas of focus toward establishing a firm building block for the next level of integration.

Figure 9. LG Chem pouch cell prior to cell test

New cells were subjected to temperature controlled Hybrid Pulse Power Characterization (USABC version), where response tables could be forwarded to dynamic models and scaled simulations could provide early estimates on vehicle capabilities. These tables also supported algorithm development where efficient utilization of onboard memory would be realized within battery monitoring microcontrollers. Vehicle models could now provide the estimated pack level contribution during drive scenarios, where the power vs. time was considered an output. Utilizing this format as stimulus for battery cyclers (either cell or pack) allowed for empirical data to further support model(s).

A large design of experiment (DoE) had been performed with cells at various temperatures, State of Charge, different depth of discharge cycles and power levels to gather data for battery management calibration and to develop a battery life model.

Module Testing

During evaluations of stand, charge, and drive profiles, thermal signatures of the cells and their interaction of support hardware were also recorded. Infrared (IR) camera scans provided for a rapid evaluation of cell gradients, tab to welding, and interactions with electrical connections for I^2R losses. Cell core temp, and its resultant relation to skin temp and how they correlate to onboard thermistors are studied at this level.

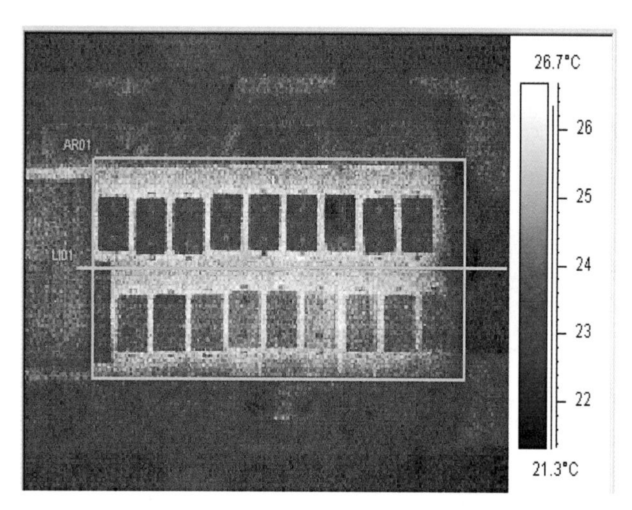

Figure 10. Infrared scan of development module

Designed best placement of temperature sensing devices is also supported by the IR scans. Multiple points provide feedback to the thermal controls as inputs as well as diagnostics.

Although multiple cells make up the series parallel configuration (96series-3 parallel), loss of a single contributor can be detected. Cells tab/welds that segregate from their parallel groups ultimately impact the group capacity, hence limiting the performance during discharge, and constraining regeneration during charge. Eventually a significant cell to cell delta Voltage will be encounter and the appropriate diagnostic triggered.

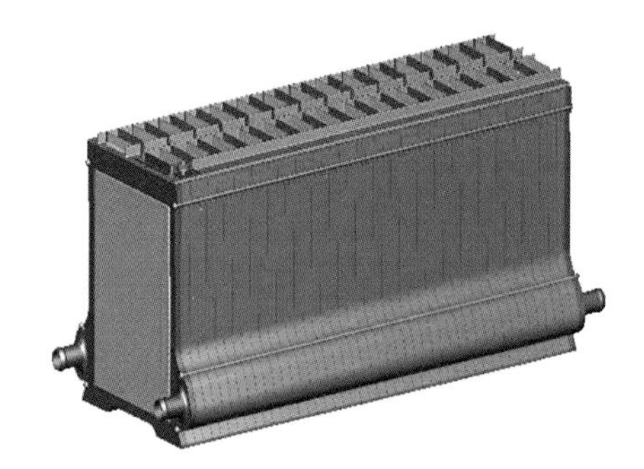

Figure 11. Math model of production design battery module

Deciphering good cells from bad hardware was best practice associated by tracing the origin of voltages (cell tabs), correlating with the onboard data-acquisition-system, and following the digital stream throughout the system. Mis-wires, redundant voltage measurements, invalid indexed arrays, and loose connections all play a part in providing an inaccurate summation of pack status and health.

Pack testing

Development involves testing to the edge of specification, looking for min/max tolerances and sensitivities. Before any significant power was exchanged, the VOLTEC batteries were evaluated for robustness of the cell/pack electronics. A halt test was established to either map operation boundaries at low and high ambient temperatures, or disclose threshold(s) of non-compliance. Hardware signatures were characterized for normal operation (household) currents, as well parasitic drain on cells for stand time compliance.

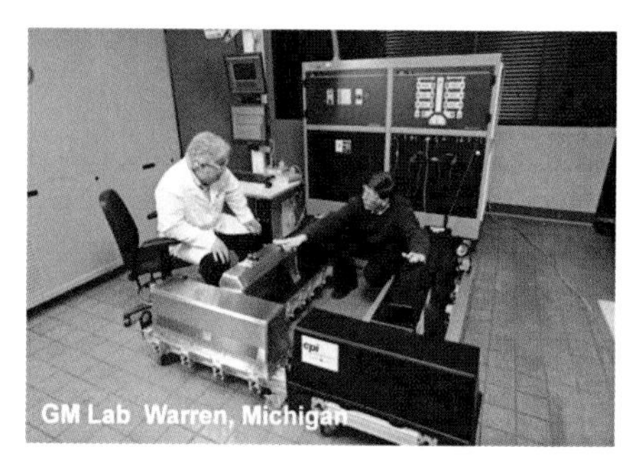

Figure 12. Proof of concept battery systems and battery cycler

Utilizing a test chamber, liquid thermal cart, and battery cycler, the system was closely monitored for reaction/ susceptibility to any non-intended conducted emissions. The battery should float, when attached connected to power processing equipment. Potential leak paths or ground loops had to be dealt with, prior to further studies. CAN message requirements depict a specific bit resolution, scaling, and periodic transmit times. Each signal is deciphered via data base file and protocol interface tool. These signals were then verified against the cycler interpretation, as they would also provide for closed loop variables and constraints toward safe operation.

Figure 13. battery systems during development test

Step response was now determined as batteries are exposed to various power pulses and ambient conditions. Current hall-effect measurements are studied for asymmetrical artifacts or gauzing offsets influenced by resident ferrite devices. CAN messages depicting both high voltage and cell voltages were correlated and compared to specifications for bit resolution, scaling, and periodic transmit times. Total transport delay was assessed from time of command vs. resulting measurement from onboard electronics.

Modeling

Configuring a high voltage battery, for the Volt, was more than just attaching multiple cells together to form a parallel/series string. Math studies, Finite Element, DFMEA, and multiple other disciplines allowed virtual components, to drive hardware requirements. Worst case scenarios, and predictions for end of life behavior, drove material selections and packaging directions. The cell aspect ratio was determined horizontally for best fit of vehicle seats, and vertically for tunnel height that correlated to corporate common air conditioning components. This was determined very early in development as multiple teams generated 3d models only to argue over pixel locations. Various connection schemes and module configurations were developed with math tools. Optimization efforts were directed toward mass distribution, materials and parts re-use, durability, secondary life usage. A robust design would allow for expansion, contraction, and frame twist, while retaining seal integrity for cover perimeter and internal manifold.

Some vehicle models rely on temperature compensated resistance tables to support simulations of a battery response to power request stimuli. A Battery Scale Factor is typically engaged to attenuate this data produced by single cell testing. Power vs. Time vectors can be extracted from such a model and be configured to instruct a Battery Cycler. In the case of the volt, many test profiles were generated from multiple models (Matlab modeled US06, FTP city and hwy cycles) and re-played against real hardware to confirm accuracy of model parameter tracking (example State of Charge and Pack Temperature). Data collected allowed for rapid development of controls, energy rules, and table independent algorithms, which ultimately closed and continues to close the loop for battery life.

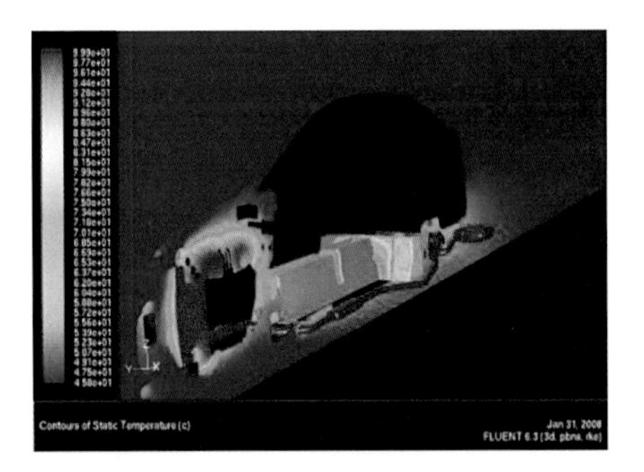

Figure 14. Thermal model of battery next to hot exhaust pipe in charge sustaining operation.

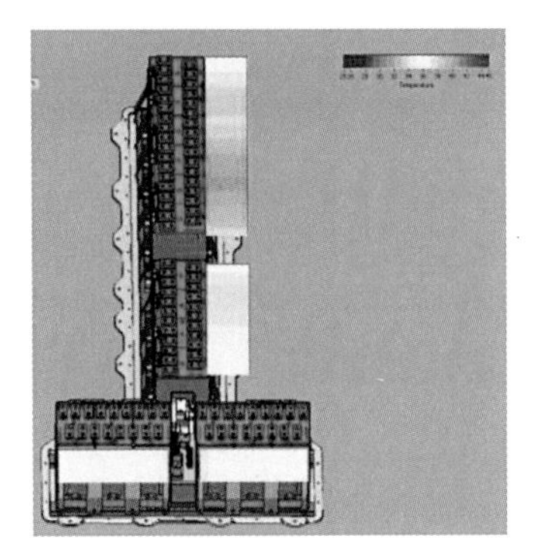

Figure 15. Measured temperature distribution in early Volt development pack.

Crush testing

To answer the question haw the battery pack would react on intrusion several crush tests had been developed and performed. For characterization the result of slow intrusion had been measured. The depth of intrusion is significant higher than worst case vehicle assumptions. Dynamic testing with sled tests confirmed the positive results.

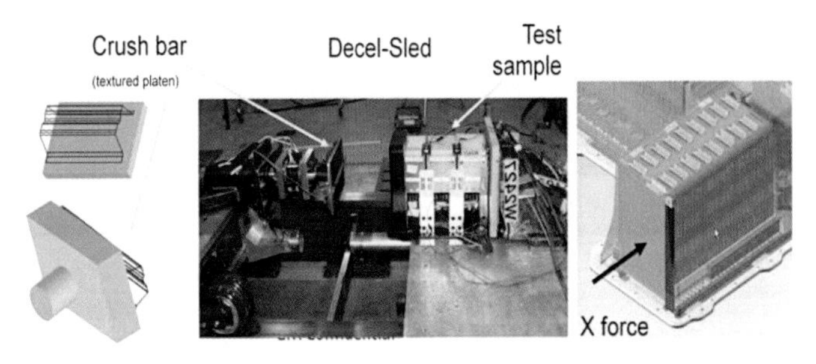

Figure 16. Static and dynamic battery module crush test.

Development of battery system in lab

To develop function and verify cycle life a number of battery systems have been cycled in a newly developed life cycle test which includes charge depletion and charge sustaining operation. The battery system is recharged, turned off for a rest period and the next cycle begins. This test represents the worst case vehicle use, deep discharge cycle and little charge sustaining operation. Four thermal regions are simulated by thermal chamber and cooling module: Hot climate based on the Phoenix data, moderate climate like in Los Angeles, warm to cold climate like in Detroit and cold simulating the Winter in Canada. To test a system for 240 000 km approximately 20 months in the accelerated cycle are required.

Figure 17. Battery system lab: Control desk and thermal chambers.

The systems is tested for capacity and resistance test every 6 weeks. The results are compared to the model based battery life prediction. The development team reviews key parameters every week. All battery data, such as all cell voltages are recorded and analyzed.

Figure 18. Battery system in production design after inspection during lab development test.

Development in Vehicle

Beginning 2008 engineering development vehicles had been equipped with battery systems followed by Mule vehicles fall 2008. The integration vehicles representing the production design had been used for development and system calibration. The battery thermal system function had been developed in these vehicles. All durability and many development vehicles are equipped with data loggers to analyze battery usage in relation to vehicle use profiles. Based on the data the battery development team can verify that the battery performs as modeled and predicted.

Figure 19. Integration vehicle on calibration drive

Vehicle testing was performed on pulsing-machines to demonstrate that all systems, including the battery system are capable to exceed one vehicle life.

Figure 20. Vehicle shaker: Vibration test with full functional battery.

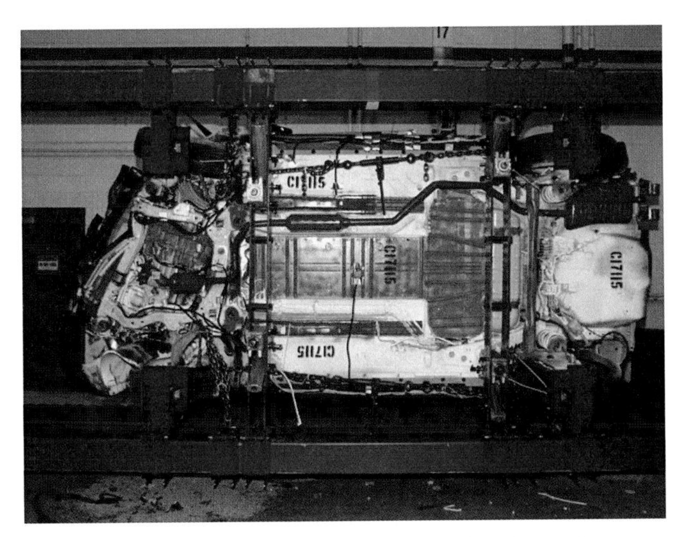

Figure 21. Vehicle Crash test with full functional battery system

To verify that the battery system is sealed for water in the severe road conditions the vehicle installed battery system had been tested.

Figure 22. Flooded road test to verify battery housing sealing

Results

The VOLTEC Battery system is meeting the requirements and has been developed ready in time for the start of vehicle production.

The battery performance had been successfully demonstrated during the final calibration rides. This included calibration in Cities with steep inclines like San Francisco, Mountains in Colorado and hot climate testing in Death Valley.

Table 7. Battery system key performance data

Technology	Lithium Ion 288 pouch cells
Voltage, nominal	370 V
Current, maximum	400 A
Capacity	45 Ah
Energy	16 kWh
Discharge Power (10 sec)	>115 kW
Regenerative charge Power (10 sec)	60 kW
Mass	190 kg
Volume	138 dm^3
Thermal Management	Liquid system providing cooling using vehicle Air condition system and specific heater
Battery Management	State of Charge estimation Cell monitoring Cell balancing Isolation measurement
Warranty	8 years

Figure 23. Chevrolet Volt on one of the final Calibration rides

SUMMARY/CONCLUSIONS

The VOLTEC battery system has been defined, designed and developed using a system engineering approach. Requirements had been generated based on vehicle targets using modern modeling tools. Requirements were broken down into subsystem and component requirements. Then components and subsystem had been developed, validated and integrated. Selection of cell technology was prioritized with safety and production readiness. An effective thermal system which allows to operate Lithium Ion cells most of the time in the best temperature range for life and performance had been invented. New electronic control cell monitoring and balancing systems, new algorithms and new test protocols had been developed specifically for the new application. Within 4 and a half years a high performance battery system for a new vehicle category, the electric vehicle with extended range had been designed and validated to get into production.

REFERENCES

1. Tate, E.D., Harpster, M.O., and Savagian, P.J., "The Electrification of the Automobile: From Conventional Hybrid, to Plug-in Hybrids, to Extended-Range Electric Vehicles," *SAE Int. J. Passeng. Cars - Electron. Electr. Syst.* **1**(1):156-166, 2008, doi:10.4271/2008-01-0458.

2. Ressler, G., "Application of System Safety Engineering Process to Advanced Battery Safety," SAE Technical Paper 2011-01-1369, 2011, doi:10.4271/2011-01-1369.

CONTACT INFORMATION

The authors can be contacted at Roland.Matthe@gm.com, Lance.Turner@gm.com, Horst.Mettlach@de.opel.com

ACKNOWLEDGMENTS

The Authors Roland Matthe, Horst Mettlach and Lance Turner like to thank the global VOLTEC Battery System team, the involved supplier teams and everyone on the greater Volt team for the great support and many early or late work hours which made the achievement of this new Volt battery system possible

ADDITIONAL RESOURCES

1. Federal Highway Administration; National Household Travel Survey, http://www.fhwa.dot.gov/policy/ohpi/nhts/index.htm

2. Southern California Association of Governments; Travel and Congestion Survey, HTTP://WWW.SCAG.CA.GOV/TRAVELSURVEY/

DEFINITIONS/ABBREVIATIONS

Ah
Ampere hour

DFMEA
Design Failure Mode Effect Analyzes

EPA
Environmental Protection Agency

EREV
Electric Vehicle with extended range

FTP
Federal Test Procedure

GM
General Motors

HV
High Voltage (Voltages above 60 V in automotive application)

HVIL
High Voltage Interlock loop

HVAC
Heating Ventilation Air Conditioning

ICE
Internal Combustion Engine

LV
Low Voltage (here: less than 60 V)

OBCM
On-Board Charger Module

RESS
Rechargeable Energy Storage System

RTS
Regional Travel Survey

SCAG
Southern California Association of Governments

SoC
State of Charge

USABC
United States Advanced Battery Consortium

VOLTEC
GM Propulsion system for electric vehicle with extended range

The Engineering Meetings Board has approved this paper for publication. It has successfully completed SAE's peer review process under the supervision of the session organizer. This process requires a minimum of three (3) reviews by industry experts.

ISSN 0148-7191

Positions and opinions advanced in this paper are those of the author(s) and not necessarily those of SAE. The author is solely responsible for the content of the paper.

SAE Customer Service:
Tel: 877-606-7323 (inside USA and Canada)
Tel: 724-776-4970 (outside USA)
Fax: 724-776-0790
Email: CustomerService@sae.org
SAE Web Address: http://www.sae.org
Printed in USA

SAE International™

High Voltage Hybrid Battery Tray Design Optimization

2011-01-0671

Published
04/12/2011

Kristel Coronado
General Motors LLC

John Lyons, Randy Curtis and Thomas Wang
General Motors Company

doi:10.4271/2011-01-0671

ABSTRACT

Hybrid high voltage battery pack is not only heavy mass but also large in dimension. It interacts with the vehicle through the battery tray. Thus the battery tray is a critical element of the battery pack that interfaces between the battery and the vehicle, including the performances of safety/crash, NVH (modal), and durability. The tray is the largest and strongest structure in the battery pack holding the battery sections and other components including the battery disconnect unit (BDU) and other units that are not negligible in mass. This paper describes the mass optimization work done on one of the hybrid batteries using CAE simulation. This was a multidisciplinary optimization project, in which modal performance and fatigue damage were accessed through CAE analysis at both the battery pack level, and at the vehicle level. The final battery tray design based on CAE analysis results and recommendations was validated through physical vehicle tests, and implemented into the vehicle production.

INTRODUCTION

With the advancement of hybrid technology, additional components are being introduced to consumer vehicles. New components have a variety of purposes, ranging from structural stability to improving efficiency. These components all have an associated mass that reduces the fuel efficiency of the vehicle. These new components must be optimized for mass to reduce the cost of materials and to pass along fuel efficiencies to consumers. In addition to optimizing the mass, the entire battery sub-system must meet multiple criteria, including safety/crash, NVH, durability, and costs.

This paper will describe the CAE analysis performed on the battery pack support system (tray), a structural component that ensures that the battery pack is securely fastened to the vehicle. In general, vehicle level analysis and safety/crash simulations are time consuming and beyond the scope of this paper. Thus, the approach taken for the project was to optimize the tray for NVH (modal) and durability performance at the pack level.

NVH/MODAL PERFORMANCE OPTIMIZATION

The modal performance of the HV battery and battery supports were assessed as a part of the HV battery NVH structural analysis. The battery modal analysis was performed in parallel with optimization trials in an effort to reduce mass and improve the structural performance of the battery support system.

The battery tray and supports were modeled using about 23,000 shell elements, 5 mm in size.

The modal performance assessment of the battery support system was performed on five design iterations (see Figures 1,2,3,4,5, below). The original design (Figure 1) is used as the baseline throughout this study. The second design (Figure 2) combines the front 3 tabs into one panel in addition to forming a different main tray design. Figures 3 and 4 are a variation of one another. Figure 3 has "open" cross members, while Figure 4 closes out the top of the two cross member supports. Figure 5 integrates the front bar of variations #3 and #4 into the mainstream tray design shown in Figure 2. Each design was evaluated using a standard free-free modal load case. The solver software used for both the analysis and the optimization was Genesis by Vanderplaats R&D.

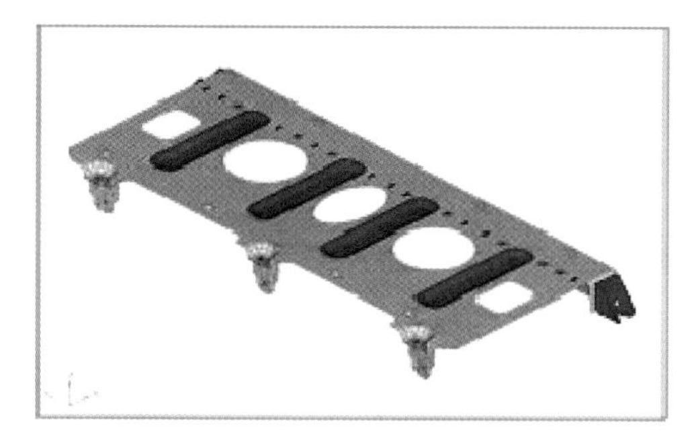

Figure 1. Original Design

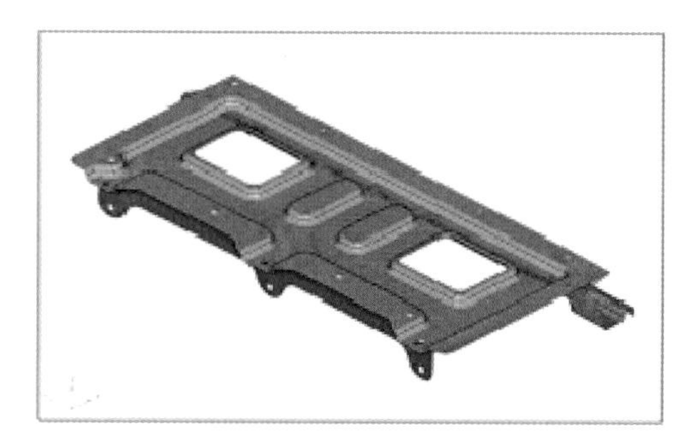

Figure 2. Second Design

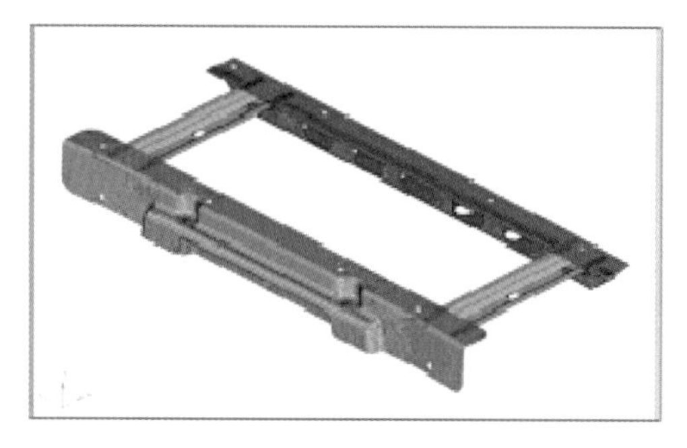

Figure 3. Third Design

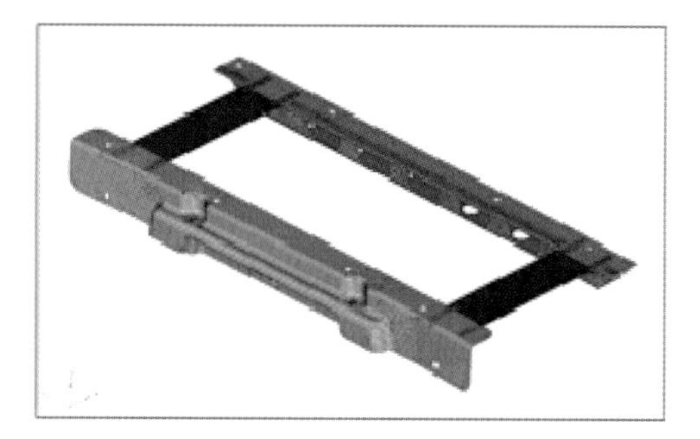

Figure 4. Fourth Design

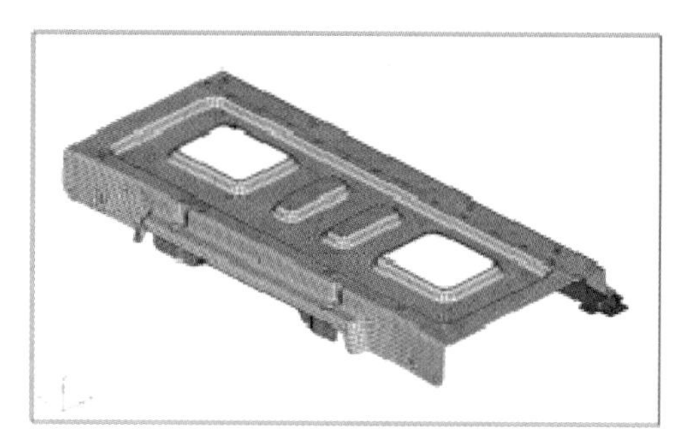

Figure 5. Fifth Design

Modal Performance Assessment of Four Battery Support System Designs

The performance of the first four designs (Figures 1, 2, 3, 4) is outlined in Table 1. In addition to a known frequency target, the modal frequency of the original design #1 was used as the baseline for performance evaluations and for design improvement. Of the four designs, none of them met the target frequency; however, design #2 was the closest, falling only 4% below the target frequency. The mass savings was the greatest with designs #3 and #4 when compared to design #2. However, only design #2 showed improvement in frequency over the baseline (design #1). Thus, design #2 was recommended as the optimal support system design for this stage of analysis and for further improvement through optimization.

Sizing and Topology Optimizations

Two optimization methods were used to evaluate the potential of design #2, sizing optimization and topology optimization. Sizing optimization was first used to determine the ideal thickness for each of the three components that make up the battery support system. Topology optimization was used secondly to provide critical load paths for potential design changes and/or adjustments. Each optimization was intended to aid in mass reduction and structural performance improvement of the design.

The sizing optimization results are shown in Table 2, in the form of normalized percentages. These results proposed that mass removal be focused on for the rear bar, while the front bar and main tray be focused on an addition or re-distribution of mass in order to improve the modal performance.

Topology optimization provided insight on the areas that are critical in strengthening the system. Figure 6 displays the battery support system as a whole. The red areas are those that are more critical to focus on improvement, while the blue areas are less critical to improving modal performance. Thus, the blue areas were used as areas for mass reduction. Figure 7 shows the load paths for each of the three support system components individually.

Table 1. Mass Analysis and Modal Performance of Designs #1-4

Tray Design	% Change in Mass from Design #1	% Change in First Mode Frequency from...	
		Target	Design #1
Design #1 (Original Design)	--	-36%	--
Design #2	**-13%**	**-4%**	**50%**
Design #3	-34%	-39%	-5%
Design #4 (Close-Out Variation of #3)	-28%	-39%	-5%

Table 2. Percentage Change in Thickness from Initial Design #2 Thickness to Final Optimized Thickness

% Change in Thickness from Initial Design #2		
Front Bar	Main Tray	Rear Bar
49%	101%	-30%

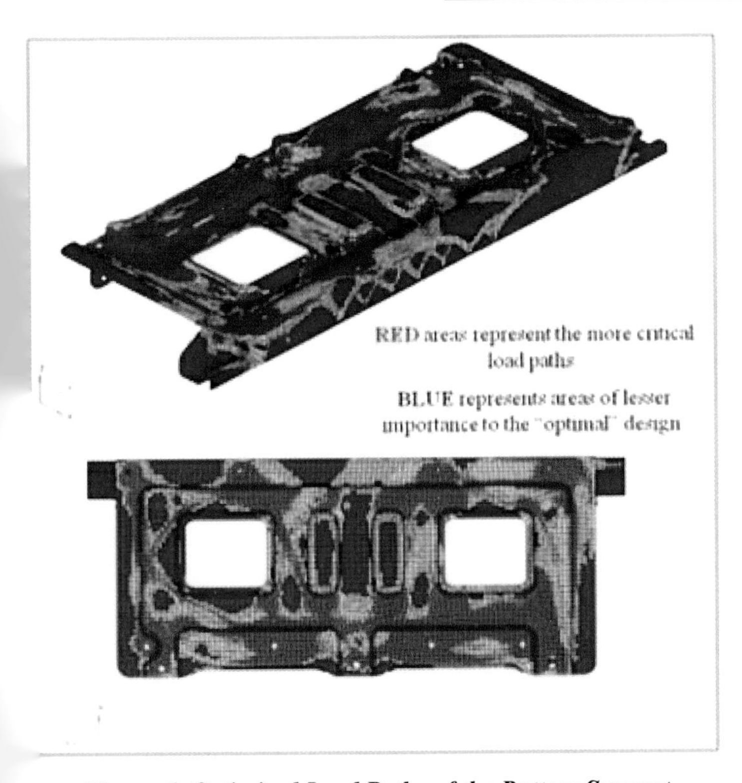

Figure 6. Optimized Load Paths of the Battery Support System

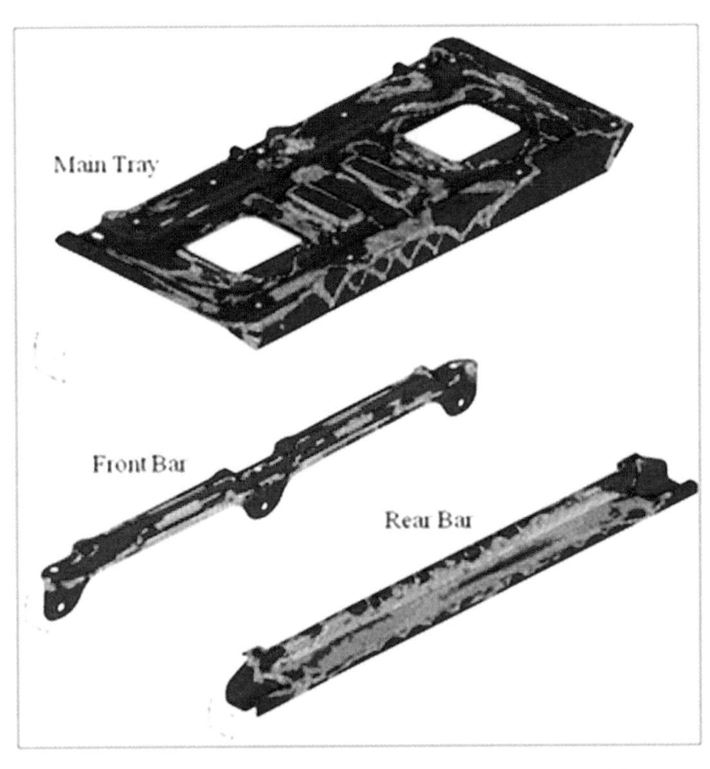

Figure 7. Optimized Load Paths for the Front Bar, Main Tray and Rear Bar.

Modal Performance Assessment of a Fifth Battery Support System Design

These load paths correlate with the sizing optimization conclusions - the rear bar is a potential area for mass reduction, while the front bar and the main tray can be focused on an addition or re-distribution of mass.

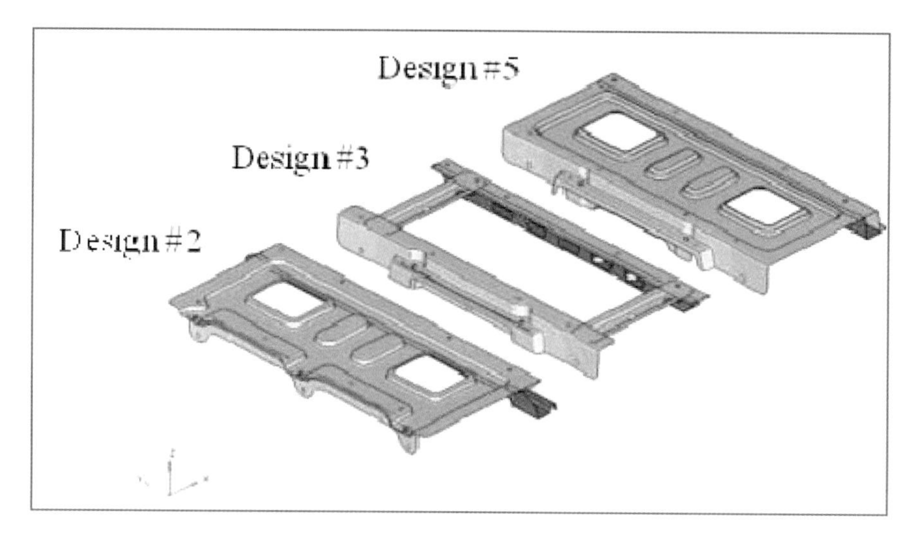

Figure 8. Design #5 is an Integration of Designs #2 and #3.

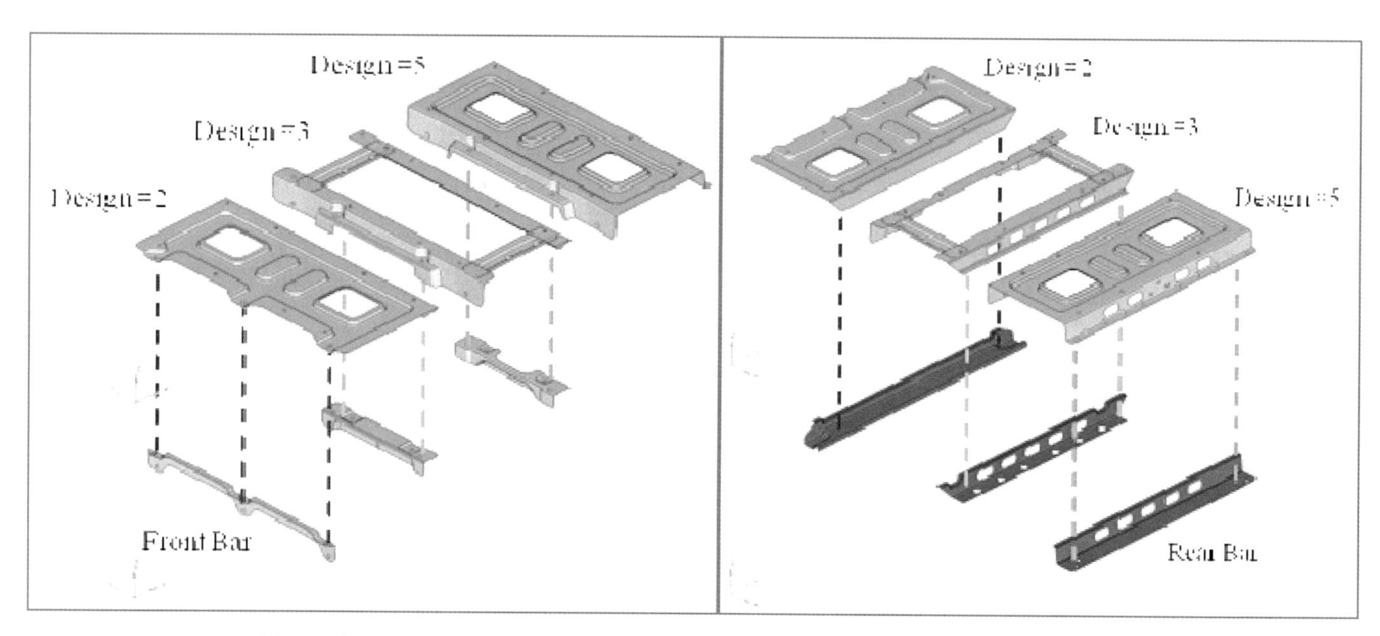

Figure 9. Design History of the Front Bar, Rear Bar and Main Tray Components.

The fifth design integrates the front bar area of design variations #3 and #4 into the mainstream tray from design #2 (shown in Figures 8 and 9). Mass was also removed from the rear portion of the tray and the rear bar, as instigated by the previous optimizations.

Multiple analysis trials were performed on design #5 to find the optimal combination of variables (thickness, mass, and attachments). First, a mini design of experiments (DOE) was performed to find an optimal thickness combination. The results are outlined in Table 3. Combinations 1 and 5 had a 19 and 15 percent mass increase from the original design, which was too large to move forward with. Of the remaining combinations (#2, 3 and 4), combinations #3 and #4 performed the best. Combination #3 decreased the mass by 1% from the original design, and although the first modal frequency was 12% lower than the target, there was a 4% improvement over the original design. Combination #4 added a minimal amount of mass (0.3%), but however, was closer to the target frequency and had as higher percentage improvement over the original design #1.

The performance analysis trials moved forward with a variation of combination #1. This design had the center of the main tray removed, as shown in Figure 10. This design has a 12% mass savings over the original design and a 26% mass savings over the initial fifth design with the same thickness combination.

Table 3. Mass Analysis and Modal Performance of Design #5 Thickness Combinations

Design #5 Thickness Combinations	Tray (mm)	FrtBar (mm)	RrBar (mm)	% Change in Mass from Design #1	% Change in First Mode Frequency from... Target	Design #1
Combination 1	1.20	1.20	1.20	18.8%	Not feasible	Not feasible
Combination 2	1.20	0.80	0.80	-18.0%	-23.4%	-9.7%
Combination 3	**1.00**	**1.00**	**1.00**	**-1.0%**	**-12.0%**	**3.8%**
Combination 4	**1.20**	**1.00**	**1.00**	**0.3%**	**-10.0%**	**6.1%**
Combination 5	1.20	1.20	1.00	14.7%	Not feasible	Not feasible

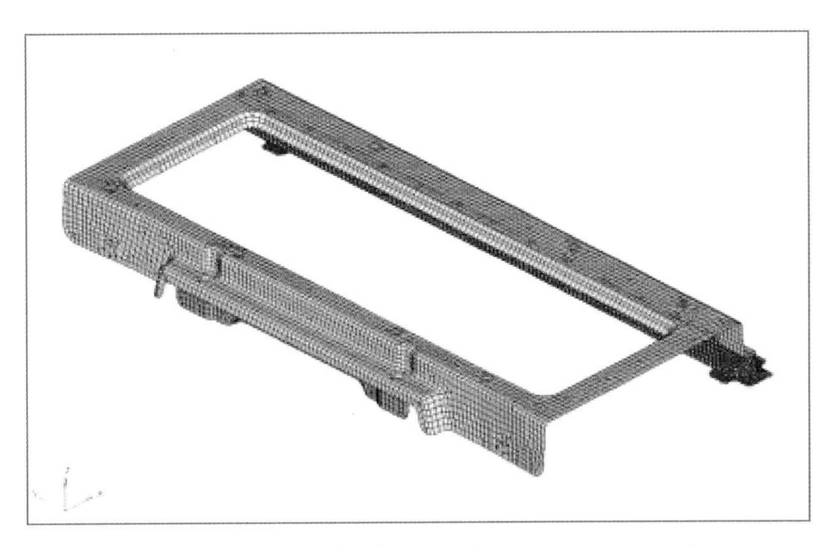

Figure 10. Variation of Combination #1, with Center of Main Tray Removed

Table 4. Mass and Performance Percentage Results of Center Removed from Design #5, Combination #1

% Change in Mass from...		% Change in First Mode Frequency from...	
Design #1	Design #5-1	Target	Design #1
-11.8%	-25.8%	-4.5%	12.6%

The combination #3 was then used to assess if the addition of more attachments could help to improve the modal performance. Two additional bolt attachments were placed from the front bar to the 5-bar location, as shown in Figure 11. The response from the bolt attachment additions was positive, as shown in Table 5. The modal frequency moved 7.3% closer to the target and improved 8.3% compared to the same design combination without these two additional attachments. However, due to limitations, it was more probable to vary the thickness and design than adding bolts to the manufacturing assembly process.

Modal Performance Assessment of Reinforcement Additions

The final trial of modal performance used a variation of design #5, combination #4. Four reinforcements were added to the system; two crossbars that flow beneath the main tray and two bulkheads that reside between the rear bar and main tray (Figure 12). Three modal analysis trials were performed; crossbars and bulkheads both attached, only the bulkheads attached, and only the crossbars attached to the battery support system. The percentage results are presented in Table 6.

The highest performance improvement is with the use of both, crossbars and bulkheads. However, it adds 6.4% mass to the baseline design (design #5, combination #4). Use of the only the bulkhead reinforcements proved to be the least effective of the three combinations assessed. The conclusion from this analysis is that the crossbars are the reinforcements of focus. The addition of these lifted the modal performance

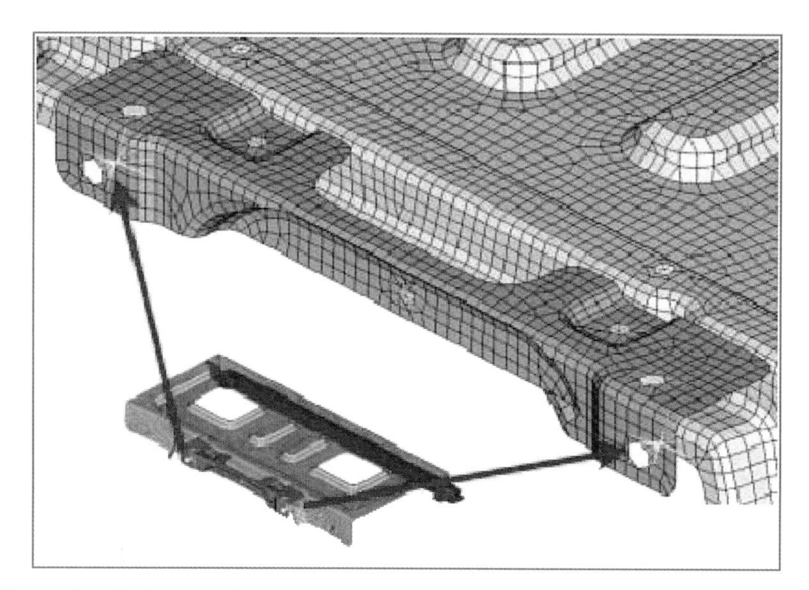

Figure 11. Variation of Combination #3, Two Additional Attachments to 5-Bar

Table 5. Performance Percentage Results of Center Removed from Design #5, Combination #3

Tray Design	% Change in First Mode Frequency from...		
	Target	Design #1	Design #5-3
Combination 3 (Design #5-3)	-12.0%	3.8%	n/a
Combination 3 : More 5-Bar Attach	-4.7%	12.4%	8.3%

to well above the target frequency, as well as surpassing the performance of the original design and the previous iteration without reinforcements.

Based on the conclusion of this optimization work, the packaging constraints, manufacturing process and cost, the design release engineer and team formed the design. The design then was assessed and further improved the modal performances at both battery pack level and at vehicle level. There was a very close correlation was found between the battery pack level modal performance and the one at the vehicle level. In other words, one could almost conclude that the better modal performance at the pack level, the better at the vehicle level. The outcome of the battery tray was then analyzed for its durability performance and optimized for improvement.

DURABILITY PERFORMANCE OPTIMIZATION

Methodology

The method used to evaluate the durability performance of the battery pack and battery support system was the shaker table test simulation. This simulation uses the method of random vibration fatigue. Figure 13 displays the CAE simulation setup. The battery pack is attached atop the support system, which then rests on the shaker table (which is considered to be a rigid).

As previously mentioned, this paper only focuses on the battery support system, as shown in Figure 14. The immediate surrounding elements were not selected for general fatigue life assessment due to stress discontinuity caused by the rigid elements representing bolts. There was a separate study was performing in a parallel process to study the fastener strategy and the areas that were excluded from this analysis using another more advanced solver which is beyond the scope of this paper.

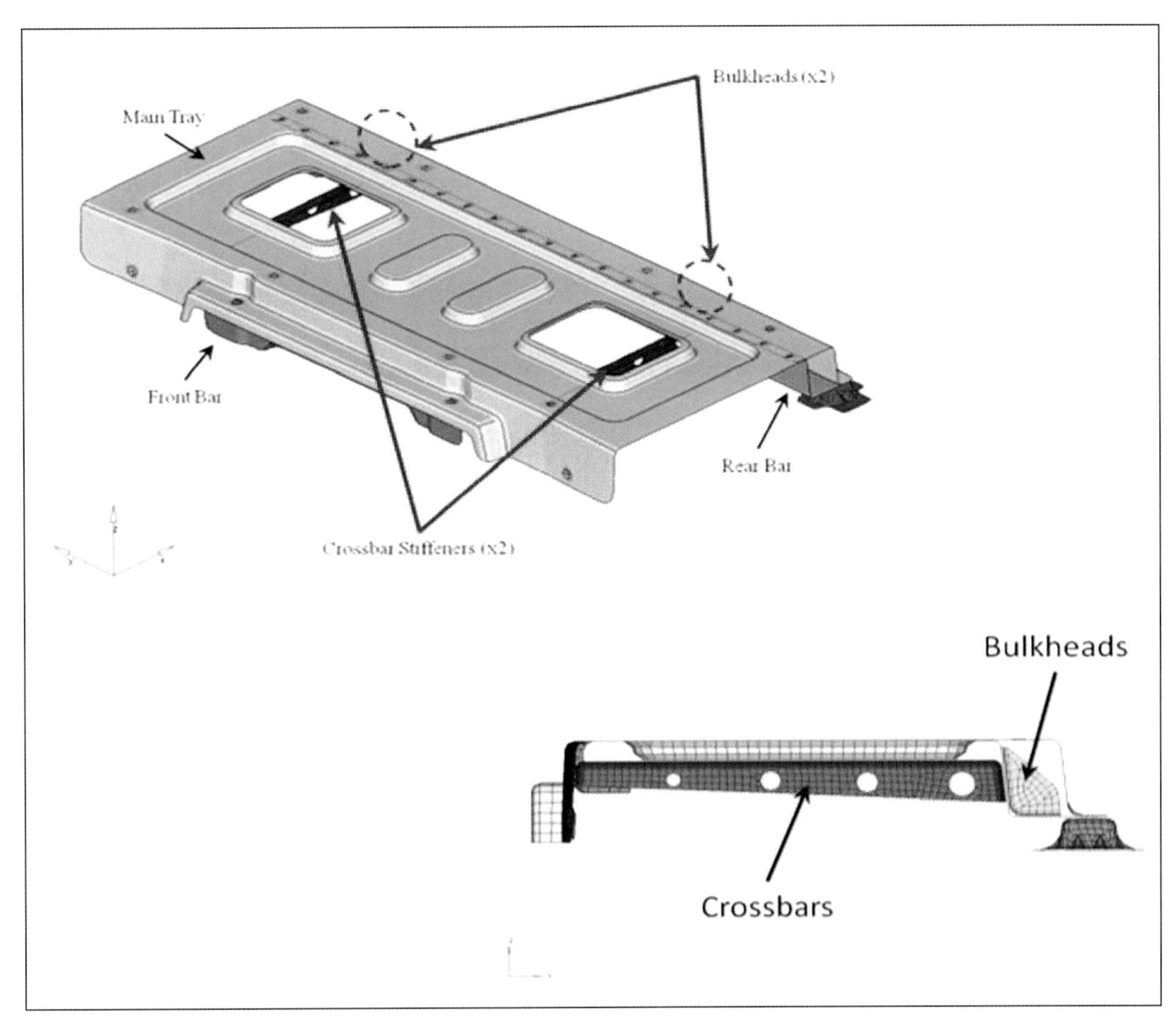

Figure 12. Support System Design #5, Thickness Combination #4 with Additional Reinforcements

Table 6. Mass and Performance Percentage Results of Reinforcement Additions to Design #5, Combination #4

Design #5, Combination #4 Reinforcements	% Change in Mass from…		% Change in First Mode Frequency from…		
	Design #1	Design #5-4	Target	Design #1	Design #5-4
Crossbars + Bulkheads	7.5%	6.4%	22.5%	44.4%	26.8%
Bulkheads Only	2.8%	1.8%	-2.8%	14.6%	0.6%
Crossbars Only	5.7%	4.6%	14.3%	34.8%	18.3%

Figure 13. The Battery Pack Sitting on a Simulated Shaker Table

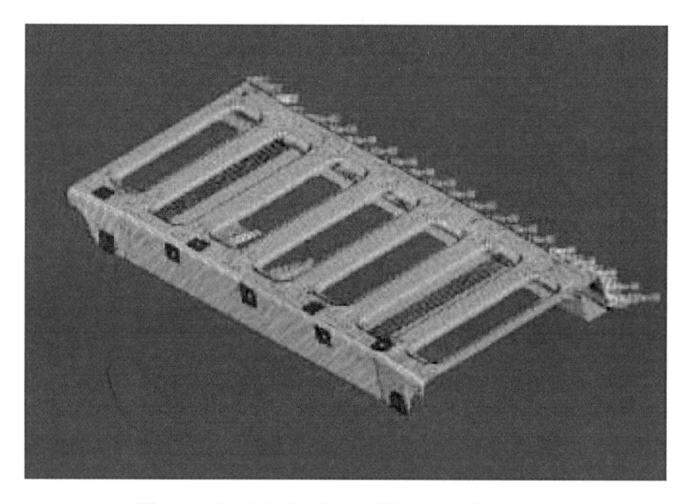

Figure 14. Main Tray Element Selection

Figure 15. Rear Bar

Figure 16. Crossbar Reinforcements

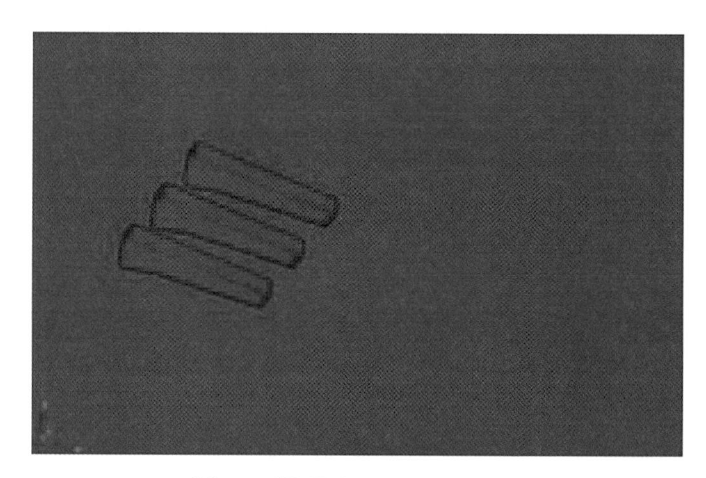

Figure 17. Tab Reinforcements

Figure 18. Rear Bar Vertical Reinforcements

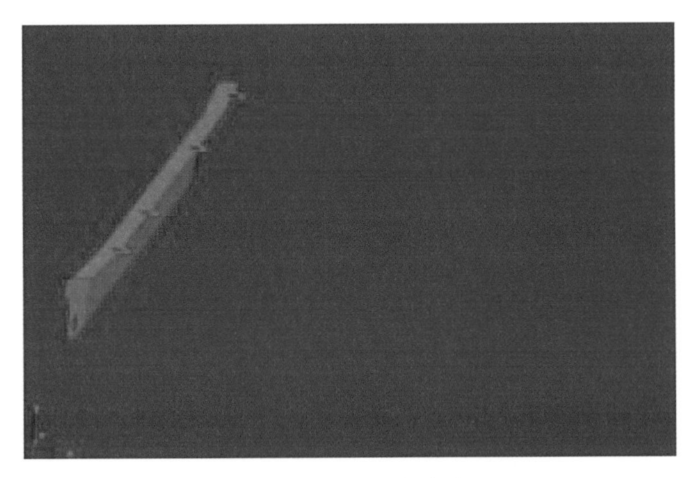

Figure 19. Front Inner Reinforcement

Figure 20. Rear Inner Reinforcement

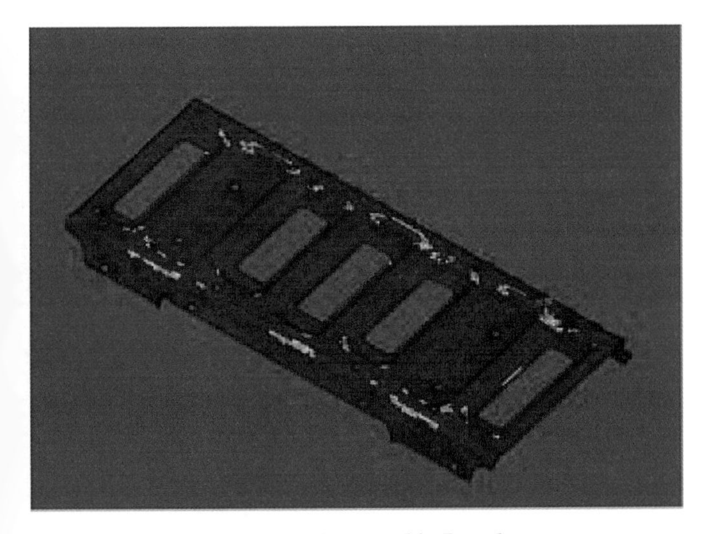

Figure 21. Tray Table Results

Three PSD profiles, which were generated from measured data, were applied to the battery pack CAE model. Optistruct was used as the solver software and nCode was used to assess the fatigue life. Figure 21 shows an example contour plot of the fatigue damage.

Durability Assessment and Optimization

A DOE was created and conducted for the mass and durability performance of the battery support system. The support system consists of the parts shown in Figure 22 in different colors.

Figure 22. Battery Support System

The thicknesses chosen for the DOE were based on typical thicknesses of sheet metal provided by the suppliers. Specifically, the thicknesses used were between 0.75 mm and 2.00 mm. The DOE varied the thickness of different tray components independently and simultaneously to determine the affect each component had on the support system's overall durability.

Next, the materials were investigated. The materials used in this DOE were typical materials provided by our suppliers, including standard steel (SAE 1008), cold rolled steel, and hot rolled steel sheet metals. Hot rolled steel has the highest durability, followed by cold rolled, and then standard steel. However, durability is not the only factor that must be considered when choosing a material grade, cost and manufacturability are also important. Fortunately, the cost of each steel is approximately the same, however, it is much more difficult to bend hot rolled steel to the design specs than it is for cold rolled and SAE 1008.

Attaching the battery support system to the vehicle body may be difficult; therefore a few modified support system designs were created. These modified designs were also included in the DOE.

Table 7. The Results of the Best/Optimal Case from the DOE

	SAE1008	Cold Rolled	Hot Rolled
Rear Rein. ΔT (mm)	0.75	0.75	0.75
Vertical Rein. ΔT (mm)	0.00	0.00	2.00
Front Inner Rein. ΔT (mm)	0.75	**1.50**	**1.50**
Rear Inner Rein. ΔT (mm)	0.75	**1.50**	**1.50**
ΔMass (kg)	1.25	1.83	2.08

The DOE showed that the battery support system components varied in the level of contribution to the durability of the entire support system. The components contributing less could have significantly reduced thicknesses or be removed altogether without much effect on the durability. More important components could have reduced thicknesses but would need to have a higher grade of steel to be structurally sound. Overall, the DOE showed that there is much to learn about the design of advanced battery components. Through proven engineering techniques, designs can be optimized, leading to lower mass and more efficient vehicles.

Thus, based on the analysis, the optimal design should be made out of cold rolled steel and have reduced component thicknesses that result in a mass reduction of approximately 1.83 kg. Although the mass is slightly higher, the cold rolled steel design is still preferred over the hot rolled design due to its ease of manufacturability.

The battery pack and battery support system were integrated into the full vehicle. The battery subsystem durability performance was evaluated under the vehicle condition using transient simulation. There was a close correlation between the shaker table simulation results at the battery pack level and the results at vehicle level, in terms of the high potential stress areas.

A physical test is necessary to verify the CAE design decision. The design review engineer created a prototype of the optimal battery support system design. The prototype was installed into a mule vehicle and run on the 4-poster test rig. No cracks were found over "2 lives". Thus the new design was verified in a complete cycle of durability performance assessment.

SUMMARY/CONCLUSIONS

A hybrid battery tray design was developed based on CAE simulation, analysis, and optimization at both battery pack level and at the vehicle level. Its performances of safety/crash, NVH, and durability were assessed while being optimized. The outcome of the optimization resulted in the best balance of mass reduction and meeting the requirements of safety/crash, NVH, and durability. The optimized battery tray design based on CAE analysis recommendation was validated by having a batch of prototype battery trays integrated in the full vehicle for both safety/crash and durability tests. The test results showed excellent correlations between the CAE analysis results and physical tests. The battery tray design was then implemented into the hybrid vehicle production.

The entire battery tray analytical process and procedures have been captured, will be documented and standardized for future HV battery programs.

CONTACT INFORMATION

Kristel Coronado
General Motors Tech Center
30001 Van Dyke, Warren, MI, 48090
kristel.coronado@gm.com

John Lyons
General Motors Tech Center
30001 Van Dyke, Warren, MI, 48090
john.lyons@gm.com

Thomas Wang
General Motors Tech Center
30001 Van Dyke, Warren, MI, 48090
thomas.j.wang@gm.com

Randy Curtis
General Motors Tech Center
30001 Van Dyke, Warren, MI, 48090
randy.c.curtis@gm.com

ACKNOWLEDGMENTS

Thanks to Simon Wang of Robust Synthesis, General Motors for his supports on many fatigue assessment runs made for the optimization work.

DEFINITIONS/ABBREVIATIONS

BDU
Battery Disconnect Unit

CAE
Computer Aided Engineering

DOE
Design of Experiments

HV
High Voltage (batteries)

NVH
Noise, Vibration, and Harshness

The Engineering Meetings Board has approved this paper for publication. It has successfully completed SAE's peer review process under the supervision of the session organizer. This process requires a minimum of three (3) reviews by industry experts.

ISSN 0148-7191

SAE Customer Service:
Tel: 877-606-7323 (inside USA and Canada)
Tel: 724-776-4970 (outside USA)
Fax: 724-776-0790
Email: CustomerService@sae.org
SAE Web Address: http://www.sae.org
Printed in USA

High Voltage Connect Feature

2011-01-1266
Published
04/12/2011

Trista Schieffer, Suneet Katoch and Richard Marsh
General Motors Company

doi:10.4271/2011-01-1266

ABSTRACT

Extended Range Electric Vehicles (EREVs), which are Off board charging capable Electric Vehicles (EV) with an on board charging generator, rely on very complex Rechargeable Energy Storage Systems (RESS) and High Voltage (HV) distribution systems to enable operation as both an EV and an EREV.

The connect feature manages the connection and disconnection of a High Voltage (HV) Rechargeable Energy Storage System (RESS) to and from the high voltage components in the vehicle. The RESS is connected to the vehicle's high voltage system to enable vehicle operation.

The HV connect feature is a part of occupant, service personnel and first responder safety for all General Motors vehicles that contain high voltage systems. Implementation of the connect feature is the method deployed in GM vehicles to meet high voltage FMVSS requirements.

INTRODUCTION

The vehicle HV system relies on the connect feature as part of the energy distribution strategy.

Functionality

• The connect feature manages the connection and disconnection of a High Voltage (HV) Rechargeable Energy Storage System (RESS) to and from the high voltage components in the vehicle. The connect feature is responsible for verifying the integrity of each portion of the HV system prior to and during use, as well as ensuring the system has been properly shut down at the end of use.

Safety

• The connect feature is used as the primary mechanism to render the HV energy storage and distribution systems safe in the event of an internal system fault or an external impingement such as a vehicle crash.

Efficiency

• The connect feature allows for complete disconnecting of the RESS from the vehicle high voltage bus to eliminate parasitic drains on the RESS.

Reliability

• The connect feature has been engineered and validated to extremely high levels of reliability as it is part of the HV Safety System and it is also a primary element of providing vehicle propulsion.

Regulatory

• The connect feature is a portion of the RESS thermal management system and thus is covered by both US EPA and California Air Resources Board requirements for On Board Diagnostics.

The Connect feature consists of physical hardware and associated control logic. The physical hardware implementation of the Connect strategy is through use of multiple HV contactors (relays) and Field Effect Transistors (FETs).

This paper will cover the functions utilized to perform the Connect feature portion of the HV energy management system, but not the physical requirements of the various contactors and FETs.

FUNCTIONALITY OF HIGH VOLTAGE CONNECT FEATURE

The connect feature manages the connection and disconnection of a High Voltage (HV) Rechargeable Energy Storage System (RESS) to and from all the high voltage components in the vehicle. The RESS must be connected to the vehicle's high voltage system to enable vehicle operation.

The connect feature utilizes a typical set of High Voltage contactors and FETs to manage the HV energy flow throughout the vehicle in all operating modes. (Fig. 1) The contactors and FETs are arranged to provide maximum flexibility for choosing operating modes while at the same time protecting the vehicle and occupants from failures that might be hazardous.

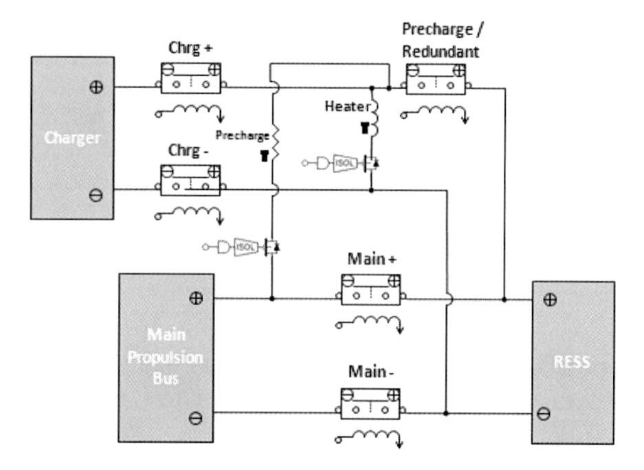

Fig 1. General layout

When the driver requests the vehicles propulsion system be available, the connect feature goes through the following start-up sequence (Fig. 2). The initialization step covers startup and communication of modules required for the connect feature. Additionally, during initialization several safety checks are performed to ensure that all necessary safety systems are fully functional. The precharge function is utilized to equalize the voltage on either side of the main contactors with a controlled current flow (Fig. 3). Diagnostics are run to further verify the safety of the HV distribution system and the contactors. The connection complete step is achieved when energy can be drawn from the RESS.

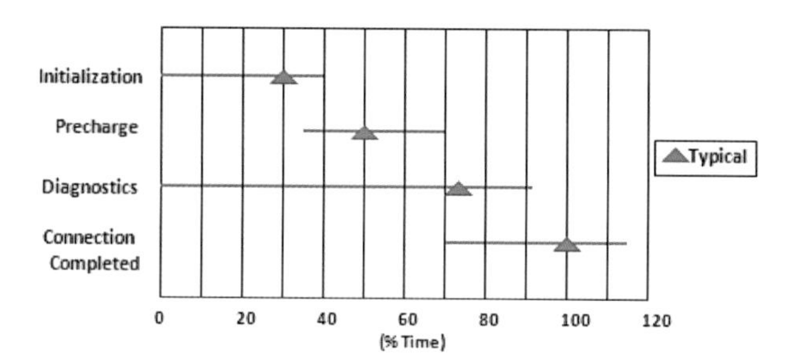

Fig 2. Start-up sequence:

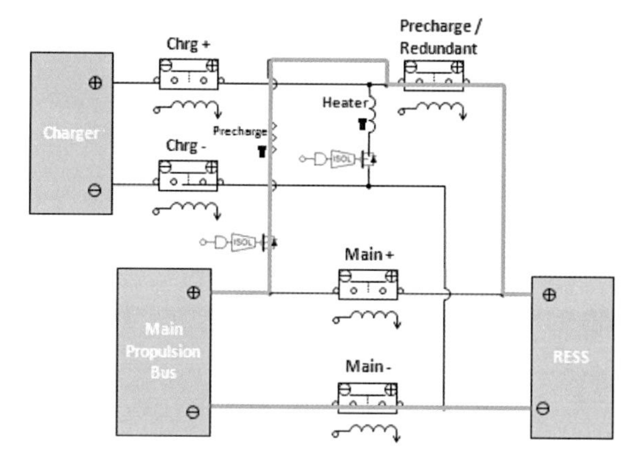

Fig 3. Precharge loop

When the voltage on both sides of the main contactors is balanced, the positive main contactor is closed, completing the main loop. An overlap when both the main loop and the precharge loop are closed is used to confirm the physical closing of the positive main contactor in the main loop. Once the positive main contactor is determined to be physically closed, the precharge loop is opened. At this point, energy can be drawn from the RESS. (Fig. 4)

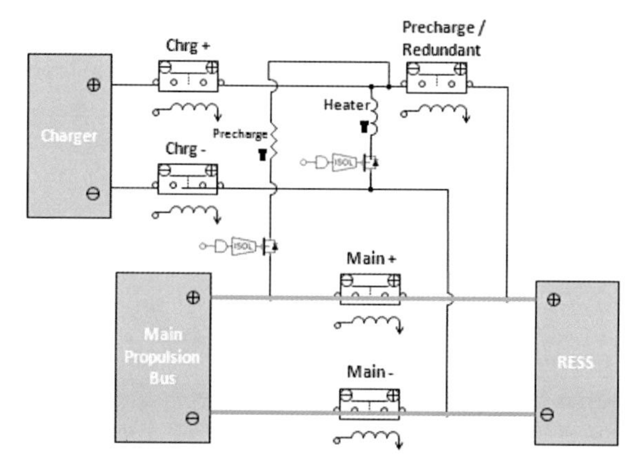

Fig 4. Main loop

Disconnection of the RESS from the high voltage components is done by a normal, a controlled or an emergency shutdown. Each shutdown has a specific time delay before the main loop is opened to allow high voltage components an opportunity to stop all high voltage use. Minimizing HV power across the contactors just prior to opening the main loop lowers the risk of damaging a contactor by welding.

When the vehicle is plugged in to an Off Board power source the charger can begin to charge the RESS. Both sides of the charging contactors must have balanced voltage prior to final closing of the charging loop and start of charging (Fig. 5). For the charging loop a startup sequence similar to the main loop is used.

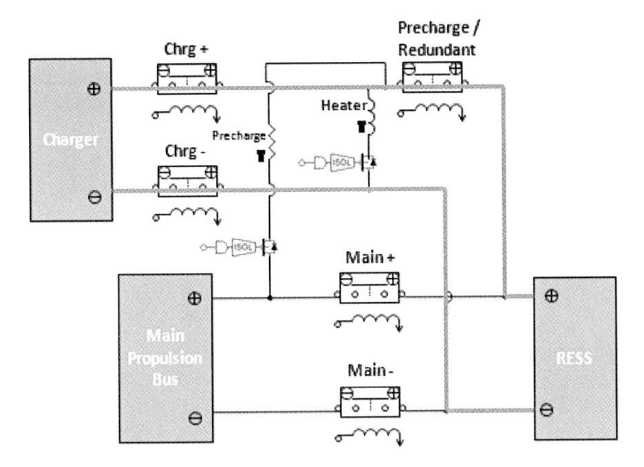

Fig 5. Charging loop

The RESS heater can be powered from either the charger or the RESS itself. Depending on the power source selected, the connect feature will engage the correct contactor(s) and then the FET is controlled to obtain the desired amount of heat. (Fig. 6)

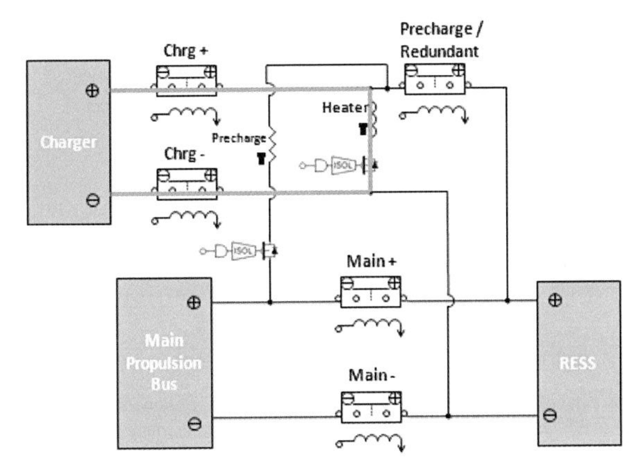

Fig. 6. Heater / Charger loop

SAFETY

EREV vehicles that utilize High Voltage energy to provide propulsion power have a sophisticated high voltage safety system to ensure safe operation during all stages of the vehicles life. This safety system operates from the early stages of vehicle assembly through normal operation, service, crash events and final disassembly at the vehicle's end of life.

The connect feature serves several purposes as part of the high voltage safety system.

- The connect feature prevents the RESS from being connected to the charger or the Main Propulsion Bus until all safety checks have been run.

- One safety check is to ensure that the main and charging loops are capable of being opened in an emergency. The primary failure mode of interest in this case is welding of the contactors, which could prevent one or both of the HV buses from discharging when required.

- Another safety check verifies the condition of the various components connected on the buses and once they have been proven safe, the power up sequence is permitted to begin.

- After the loops are connected, the connect feature disconnects the RESS from the charger and / or the main propulsion bus if the safety system detects a potentially hazardous situation.

As part of the HV Safety system, the connect feature is engineered to meet the requirements identified in ISO 26262 for safety critical control systems

- The connect feature is controlled by a safety critical OBD compliant control module. This control module is designed with appropriate redundancies and internal microprocessor checks to ensure safe operation.

- The connect feature is designed so as to not inhibit the HV Safety System's ability to determine high voltage bus isolation and detect loss of high voltage bus isolation. Isolation detection is essential to GM's strategy to comply with FMVSS 305.

The connect feature employs numerous strategies to ensure vehicle operation is shutdown only when absolutely necessary. Redundancies in specific input and output systems and continuously monitoring of the rationality of the inputs is used for safety critical control decisions.

EFFICIENCY

The connect feature allows for complete electrical disconnection of the RESS from the vehicle high voltage bus to eliminate parasitic drain on the RESS. While the RESS typically has low self discharge properties, other systems on the HV bus have sufficient parasitic losses to reduce the stand time. It is more efficient to do a full shutdown of the HV buses and open the contactors and then reenergize the HV bus when the vehicle is started than to leave the buses energized over long stand times.

RELIABILITY

The connect system has been engineered and validated to very high levels of reliability as it is part of the HV Safety System and it is also a primary element of providing vehicle propulsion. Connect system failures have the potential to lead to a walk home event and must be kept to an absolute minimum.

Based on the various functions the contactors are required to perform, the following reliability requirements have been established (Table 1):

Type 1. Safety requirement specifying the ability to open

- Must meet ISO 26262 requirements for an ASIL D system

Type 2. Availability requirement specifying the ability to close

Type 3. Emission Component requirement for ability to close

Table 1. Contactor Reliability Summary

Contactor	Disconnect	Connect
Main +	Type 1	Type 2
Main -	Type 1	Type 2
Precharge	Type 1	Type 2
Charge +	Type 1	Type 3
Charge -	Type 1	Type 3

Additionally, each contactor has a life requirement for the number of times it is required to open under full load and still maintain full capability.

REGULATORY

Plug in Electric Vehicles with an On Board Generating capability (EREVs) powered by an internal combustion engine and Hybrid Electric Vehicles are required to be compliant with EPA and CARB regulations for On Board Diagnostics. (OBD)

The connect devices used with a Rechargeable Energy Storage System (RESS) may be considered OBD if their failure could result in altered vehicle performance. For example, if RESS charging or heating or cooling are impacted by failures of the connect feature, the system and its components are considered OBD. If the hybrid operation of the vehicle is changed by certain failure modes of this system, the contactors would be considered OBD.

The connect feature on GM's EREV meets the criteria and has been designed to be OBD compliant

REFERENCES

1. ISO 26262.
2. FMVSS.

The Engineering Meetings Board has approved this paper for publication. It has successfully completed SAE's peer review process under the supervision of the session organizer. This process requires a minimum of three (3) reviews by industry experts.

ISSN 0148-7191

Positions and opinions advanced in this paper are those of the author(s) and not necessarily those of SAE. The author is solely responsible for the content of the paper.

SAE Customer Service:
Tel: 877-606-7323 (inside USA and Canada)
Tel: 724-776-4970 (outside USA)
Fax: 724-776-0790
Email: CustomerService@sae.org
SAE Web Address: http://www.sae.org
Printed in USA

CHAPTER THREE:

Engineering with a maniacal focus

A dedicated, cohesive team and a conservative engineering approach put this innovative vehicle into production at moon-shot speed.

Chief Engineer Andrew Farah, an EV-1 veteran, kept the program on schedule while managing to keep his sanity intact.

Volt's many learnings are now part of GM's "tool kit" for electrified vehicle development, said VLE Tony Posawatz.

Former Global Chief Engineer Frank Weber directed the program's critical early development before leaving to join Opel.

Not long after he became General Motors Vice Chairman for Product Development, Bob Lutz spoke with *AEI* about what he called "decoupled development." By disconnecting development of the most engineering-intense subsystems (*i.e.*, a new powertrain) from a vehicle program's critical path, overall efficiencies can be realized and a higher-quality end product can result.

But Lutz gave us his dissertation before the Volt program began. Few vehicle development efforts in modern times have required such an intense engineering effort, not to mention inventions, on the critical path. Before he retired, Lutz called Volt a "moon shot" and equated it to that mother of all critical-path projects, NASA's space program of the 1960s.

Reflecting on the four years of Volt development during the car's media launch in early October, Vehicle Line Executive Tony Posawatz said the program's success was due to one critical element.

Engineering
with a maniacal focus

Interior Design Manager Tim Grieg's group worked with the body and powertrain development teams to wrap the four-seat cabin around the large battery pack.

"It really came down to how well the overall team—from those dedicated to the Volt program, to the engineers and scientists in GM R&D, to our key supplier partners—worked together. Looking back, the team made many, many smart decisions and maintained a maniacal focus on developing this vehicle."

Posawatz joined Volt from the start. A superb communicator with a comprehensive understanding of the vehicle, its subsystems, engineering challenges, regulatory issues, and the charging environment, he played a key role in connecting key constituent groups and keeping them informed on Volt's progress.

When a reporter noted that GM kept the Volt program alive throughout its financial meltdown, bankruptcy, and government-funded resurrection, Posawatz joked that "regardless of which CEO had recently fallen, the next one always seemed to say he liked Volt more than the previous guy!"

The team's strength also proved itself after losing a few key engineering executives—including Global Chief Engineer Frank Weber (now at Opel); Director of Energy Storage Systems Denise Gray (at a California battery start-up); Executive Director of Global Hybrid and Electric Vehicle Engineering Bob Kruse (now at battery R&D company Sakti3), and Assistant Powertrain Chief Engineer Alex Cattelan, who left to join AVL North America.

By making "smart decisions," Posawatz refers to not reinventing technology wherever possible and always erring conservatively on the side of robustness, reliability, operating safety, and customer delight.

"There are many, many examples of our team doing things intelligently," he noted. "Instead of developing a new blended braking system, we leveraged what we had done on the hybrid trucks (T900) and Saturn Vue Hybrid. We used the same suppliers, same everything, but with a few subtle enhancements." (See sidebar interview with TRW's Dan Milot, page 45.)

Volt's electric power steering system is very similar to what GM is applying across its global portfolio of conventional vehicles, he added. But the battery, its controls and management, and the car's super-efficient HVAC system required "enormous effort" to develop, test, and validate.

"We opted to over-engineer everything—to make all subsystems so safe. We lavished loving care on the battery and its thermal management, and on the charger and power electronics. They're all liquid-cooled, so we can keep them in a bandwidth that's easily controlled, and maintain behaviors that we fully understand," said Posawatz.

The team "engineered-in multiple [coolant and control] loops" to manage the battery and electric powertrain, a strategy that ended up costing more than if GM had chosen air cooling or a less conservative design approach, Posawatz explained. He added, however, that being "too conservative" in the early stages of vehicle electrification will pay off as GM Powertrain continues to refine and optimize the Voltec system.

"I believe the one that's the fastest learner in this game has the greatest advantage," he said. "The regulatory requirements, the customer-delight issues, the warranty-cost issues, all had to be managed. But basically we had a great team and we were on a mission."

Was there ever a point in the program where the team faced hurdles that appeared insurmountable—

GM Director of Global Battery Systems Engineering Ronn Jamieson worked closely with his counterparts at LG Chem's Compact Power group in tuning, testing, and validating the new prismatic Li-ion cells.

that may have caused them to question the effort's chance of success?

"There was never a point where we went back to senior leadership and said, 'We're stumped.' There was never a point of anyone even considering that," Posawatz stated. "But there were many heroic efforts by team members to accomplish this project. We didn't hit a single huge pothole or sinkhole in the road, but we hit a lot of little ones. And the team responded incredibly well and kept moving forward."

Solving the BSE riddle

One of the program's most daunting technical challenges, one whose solution caused a "hurrah" moment, was developing the car's BSE (battery-state estimator) program. The BSE is an enormously complex control algorithm on which each powertrain subsystem's operation and performance is dependent.

A BSE that's not properly developed, whose calculations are off by mere percentage points, can cause the ICE to engage at inopportune times and decrease the car's EV operating range. NVH and driveability also can suffer.

"Talk about an algorithm that's really, really hard to develop and get absolutely right, the BSE was a big one for us," Posawatz recalled. "But now that our guys know how to do it, we have a major competitive advantage that opens the door in a lot of technology areas going forward." Volt's BSE is key intellectual property for the company, he added.

The engineers who found the BSE solution worked for Scott Miller, the Vehicle Performance Manager at the Milford Proving Ground, in charge of controls. He told *AEI* there were times where it seemed the team had run up against a wall.

"This is the first production extended-range electric vehicle with a lithium-ion battery that does plug-in charging," Miller explained. "Monitoring what's going on in the battery when it's in charge-sustaining mode, with the engine running, is different than when the car's charging plugged in. The BSE predicts this activity. It's a major algorithm and it has to consider a lot of variables. A small error in monitoring any one of the 288 battery cells can create a performance issue."

The team "significantly beat" its extremely aggressive original target for error, he noted.

Developing the architecture, the hardware, and the propulsion modes put a lot on each engineer's platter. As Volt's launch arrives, Posawatz and Miller quipped that most of GM's engineering staff will be putting in for the vacations they missed during the crushing 40-month program.

"Volt is all about balanced operation—delivering an efficient point of running the engine, staying within the bandwidth that the regulators like, and making the car pleasing to the customer," acknowledged Posawatz. "There were a lot of struggles in solving this three-legged stool.

"But now it's all part of our 'tool kit.' We plan to do a few more electrically powered vehicles, so we think our learnings may come in handy!"

Chevrolet Volt
Development Timeline

January 2007 – GM reveals the Volt concept car at the North American International Auto Show.

June 2007 – Advanced Li-ion battery-development contracts awarded to LG Chem and Continental Automotive Systems.

August 2007 – GM and A123 Systems announce partnership to co-develop battery cells with A123 Systems' nanophosphate battery chemistry.

November 2007 – "Decoupled development" phase including proof of concepts, engineering development vehicles, and mules begins. Design of Volt's electric drive system incorporates elements of the Two Mode hybrid transmission. First mules use Malibu bodies (called MaliVolts).

May 2008 – GM Vice Chairman Bob Lutz test-drives the first pre-production Volt at the Milford Proving Grounds. "Decoupled development" phase for subsystems ends.

June 2008 – GM's Board of Directors approves the Chevrolet Volt and "Voltec" electric propulsion system for production starting in late 2010.

Sept. 2008 – Many aerodynamic changes to the Volt body are revealed at GM's Centennial celebration. Second-gen mules using Cruze bodies arrive.

November 2008 – Product development phase begins.

January 2009 – GM announces it will develop and manufacture Volt's battery pack in the U.S. It also selects LG Chem as the Li-ion battery cell supplier for production.

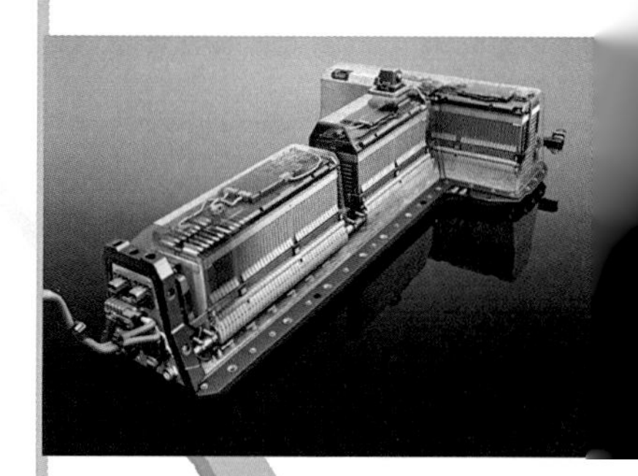

March 2009 – Volt's European cousin, the 2012 Opel Ampera, is revealed at the Geneva Motor Show.

June 2009 – GM opens its new Global Battery Systems Lab at the Warren Technical Center. The facility is four times larger than GM's previous battery lab. It employs >1000 engineers.

August 2009 – As part of the U.S. Recovery Act, the DoE awards GM over $240 million for the high-volume assembly of battery packs and electric motors, and for testing hundreds of Volts.

October 2009 – Volt road testing moves to Pikes Peak, where the car's performance on long grades and at high altitudes, plus brake regeneration and other aspects, are evaluated.

November 2009 – 80-car PPV (prototype production vehicle) build completed. 300 Volt battery packs under test. PPV mules have racked up 250,000 mi in all conditions. Systems calibration is 65% complete.

December 2009 – Full-vehicle simulation tests (300,000-mi. equivalent) completed. The Detroit/Hamtramck plant begins building Volt parts using production tools and processes.

January 2010 – Chevrolet and OnStar unveil the industry's first working smartphone application. The Volt mobile app works on the Apple iPhone and Motorola Droid. It allows 24/7 remote connection and control of various vehicle functions and OnStar features.

February 2010 – GM's new Brownstown Twp. (MI) Battery Plant builds the first Volt pre-production battery pack.

March 2010 – The first production-spec Volts are built on the Detroit-Hamtramck assembly line. These vehicles are used for final validation and testing prior to on-sale.

April 2010 – GM announces $8 million expansion of the Global Battery Systems Lab, doubling the facility's size to 63,000 ft^2. The first pre-production 2012 Opel Ampera is built at the pilot plant in Warren. GM unveils Volt MPV5 crossover concept at Bejing Show.

May 2010 – Final major calibration test drives are under way in the western U.S. Systems calibration is 99% complete. GM unveils Volt's "Mountain" driving mode for extreme duty cycles.

July 2010 – Chevrolet announces Volt's MSRP: $41,000. Buyers are eligible for U.S. income tax credits up to $7500. A $350/month 36-month leasing scheme also debuts.

August 2010 – First production battery packs are built at the Brownstown Twp. plant. President Barack Obama drives the Volt at Detroit-Hamtramck. Chevrolet ups its planned Volt production capacity in 2012 to 45,000 units.

October 2010 – The global automotive media drives production Volts in Michigan. GM reveals hybrid-drive elements of Volt's electric propulsion system.

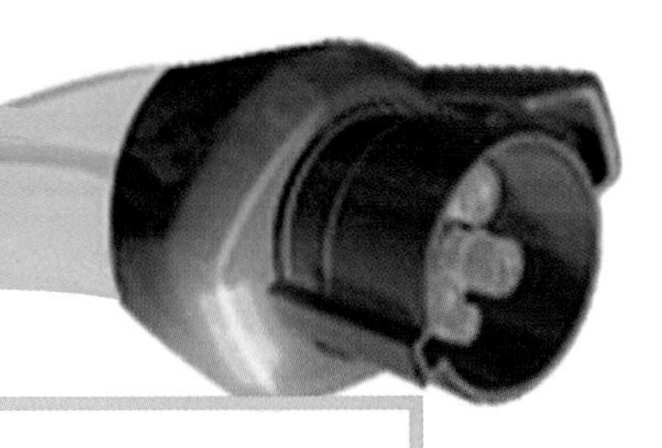

November 2010 – Volt series production begins.

OnStar deployed as a tool for Volt engineers

Engineers working on the 2011 Chevrolet Volt are using OnStar to pull data from various modules as the car is test-driven.

The Chevrolet Volt's engineering team is using OnStar's wireless connectivity to get real-time data from preproduction Volt vehicles accumulating test-drive miles.

"We are literally updating, refining, and making more robust every module in the car all in parallel," said Dana Fecher, Group Manager of the OnStar EV Lab in Detroit.

Nick Pudar, Vice President of Planning and Business Development for OnStar, said that as of Dec. 17—118 days since data collections began—24 Volt development vehicles had accumulated a total of 86,405 mi (139,055 km). "We're pulling data from 20 different modules on each of those vehicles, and we've reached out to those vehicles 1741 times to pull information," said Pudar.

According to Fecher, those 1741 information pulls netted approximately 200,000 individual pieces of data.

Volt Vehicle Line Director Tony Posawatz said that OnStar's connectivity helps engineers make the Volt's lithium-ion battery pack more robust. For example, engineers can obtain data that indicates battery temperature or battery voltage at given points in time.

Access to the real-time data is helping engineers diagnose problems quicker.

"The speed of getting the feedback and the ability to crunch and manipulate and use the data for diagnosis is really, really helpful. We're feeling very good about the progress we're making," said Posawatz.

Volt owners will get an assist from a forthcoming Volt/OnStar mobile smartphone application, which will enable remote vehicle communication via Motorola Droid, Apple iPhone, or Blackberry Storm. The Volt/OnStar application also will be available on a mobile browser for other Internet-capable phones.

A Volt owner can use the smartphone application to communicate with the vehicle, for example, to schedule a time to charge the battery or initiate an immediate battery charge. Other functions include receiving text or e-mail notifications for charge reminders/charge interruption/full charge as well as starting the vehicle remotely to precondition the interior temperature, thus preserving the battery charge for electric-mode driving.

"We've accelerated the implementation of this technology, but we've done it in a manner that's very prudent and very robust so customers should say, 'Wow, these guys are thinking about everything,'" said Posawatz, adding, "The Volt has always been about an EV that can become your primary car—no range anxiety, all the features, all the amenities, and fun-to-drive characteristics. And we're going to deliver it November 2010."

– Kami Buchholz

CHAPTER FOUR:

A unique electrified transaxle

The view inside GM's new 4ET50 electrified transaxle (at left) shows the dc generator motor, ac traction motor, a trio of hydraulic clutches, and the planetary gearset. The control module is integrated into the main case.

Hybrid or not? Definitions aside, what really matters is GM wisely leveraged its next-generation Two Mode propulsion technology to give Volt greater overall efficiency.

A unique electrified transaxle

Volt's combination of 1.4-L four-cylinder ICE and 4ET50 is a compact unit.

Leave it to the Volt to cause the telephones at SAE International to ring almost non-stop. Since the car's media launch in early October, reporters have been looking for answers as they try to apply a standard technical description to the Volt's cleverly engineered powertrain.

GM has called it an "E-REV" (extended-range electric vehicle) ever since the concept was unveiled in January 2007. Over the course of the next 40 months, simply mentioning the word "hybrid" around any of Volt's senior development team members netted plenty of discourse as to why their baby is a new breed of propulsion system. Indeed, *AEI*'s first story covering the original show car described Volt as a series-type plug-in hybrid, drawing fire from two ranking executives on the program.

So what type of powertrain is Volt's? SAE's definitions for the main types of hybrid vehicles define "hybrid" as "a vehicle with two or more energy storage systems, both of which must provide propulsion power—either together or independently."

We describe a series-type hybrid as a vehicle "in which both sources of energy go through a single propulsion device." And a plug-in hybrid (PHEV) is defined as "a hybrid vehicle with the ability to store and use off-board electrical energy in the RESS (rechargeable energy storage system).

Currently there is no standard definition for an E-REV, but *AEI* continues to see Volt as primarily hybrid. It combines series and PHEV capabilities, and delivers battery-EV performance for those owners who travel less than 40 mi in their daily driving cycles. It can't deliver an EV's zero tailpipe emissions, however, because as explained in Chapter 8 the 1.4-L internal combustion engine (ICE) is calibrated to start up every 45 days, regardless of whether it's called on for generator duty, and run for about 10 min for various checks and maintenance.

Battery EVs, however, are currently incapable of matching Volt's go-anywhere, anytime, capability, made possible by its liquid-fueled generator in the powertrain support role.

As a relevant aside, the car's customers won't care about technology labels. They'll only care that Volt delivers as promised—as it did during *AEI*'s recent 150-mi (241-km) test drive of a production-spec Volt in southeastern Michigan. Under 70°F (21°C) ambient conditions our car with driver and two passengers netted 44.5 mi (71.6 km) of EV range in varied traffic conditions and terrain.

The four-cylinder engine engaged remarkably quietly and seamlessly, chiming in at its governed 4800-rpm maximum but effectively muffled through great attention to NVH details.

The car had no trouble accelerating up to and sustaining its 101 km/h (63-mph) governed maximum speed. The average "fuel economy" in electric and extended range (with generator) modes combined was 51 mpg. It will be interesting to observe the vehicle's performance in daily use under thermal extremes. Your mileage, as they say, may vary.

GM Powertrain engineers draw distinctions between Volt and a classic PHEV, noting their

Providing the tools for fast, accurate simulations and testing

Mahendra Muli

Speed, efficiency, and engineering precision were critical to all aspects of Volt's development. General Motors engineers explained that only with the latest simulation and testing tools were they able to bring the car to market in such a compressed timetable, given the high level of electrical and electronic systems developed on the critical path. *AEI* asked Mahendra Muli, Manager of New Business Development for dSpace, about his company's involvement.

Q: What specifically did dSpace contribute to the 2011 Volt program?

Muli: We provided rapid controls prototyping and hardware-in-the-loop systems for electronic control unit software development, validation, and verification for technology development of powertrain, vehicle dynamics, body electronics, safety, and, finally, integration testing.

Q: How did dSpace collaborate with the Volt/GM engineers?

Muli: We provided engineering support for our products for various product development and testing departments at GM. Our engineers helped ensure that the development and testing platforms performed to the satisfaction of GM engineering teams involved in meeting complex challenges of the technology development under tighter timelines.

Q: What challenges did the Volt program present in terms of technology development/systems integration?

Muli: In general, vehicle electrification has posed challenges to the simulation systems required for development and testing of these complex vehicle architectures. The fast dynamics of the electrified vehicle powertrain systems require high computational power. dSpace responded by providing test systems based on the latest processor and FPGA technology and high-speed interconnection between processor boards. This resulted in powerful multicore, highly scalable simulation platforms used for component and integration testing of vehicle systems.

Q: What learnings did your team take away from your company's involvement with the Volt program?

Muli: Automotive engineering is taking a huge leap in evolution with electrification, led by automotive OEMs such as GM. The tasks and challenges are enormous both at the OEM level and at supplier levels. The challenges demand every player to put their best foot forward and more. Individuals and companies have to be highly adaptive in learning newer technologies, and to respond with solutions to various problems that arise. In summary, we learned that it takes the entire community to collaborate and move at a rapid pace in this evolution of vehicle propulsion.

Q: What opportunities does vehicle electrification present to your company?

Muli: It has expanded the industry's scope in both energy storage systems such as batteries and ultra-capacitors, as well as semiconductors, battery management systems, and components for electric drives and related electronics. Also, it has increased the complexity of the overall vehicle electronics, resulting in the need for extensive validation and verification of software in production programs.

decision to equip the car with a relatively large battery (compared with plug-in hybrids) for greater EV-mode range. This allowed them to downsize the combustion engine, although not to the extent originally envisioned. First, they admit Volt resembles a series hybrid in its basic configuration. They acknowledge its operating characteristics are similar to those of a PHEV in Initial EV drive mode.

However, they also note an E-REV must maintain this mode of operation during all operating schedules when energy is available from the battery. Unlike the PHEV, it doesn't need to transition to a blended-operation strategy when battery energy is available.

The Volt's battery and electric drive system are sized so that the ICE never is required for vehicle operation when battery energy is available. This configuration doesn't require a specified operating cycle, while PHEVs require discussion of the urban schedule, the engineers argue.

The accompanying bar graph from GM shows that as greater electric-only operation is required, motor size and overall electric-propulsion capability must be increased. Hybrids and PHEVs are able to blend electric and ICE power to propel the vehicle, so they require less total onboard power than Volt.

"Unlike any hybrid, Volt delivers full performance on electric power alone," said the car's "godfather," GM Ventures President Jon Lauckner. "It only draws energy from the liquid fuel in the tank after the initial battery SOC is depleted. This is an electric vehicle with full vehicle performance available as an EV, but with the onboard electric generator to extend driving range and take advantage of the liquid-fuel infrastructure that will be in place for many years globally."

Three operating modes

Volt's drive system can be operated in three distinct driving modes, via the center console mounted selection lever. Normal Mode is what most owners will likely use for their daily commutes. It biases a moderate energy draw from the battery to the traction motor for maximum range. Sport Mode changes the accelerator map to increase torque response.

Mountain Mode helps optimize the car's performance under long, steep-grade conditions.

HEV and E-REV peak motor and engine power as a percentage of peak vehicle power demand

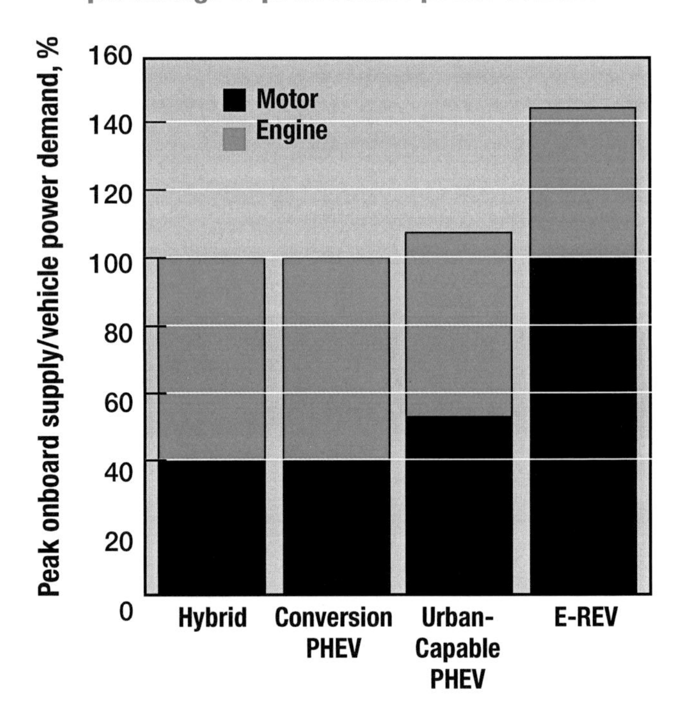

GM bar graph shows that as greater electric-only operation is required, motor size and overall electric-propulsion capability must be increased. Hybrids and PHEVs are able to blend electric and ICE power to propel the vehicle, so they require less total onboard power than Volt, which offers full-electric-propulsion capability.

According to Larry Nitz, GM Executive Director of Hybrid and Electrical Powertrain Engineering, Mountain mode enables the power controllers to "dip deeper into battery" while under range-extended operation.

When the driver selects Mountain Mode, the controllers place more of the battery's energy capacity in reserve than during Normal Mode operation. This causes the ICE's revs to rise a few hundred rpm, providing supplemental power and boosting the charge-sustaining capability of the ICE.

"When you have a substantially downsized engine as we do—we call it "half-sized" in terms of its power-delivery capability of the electric side of the vehicle—you need to be able to reach into the battery and pull that extra performance," Nitz said during a Western states test session last May.

Micky Bly, GM's Executive Director, Electrical Systems, Hybrid & Electric Vehicles and Batteries, said future generations of the Voltec propulsion system are well into development.

The Specifications section in this publication provides details of Volt's powertrain component set. But it wasn't until the car's media launch that GM officially revealed the internal workings of the new 4ET50 electrified transaxle. Various mechanical and electronic-control aspects of this cleverly designed, slick-operating device were first implemented in GM's Two Mode drive developed for C/D-segment front-drive applications. Its initial application was to be the 2010 Saturn Vue Green Line prior to Saturn's closing.

The Two Mode Vue, a PHEV, was configured for EV-only range of more than 10 mi (16 km). Higher speed and load demands would engage the vehicle's ICE and/or electric power to propel the vehicle.

4ET50 trickery yields benefits

GM's recent surprise revelation with Volt is the capability to clutch-in the 55-kW generator motor and connect it to the ring gear of the planetary gearset for greater operating efficiency. Under most road speeds and loads, the engine/generator is waiting to be deployed when battery state-of-charge drops below 35-30%. It's basically along for the ride unless the ride is longer than Volt's 25-50 mi (40-80 km) of battery range.

When the battery is depleted, the generator engages to maintain minimum SOC until the car can be plugged in for cheaper and more efficient static charging. Torque to the drive wheels is provided by the 111-kW traction motor coupled to the planetary's sun gear.

But in a few narrow operating conditions, typically accelerating from 70 mph (113 km/h) while in extended-range mode, the shaft speed of the 111-kW traction motor begins to exceed its peak efficiency. So the system's designers took advantage of the more direct and efficient mechanical connection to the car's wheels offered by coupling the generator and ring gear. In doing so, the generator brings the traction motor's shaft speed down to a more optimum rpm.

GM engineers claim the resulting power flow provides a 10-15% improvement in highway fuel economy.

The engineers set up this subtle technique because the 4ET50's three hydraulic clutches, two electric motors, planetary gearset, and ample control software gave them the right stuff to do it. "If the hardware and controls are there, why not use them," commented Volt VLE Tony Posawatz.

So is the Voltec drive system a hybrid or something else? The car simply cannot move without power from the electric motors. Remove the traction motor from the equation and the ICE generator cannot propel the vehicle alone. Remove the ICE, and Volt remains an EV.

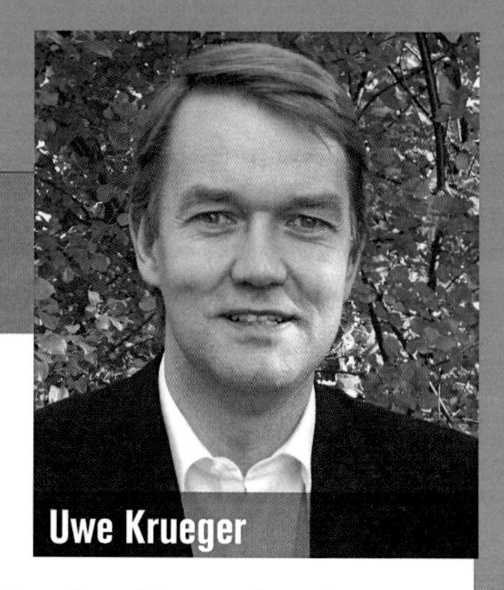

"Chilling out" **Volt's battery pack**

Behr America engineers collaborated with their General Motors counterparts to create an extremely sophisticated thermal management system for Volt. Behr's Vice President of Engineering, Dr. Uwe Krueger, talked about his company's technical contributions.

Q: What did Behr contribute to the Volt program?

Krueger: We contributed two major parts. One is we're doing the full front-end cooling module. The second is what we call the 'chiller.' [This is a highly compact, purpose-built heat exchanger designed to efficiently transfer waste heat from the secondary battery-cooling loop to evaporated refrigerant from the car's A/C circuit. The evaporated refrigerant is used to re-cool the secondary loop in extreme-temperature conditions.] In designing the chiller we also had discussions with **CPI** so that we understood each other and to ensure the systems integration works well.

Q: How far up front did Behr work with the GM Volt engineering team?

Krueger: We've been involved since the very early stage of the project. We were part of a concept consortium which brought together the basic concept of how the thermal management system of this very unique range-extended car could work. We started obviously with a lot of design studies and thermal analyses. Then we got the production contract. We also helped optimize the thermal management technologies when GM revised the car's aerodynamics during body development.

Q: What learnings did you and your team take away from the Volt program?

Krueger: First, that early involvement is key to the program's success. If we had not been involved with the program from the beginning, it would have been difficult to pull it together in the end. Cooling of the electric drivetrain is in some ways more challenging than cooling conventional IC vehicles. By working with GM on Volt we learned a lot about the needs of advanced Li-ion batteries and how to optimize the next generation of EV thermal systems. We have great ideas on how to make them even more robust and help to lower costs.

Battery Use	2010	Around 2015	Around 2020
HEV-focused batteries	>32 kW	>32 kW	>32 kW
	260-300 V	260-300 V	260-300 V
	$45-60/kW	$30-40/kW	$20-30/kW
E-REV-focused batteries	>120 kW	>100 kW	>100 kW
	>8 kW·h usable	>8 kW·h usable	>8 kW·h usable
	$500-2000/kW·h	$300-400/kW·h	$200-300/kW·h
Data derived from industry targets. Source: GM			

GM and other OEMs expect advanced lithium battery costs for automotive use to fall while battery performance rises—but neither will likely happen as quickly as product planners would prefer.

The Voltec 4ET50 Electric Drive System

2011-01-0355
Published
04/12/2011

Khwaja Rahman, Mohammad Anwar, Steven Schulz, Edward Kaiser, Paul Turnbull, Sean Gleason, Brandon Given and Michael Grimmer
General Motors Company

doi:10.4271/2011-01-0355

ABSTRACT

General Motors' Chevrolet Volt is an Extended Range Electric Vehicle (EREV). This car has aggressive targets for all electric range with engine off and fuel economy with the engine on. The Voltec 4ET50 transaxle has gears, clutches, and shafts and controls that execute two kinematic modes for engine off operation or Electric Vehicle (EV) operation, and two additional kinematic modes for extended range (ER) operation. The Voltec electric transaxle also has two electric motors, two inverters, and specialized motor controls to motivate to execute each of those four driving modes. Collectively these are known as the Voltec Electric drive. This paper will present the design and performance details of the Chevrolet Voltec electric drive. Both the machines of the Voltec electric drive system are permanent magnet AC synchronous machines with the magnets buried inside the rotor. The motor has distributed windings. However, as opposed to a conventional stranded winding the Chevrolet Volt motor has bar-wound construction to improve the motor performance, especially in the low to medium speed range. At higher speed the skin effect and proximity effects in the stator bars lead to increased stator winding losses but are addressed in the design. The bar-wound construction also has excellent thermal performance in both the steady-state and transient conditions necessary for full EV driving. The generator uses concentrated windings. The concentrated winding construction has good slot fill and extremely short end-turn length. These features resulted in good performance in the intended operational region and were an enabler for machine packaging inside the transmission. Both the machines exhibit excellent efficiency and exceptionally smooth and quiet operation. Machine design and construction details as well as the measured thermal, electromagnetic and acoustic noise performance are presented in the paper. The Traction Power Inverter Module (TPIM), which is designed to commutate these motors and uses state-of-the-art power electronic components, adopts flat and compact structure to achieve high current density and exhibits improved manufacturability., Hardware and software design tradeoffs were considered to meet TPIM reliability, efficiency, performance and overall security requirements. These features will be discussed in this paper. To address the aggressive targets for fuel economy and EV range of Chevrolet Volt, engineers were forced to closely examine all loss mechanisms involved in the electric drive system and optimize motor controls to provide the best drive system efficiency. While optimizing the drive system efficiency, other constraining requirements such as vehicle Noise Vibration and Harshness (NVH), high voltage bus ripple and resonance, component thermal performance and controllability, etc. were addressed. This paper will further discuss the issues encountered and solutions adopted to optimize the drive system efficiency to help the Chevrolet Volt meet the fuel economy and EV range targets.

INTRODUCTION

The main goals of GM's alternative technology vehicles are the reduction of emissions, the reduction of fuel consumption and the diversification of the energy sources. One way to reach these goals is by increasing the level of "electrification" [1]. GM was first in modern days to offer complete electrification when it's Battery Electric Vehicle (BEV) EV1 was introduced in 1996 [2]. Subsequently GM introduced hybrid electric vehicle (HEV), plug-in HEV (PHEV) and most recently, the Extended Range EV (EREV). An EREV has full functionality as an EV as long as battery has enough energy and hence provides more electrification than PHEV that has blended operation of gas and electricity and the engine may be often forced on to support vehicle speed or power, even hen the battery has plenty of useable energy[1]. As explained in [1], an EREV is very similar to the vehicle which was envisioned by the expert panels of Air Resource Board (ARB) in 2007 [3]. An EREV has zero emission vehicle (ZEV) mode when it runs as pure EV. GM is first among OEMs to produce EREVs.

Figure 1 shows Chevrolet Voltec electric drive system context diagram. It has two machines. Motor-B is connected to the final drive and is mainly used as motor. Motor-A is connected with the engine and Motor-B through appropriate clutch configurations depending on the mode of control. Motor A is as motor in EV operation and mainly as a generator in ER operation. The Voltec electric drive system also contains an oil pump motor. The oil pump motor supplies the oil, used for kinematic mode control, lubrication and coolant for the entire transmission system when the main hydraulic pump is off. The inverter module, labeled as TPIM, acts as a power conversion and distribution box between the HV battery and the motors. Voltec TPIM has three power inverter modules (PIMs). Two main inverters, PIMA and PIMB, to run these traction motors and a third inverter, PIMC, to run the oil pump motor. These PIMs are voltage source inverters (VSI).

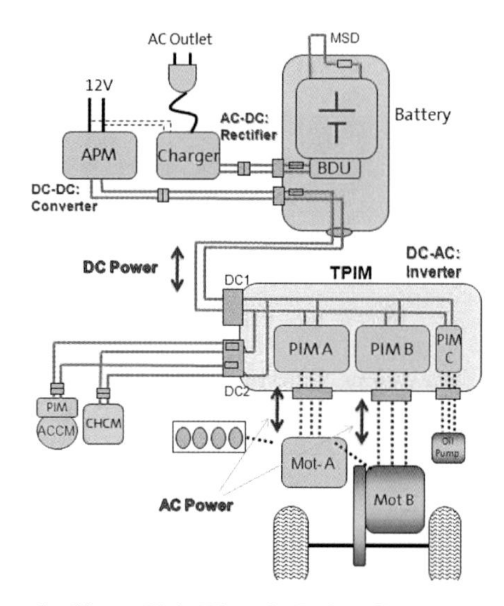

Figure 1. Chevy Volt Electric Drive Context Diagram.

Figure 2 shows a generic block diagram of a voltage source inverter (VSI) showing major building blocks. The VSI interfaces with HV battery, motor, inverter cooling system and the rest of the vehicle electrical systems and controls. The main building blocks of the VSI are the power modules, the DC bus capacitor, EMI filter capacitor, switching signals, motor controller. The motor controls algorithms are programmed into the motor controller.

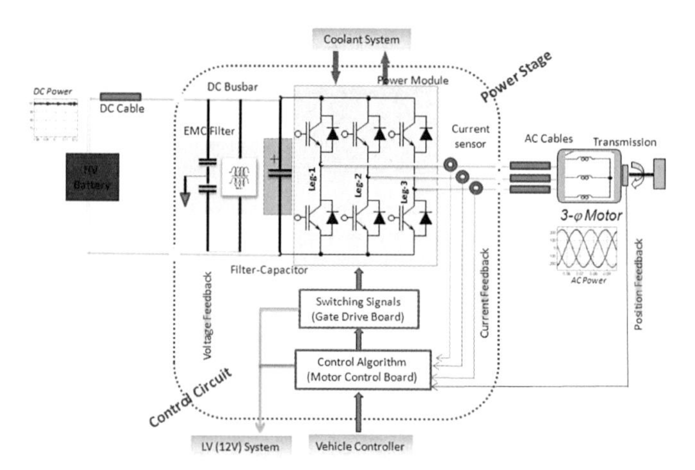

Figure 2. A voltage source inverter (VSI) showing major building blocks and system interface.

The Voltec electric system has very aggressive targets for both all electric range and fuel economy. The traction drive efficiency is a large contributor to the overall vehicle performance. To meet these targets, engineers were forced to closely examine all loss mechanisms and optimize motor controls to provide the best drive system efficiency. However, the optimization the drive system efficiency was done within the other constraints presented, such as NVH, bus resonance, controllability. Hence, a trade-off involving input from various vehicle design disciplines was necessary. This paper will discuss the issues encountered and solutions adopted to optimize the drive system efficiency to help the Volt meet the fuel economy and EV range targets.

This paper is organized as follows: Volt traction machine design and construction will be addressed first. The power electronics drives designs will be described next followed by the motor control algorithm development. The final section would present some relevant test data. Summary and conclusion are presented at the end.

ELECTRIC MACHINES

Both the machines are buried (interior) permanent magnet ac synchronous machine type. The larger of the two machines is used mostly as an electric motor (to provide electrical propulsion or tractive effort) and is of a bar wound type construction. This machine is shown in figure 3 and is referred to as motor B in this document.

Figure 3. Motor B stator and rotor assemblies

The bar-wound construction, as opposed to the more conventional stranded type has several advantages, i) higher slot fill, ii) shorter end-turn, iii) improved cooling performance, iv) fully automated manufacturing process, v) improved high voltage protection etc. Some of these attributes would be elaborated later in this section. Due to bar-wound construction, the slots have parallel sides as opposed to stranded design where the tooth hasve parallel sides. This is illustrated in figure 4.

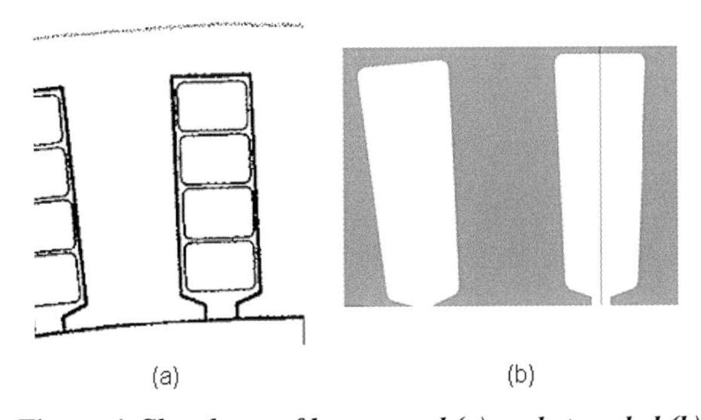

Figure 4. Slot shape of bar-wound (a) and stranded (b) design.

The winding layout of the bar-wound motor B has wave-winding construction whereas lap winding is typically used for stranded design. "Hairpin"s of the bar-wound construction are formed outside and then inserted in the slot. A twisting operation forms a frog-leg type shapes which are then welded together to form the wave-winding pattern. Figure 5 shows a hair pin shape before and after insertion and twist.

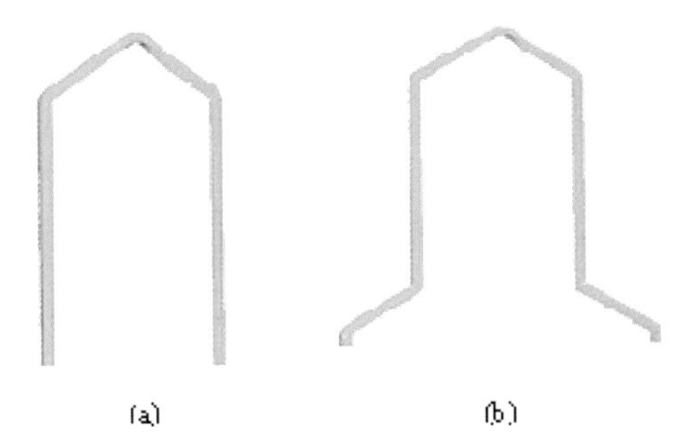

Figure 5. Hairpin shape before (a) and after insertion and twist (b)

The use of a bar-wound design on the Volt motor B allows the conductors to be designed with specific shaped coils and phases. As a result Volt motor B exhibits an excellent voltage protection. The amount hairpin width or span and the conductor width mostly dictate the end-turn height. Volt motor B has a small end-turn height which resulted in a larger active stack for the given packaging dimension. A cross section view of the Volt motor B stator and the end-turn region are shown in figure 6.

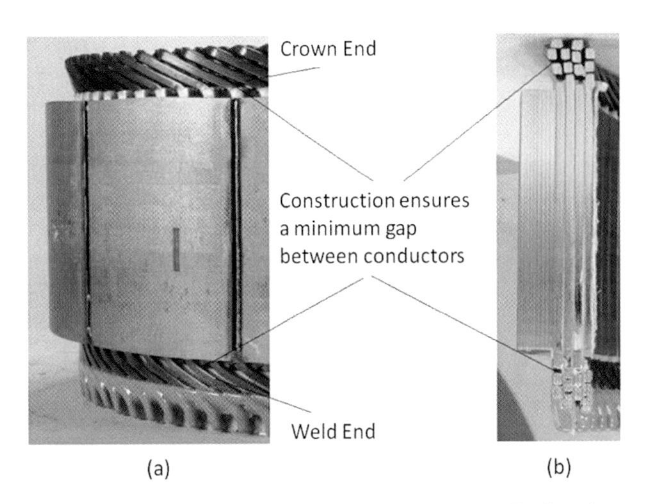

Figure 6. Cross sectional view of Volt Motor B showing the end-turn crown and the weld-end (a) and the detailed view of the conductors in the end-turn region (b).

Volt motor B has four bars (conductors) per slot as shown in figure 7 (a). The insulation system was further improved by the geometry of the slot insulation that provides isolation of the conductors from the stator laminations. This was accomplished through the use of a B-shaped insulator rather than a more conventional S shaped insulator as shown in figure 7. This is an improvement from the earlier designs, GM Rear Wheel Drive two-mode hybrid, where an S-shape insulation paper was used. The B-shape provides better insulation by eliminating the gap in the corner that exists in an S-Shaped insulator between conductor and stator laminations.

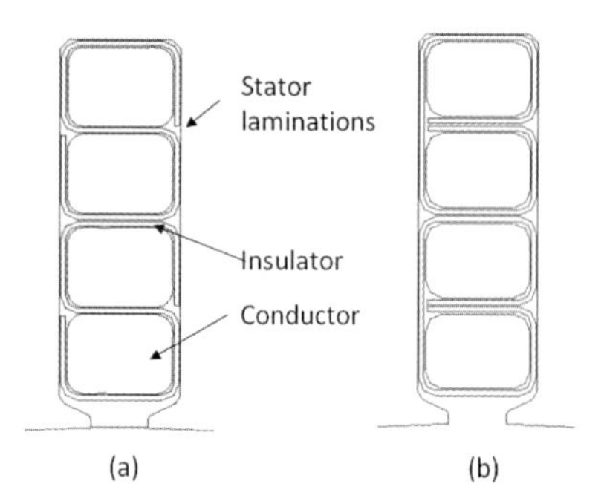

Figure 7. S-shape insulation (a) as used in GM Rear Wheel Drive 2-mode hybrid and the improved B-shaped (b) insulation as used in Volt motor B.

To provide further isolation between conductors, insulation material was inserted between layers on the weld end of the stator coils. The insulator between the two sets of conductors was made to protrude higher than the others to provide increased insulation between adjacent welds. To further make the bar-wound construction robust to manufacturing variation, the unique wire forms were coated with an insulating epoxy. This epoxy while providesing an additional insulation coating.

The dc resistance of the bar-wound machine is significantly lower than the stranded design. This results from two facts as stated earlier, i) the slot fill of the bar-wound design is significantly higher than the slot fill of the stranded design, ii) the end turns are tighter for the bar wound design as compared to the stranded design. For Volt motor B, the phase resistance is roughly 30-40% lower than the equivalent stranded design which would produce the similar torque and power. However, the bar-wound construction would have significant skin and proximity effect which would increase the ac resistance of the machine at high speed. Figure 8 shows the ac resistance of the motor B as function of machine fundamental frequency and machine phase current. The conductor was sized properly to minimize the eddy current effect at higher speed. Hence, the skin effect in this particular design is not excessively high. Figure 8 also shows the phase resistance of an equivalent stranded design. It is assumed that the stranded design has no skin or proximity effect. This assumption is quite valid for machine built with multiple thin strands. Although the skin effect is computed as function of total phase current and machine speed, the skin effect actually depends not only on phase current but also on the split of the phase current between d-axis and the q-axis. Hence, it is a three dimensional function of d-axis current, q-axis current and machine speed.

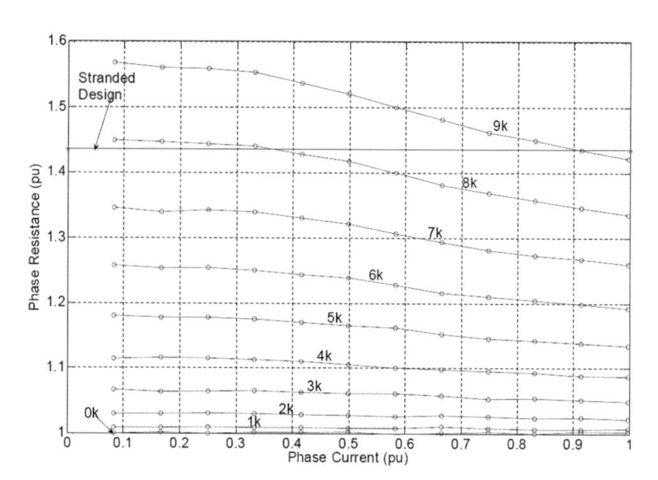

Figure 8. Machine phase winding ac resistance as function of phase current and machine speed.

It is evident from this plot that bar wound design has favorable winding resistance up to 8 krpm. Above that speed the stranded design starts to perform better than the bar wound design. Motor B usage profiles for most drive cycles show most significant motor B operation is all well below 8000 rpm above which bar wound starts to lose its advantage. Figure 9 shows the motor B usage profiles for the ftp urban 2 (a), highway (b), and US06 (c) drive cycles, where most of the motor B operations are below 5000 rpm where bar wound have its clear advantage over the stranded design.

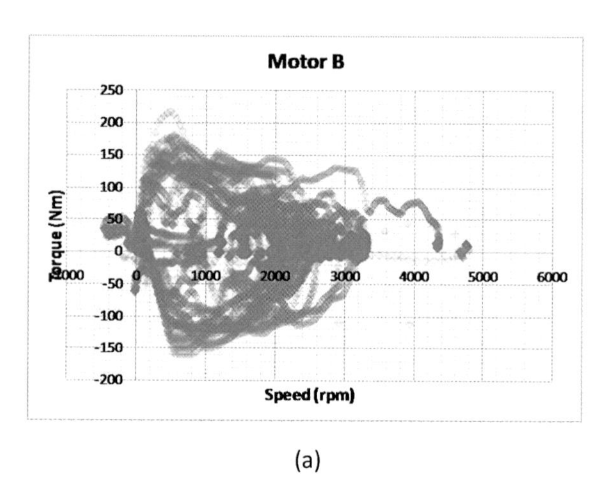

(a)

Figure 9. Volt Motor B usage profile (a) ftp urban2, (b) highway, and (c) US06

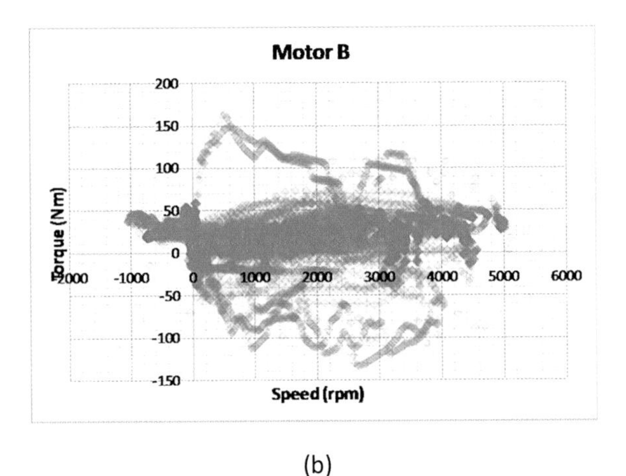

(b)

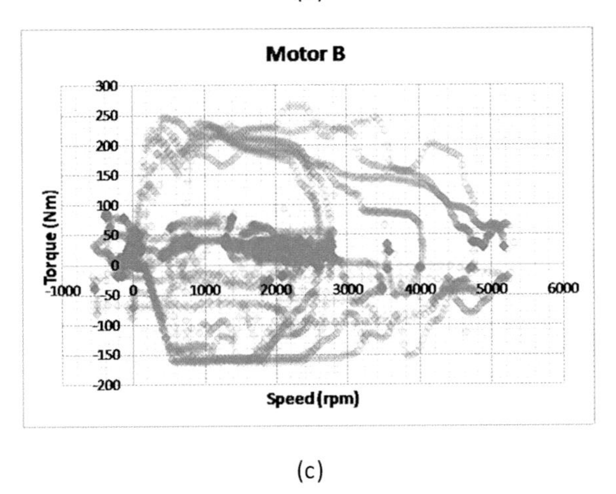

(c)

Figure 9. Volt Motor B usage profile (a) ftp urban2, (b) highway, and (c) US06

However, these usage profiles also show that there are significant motor B operations at light load areas at medium to high machine speed where the motor iron loss starts to dominate the total machine loss. Hence, reduction of machine iron loss was quite critical. This was addressed during the machine design, especially the rotor design.

Machine rotor of motor B has a V-shaped construction. This construction has favorable rotor saliency behavior, thus improving machine torque density. The rotor barriers, where the magnets are buried, are constructed with additional air pocket (barrier) on the top of the magnet. This is illustrated in figure 10. The additional air pocket on top of the magnet acts as an air-gap to the magnet. Hence, addition of this air pocket reduces machine iron loss caused by the rotating magnet due to reduction in the airgap magnet flux. This additional air pocket also reduces the harmonic in the air-gap, thus, further lowering the iron loss due to rotating magnet. The addition of this air pocket reduces the d-axis inductance and hence improves the rotor saliency. As a result, reluctance torque of the machine is increased which to some extent offsets the loss of magnet torque due to reduction of the air gap flux. The impact of the air pocket on the machine peak torque has been small.

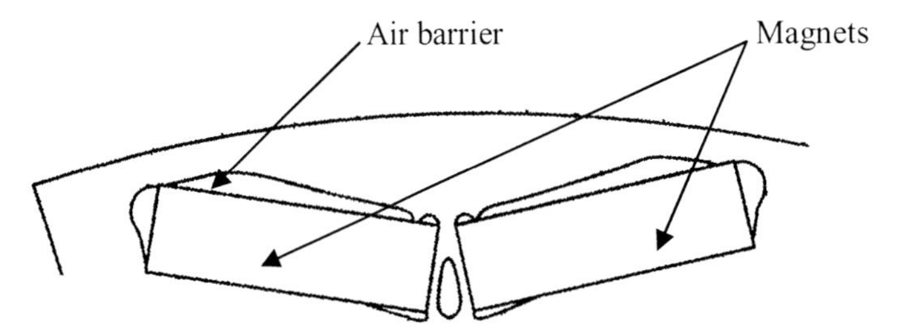

Figure 10. IPM Rotor Geometry of motor A.

The simulated unpowered iron loss of motor B is shown in figure 11 as compared to a design without the air pocket where the iron loss for the machine with air-pocket is the reference. A significant reduction of machine iron loss has resulted due to the addition of this air pocket.

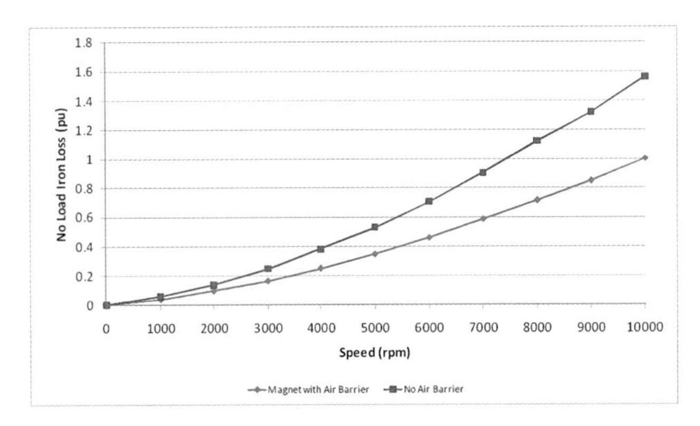

Figure 11. No load spin loss with and without air-barrier.

The other traction machine in the Volt drive system which mostly runs as a generator is a concentrated winding type. This machine is referred to as Motor A in this paper. The concentrated winding machine has significant packaging advantage and is well suited for limited space. Motor A rotor has 16 permanent magnet poles buried inside the rotor while the stator has 24 slots (teeth). An end view of motor A rotor is shown in figure 12. Noise is a concern with concentrated winding machine. There are two major components for noise. One results from the pole passing frequency and the other results from the torque ripple frequency. The distribution of tooth forces as the pole passes the teeth is very critical for noise. The 24-16 geometry has an 8 lobe forcing distribution. Figure 13 below shows the 8 lobe forcing distribution of the 24-16 geometry (which was selected for Voltec Motor A) and compares that to another geometry which has 4 lobe forcing distribution such as of a 24-20 geometry (20 pole rotor). 24-16 geometry has same peak level of tooth forces, however, more evenly distributed than the 24-20 geometry. Moreover, the eight-point pull of the 24-16 geometry as opposed to four-point pull of 24-20 geometry results in a much quieter drive performance. This was exhibited during the Chevrolet Volt vehicle NVH test which showed exceptional noise behavior. Some of these noise performance plots would be presented in the TEST DATA section.

Figure 12. Motor A rotor end view.

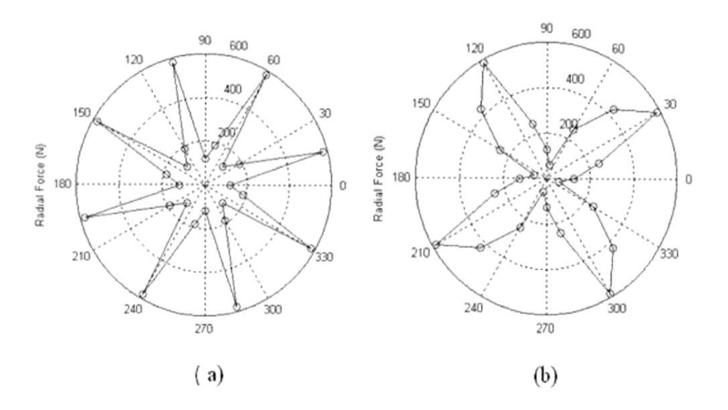

Figure 13. The tooth force distribution of 8-lobe 24-16 geometry (a) and 4-lobe 24-20 geometry (b).

Volt Motor A and B specifications are presented in Table I below. Motor B operates to a maximum speed of 9500 rpm. Motor A maximum speed is dictated by the engine speed and operates to a maximum speed of 6000 rpm.

Table I. Specifications of Volt Motor A and B

Motor B	Torque	370 Nm
	Power	110 kW
	Max Speed	9500 rpm
	Number of Poles	12
Motor A	Torque	200 Nm
	Power	55 kW
	Max Speed	6000 rpm
	Number of Poles	16

Both of the Voltec motors demonstrate superior thermal performance. This is achieved by cooling of the rotor and stator end windings by the use of ATF (automatic transmission fluid). Both motors are mounted in a manner that reduces the amount of energy that can be transferred to the transmission structure. As a result, the primary method to remove heat is through the direct cooling provided by the ATF rather than conduction to the case. Motor B with bar-wound construction specifically has improved thermal performance due to its increased surface area of the end winding which is in direct contact to the ATF. Figure 14 shows the typical ATF cooling inside the transmission for motor B. The cooling flow inside the transmission is shown by the arrows in figure 14. The cooling flow has two major paths. Coolant flow flows on the surface of the end turns on both ends by cooling tubes. The second flow is from inside the rotor which flows through the rotor and cools the rotor and additionally splashes in the end-turn thus providing further cooling for the stator. Motor thermal as well as efficiency performance data would be presented in the TEST DATA section later.

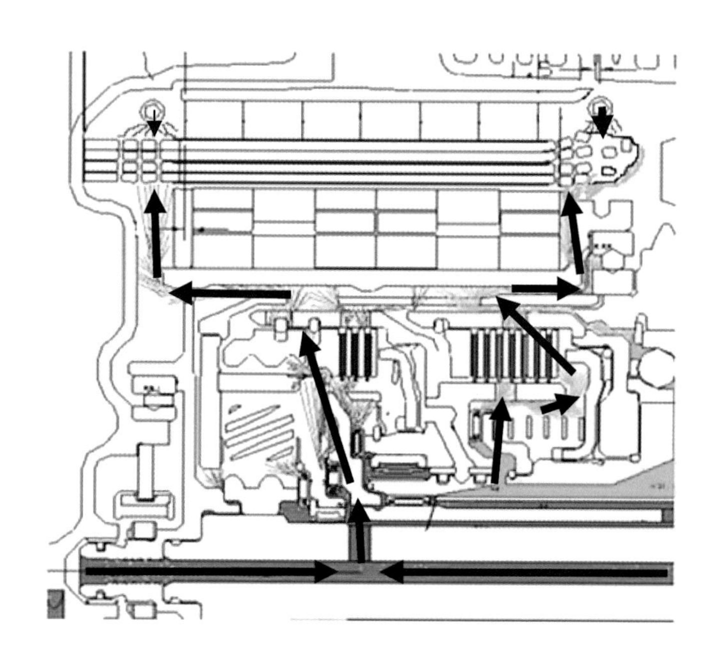

Figure 14. Motor B ATF cooling flow

POWER ELECTRONICS

Power electronics, particularly traction power inverter, functions as the crucial power distribution and conversion box between the battery and the electric motor. It plays an important role in realizing the unique performance feature for this car. This state-of-the-art power electronics design is optimized balancing the quality, reliability, size, cost, efficiency, performance and overall safety [4-5]. The Voltec system functionality, interdisciplinary nature of power electronics technology, optimal integration of different HV components were considered to synthesize and complete the design.

PRISMATIC AND SPECIFICATION

The Voltec Traction Power Inverter Module (TPIM) contains three voltage source inverters (VSI). PIMA and PIMB run two main motors (Mot-A and Mot-B) and PIMC run the oil pump motor as shown in the context diagram of Figure 1. The major building blocks for each of these inverters are as shown in the generic block diagram of Figure 2. TPIM operates in under hood environment with its own cooling system (water ethylene glycol) with nominal cooling temperature at 75 C. It interfaces with the motors motors via shielded copper cables as shown in Figure 15.

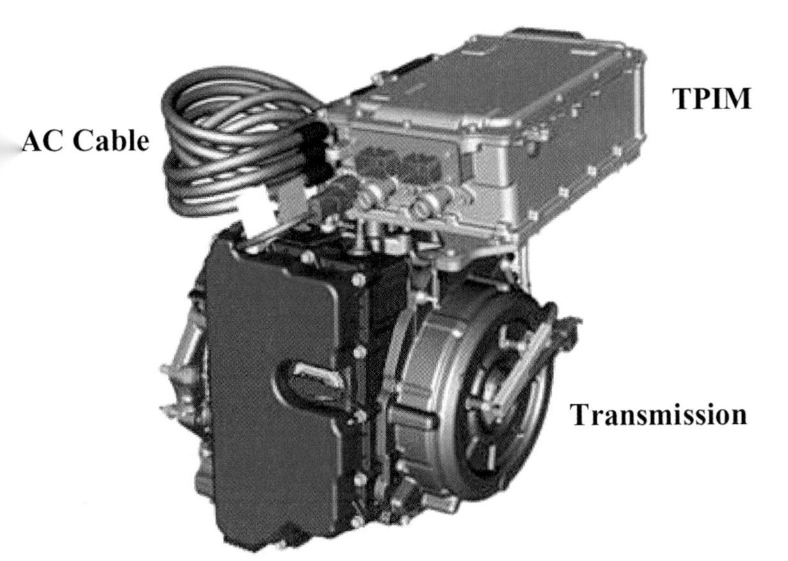

Figure 15. Chevy Volt Electric Drive System

From operational and cost perspectives, the power module, filter capacitor, EMC components, DC and AC bus bars, harness and connectors require significant design attention. Figure 16 shows a detail representative cost break down of TPIM. It shows power stage represents a large percentage, approximately 56% of the total TPIM cost. Understanding of these cost drivers help determine GMs level of system specification required to optimize the power electronics system. System level design selection tradeoffs as well as control algorithm optimization are considered during development.

The major specification for Chevy Volt TPIM is shown in Table II. Compared to General Motors Gen 1 TPIM which is used in the Chevrolet Tahoe 2-model hybrid vehicle, this TPIM has 22% lower mass and 47% lower volume despite it's 50% higher total power (kW) ratings. This is a high current, flat and compact package inverter with good productivity and miniaturization. These size and volume reduction are realized because of stat-of-the-art power electronics components, simple bus bar and terminal designs, simple coolant manifold, and reduced number of harnesses and unifying circuit boards.

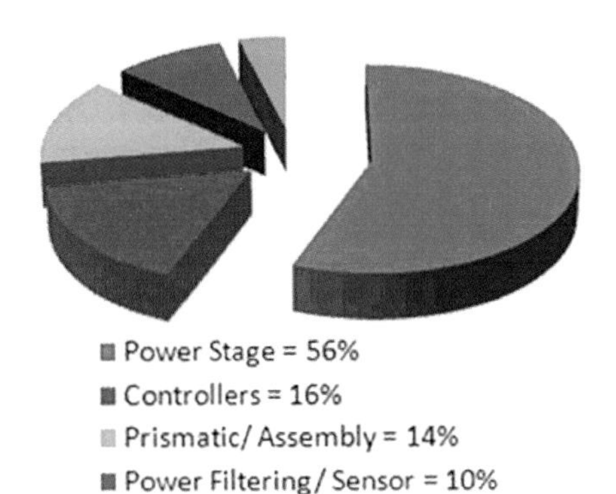

Figure 16. Power electronics cost distributions.

Table II. TPIM Major Specification and Ratings

Parameters	VOLT TPIM
Size (mm)	350x230x133 (mm)
Volume (L)/Mass (kg)	10.7 / 14.5
Incorporated main components	Motor-A Inverter Motor-B Inverter Oil Pump Inverter Motor Controller Hybrid System Controller
Power Rating (A+B+C) KW	(120+120) + (1.5 Oil Pump)
Current Rating A/B (Arms)	150 (Continuous), 450 (Peak)
Voltage Rating (Volt)	430 (Peak), 360 (Nominal)
Coolant Temp (C)	75
Power Density (kVA/L)	36.2
Gravimetric Power Density (kVA/kg)	26.7

Ease of mass production and manufacturability improvements helped reduction of unit cost. The flat structure and simple assembly process can be understood from the exploded structural diagram of figure 17. The number of assembly process is reduced by 20% and hand-operated (manual) processes is limited to nine both of which have reduced total assembly process by about 63%, as compared to Generation-1 Tahoe TPIM.

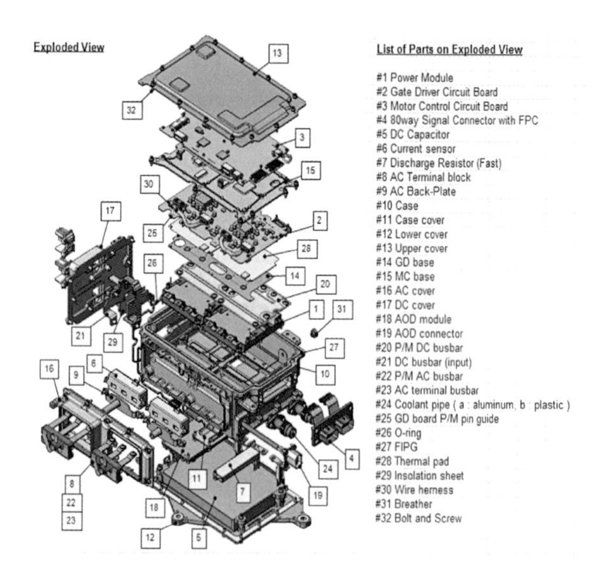

Figure 17. Exploded Structural Diagram of VOLT TPIM.

POWER MODULE

As mentioned in previous section that, Chevy VOLT TPIM has used state of the art power electronics, particularly power module, to achieve a high current density and a compact design. Figure 18 shows power module for one of the traction inverters. It is using high performance trench IGBTs (HiGT) and modified diode (U-SFD) in a low inductance bus bar layouts. This power module package is also ensuring enhanced coating and better contact with heat sink. Figure 19 shows that IGBT silicon surface area has to be increased by only 11% to handle peak phase current requirement even though the peak current is 50% higher than the previous TPIM design.

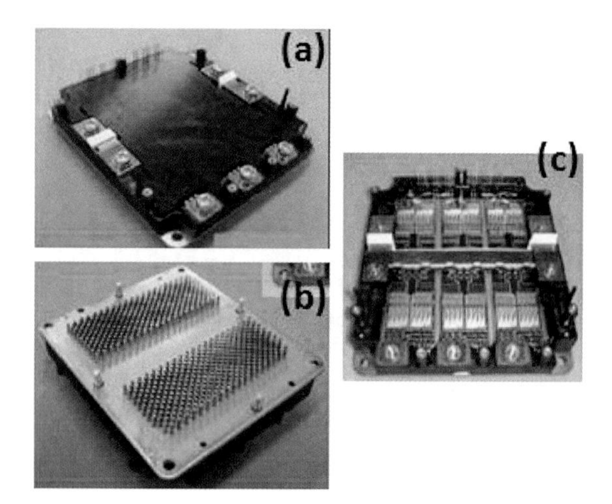

Figure 18. IGBT Power Module. (a) Top View, (b) Bottom View and (c) Silicon Structure.

Figure 19 also shows comparative improvement on silicon current density, thermal impedance, stray inductance, maximum junction temperature and silicon thickness.

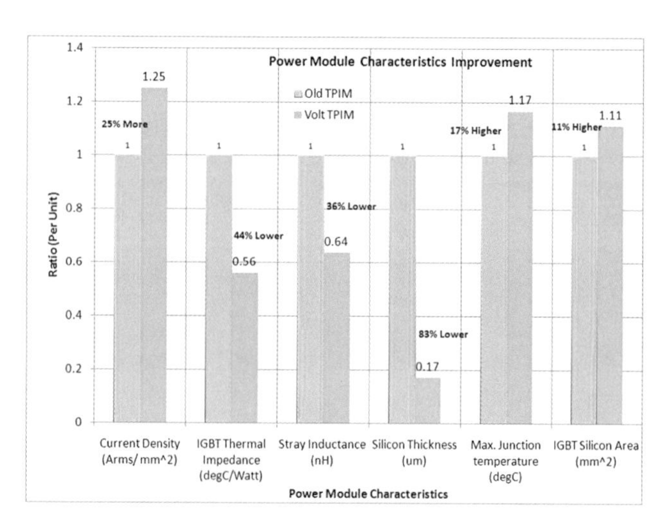

Figure 19. Power Module Characteristics Improvement.

Power module design selection follows a rigorous analysis and validation process to ensure reliability and performance. Some of the main items in these regards are module life expectancy, inverter efficiency and module switching dynamic. Module switching dynamic includes voltage overshoot, rate of change of voltage and rate of change of current. These depend on module layout, loop inductance, gate drive parameters and silicon technology. Lower switching speed (lower rate of change of voltage, 'dv/dt') means lower voltage overshoot that set up the maximum voltage rating for the power module. This voltage overshoot is also detrimental to the traction motor insulation life. Reduction of voltage overshoot thus helps extending the machine life. Lower switching speed, however, also means more switching loss that affects the inverter efficiency. However, lower 'dv/dt' is also preferred for EMI (electromagnetic interference). These tradeoffs in choosing the appropriate switching speed are part of this overall selection process.

Power module switching frequency is another important parameter that plays an important role in meeting inverter efficiency and acoustic noise. The selection of an appropriate switching frequency profile depends on HV wire mechanization, HV DC bus filter components, HV bus ripple/ resonance and dynamic response as described in [6]. More on this would be discussed in the next section.

FILTER CAPACITOR

The attributes for the filter capacitor for VOLT TPIM show superiority not only for electrical performance but also for productivity because of the flat design structure as shown in Figure 20.

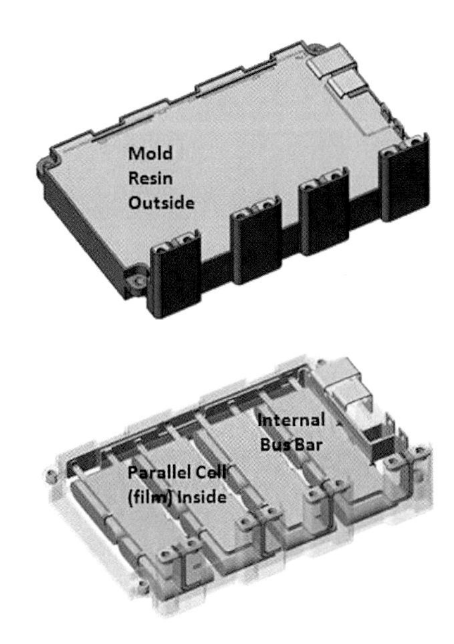

Figure 20. Filter Capacitor for VOLT TPIM.

This filter capacitor uses polypropylene film technology with thin films where the parallel cells are connected to each other and exterior HV DC terminals through low inductance bus bars. It ensures low equivalent series resistance and inductance (ESR and ESL) and higher filter current capabilities and at the same time has reduced the exterior volume by 35%.

All these design enhancement and design tradeoffs have resulted in efficient, low cost, low noise and high density power inverter modules for the Volt drive system.

MOTOR CONTROL OPTIMIZATION

For the Volt system, key motor controls algorithms were found to be largely constrained by efficiency, NVH, and bus resonance issues. For a given operating condition (torque, speed, and HVDC voltage), it was found that the motor stator current trajectory, PWM method, and switching frequency are the most critical control parameters of those which can be manipulated in order to meet the stated objectives. Since the margin with respect to each of the identified constraints (efficiency, NVH, bus resonance etc.) varies with operating condition, the best solution can be obtained by varying the key control parameters as a function of the operating condition. By thorough analysis and experimentation, the best solution was selected and implemented. This section will examine how some of the key control algorithm choices can affect these critical system behaviors, and how these were optimized for the Volt platform.

The power inverter hardware and motor controls algorithms can influence vehicle NVH by creating current harmonics on the high voltage DC input and the AC output interfaces. Please refer to Figure 2. The harmonic currents and their associated electromagnetic fields can create forces which induce vibrations on a variety of physical components. The current harmonics are typically a function of the PWM switching frequency, as well as the fundamental electrical frequency of the load motor. Additionally, the power inverter can contribute to noise and vibration due to vibration of components within the inverter box itself.

Below base speed, the impact of motor controls on the vehicle system efficiency and NVH is largely determined by stator current trajectory, PWM method, and switching frequency. For a given torque request, DC bus voltage, and motor speed, there may be multiple Id/Iq operating points which can satisfy the operating condition. For example, there exists a family of constant torque curves in the DQ current plane as shown in Figure 21. It is most common to operate along the maximum torque per amp (MTPA) trajectory below base speed. However, for a given torque command one could depart the MTPA trajectory and traverse left along the constant torque curve, as long as voltage ellipse and current limit constraints are met. In general, moving left along the constant torque curve means additional stator current and conduction losses. Also, the radial and tangential forces in the electric machine often increase significantly as the control angle (beta) is increased. This is typically associated with increased NVH. On the other hand, in some cases such as higher speeds, moving slightly to the left along the constant torque trajectory could result in lower core loss and an overall efficiency improvement even though the conduction losses may have increased. Besides this small effect, both higher efficiency and low NVH are obtained by staying as close to the MTPA line as is possible within the current and voltage constraints. In general, it was found that the stator current trajectory can be selected for maximum efficiency and not be constrained by NVH concerns.

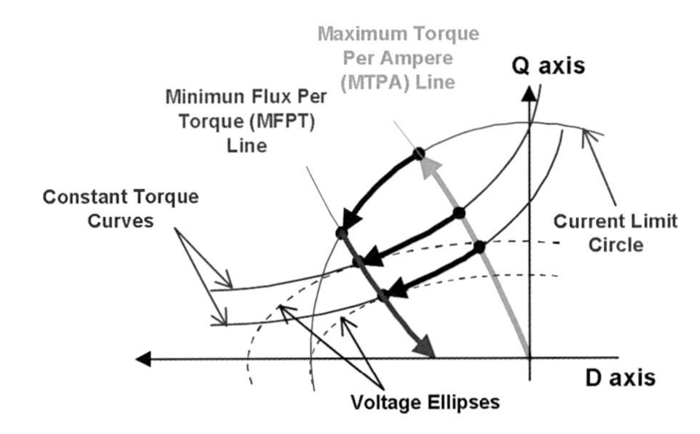

Figure 21. DQ current plane illustrating MTPA and constant torque lines. The phase current harmonic spectrum is plotted for 2000rpm, 100% motoring torque.

The contribution of PWM method to DC input and AC output current harmonics is fairly well documented in the literature [8]. At the most basic level, the controls engineer must select between a continuous and discontinuous PWM method.

Continuous PWM methods switch each phase of the inverter during each PWM period. This results in two different zero vectors being applied for each period, with zero vector duration split between the edges and center of the PWM period. During these zero vector periods, there is no current sent from the power module towards the DC bus. Hence, the DC bus harmonic current spectrum of continuous PWM methods is dominated by the second carrier group (2*fsw) and its sidebands. A similar phenomenon occurs on the AC output current ripple as well. Pushing the harmonic content out to the switching frequency second harmonic carrier group has multiple obvious benefits, such as better filtering of the DC bus capacitor and reduced motor current ripple due to increased inductive impedance of the motor at high frequencies. Along with smoother DC input and AC output currents, the content is more easily pushed above the audible range with the doubling effect provided by continuous PWM methods. However, for all the advantages of the continuous PWM methods, there is one significant drawback: high switching losses.

Discontinuous PWM methods offer significant efficiency improvements over the continuous methods by reducing the number of switch transitions per PWM period by 1/3. This is accomplished by not switching one of the inverter phases for two 60 degree periods per fundamental period. Coupled with intelligent selection of which phase not to switch (ideally the phase carrying the largest current) switching losses can be reduced by approximately ½. This significant reduction in switching losses does come at a cost. The DC and AC bus current ripple is significantly increased. In general, with DPWM the switching energy is concentrated around the first harmonic carrier group. Ripple current harmonic content is at a much lower frequency compared to continuous PWM methods. Hence, DPWM is often more noisy than its continuous counterpart. Figure 22 compares the AC output current spectrum for the two PWM methods under the same nominal operating conditions. Note that the CPWM method pushes more of the energy out to the higher frequency second carrier group centered around two times the switching frequency.

In a system constrained by inverter losses, it would be logical to select DPWM mode at high torque/current operating conditions to minimize switching losses and control junction temperatures. However, as discussed above, DPWM results in increased harmonic currents. Using DPWM at high torques is doubly troublesome since the fundamental current amplitudes are larger and machine inductances are typically lower due to saturation effects. On the other hand, the typical EV/HEV spends the majority of its operational time at lower torque levels. It is at the high duty cycle, low torque operating points that the control algorithm selection should heavily weigh efficiency as compared to NVH concerns.

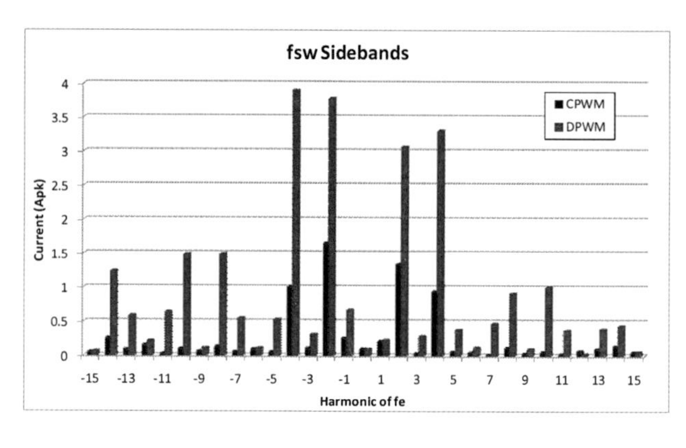

(a)

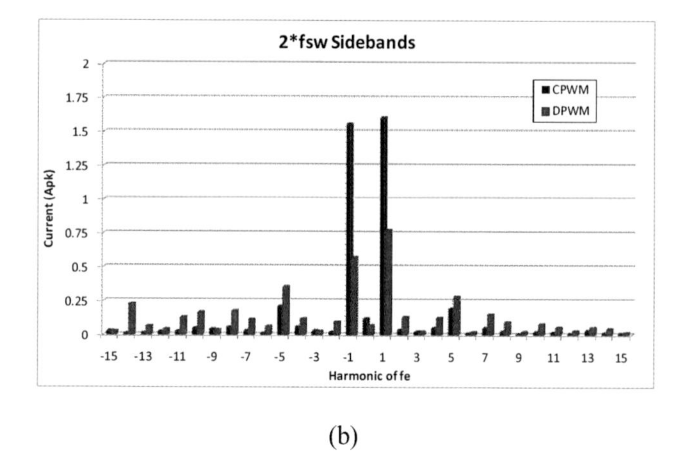

(b)

Figure 22. Phase current harmonic sidebands for CPWM and DPWM around (a) first carrier group fsw, (b) second carrier group 2*fsw

With the above considerations in mind, it was chosen to utilize DPWM methods at low torque levels, while transitioning to continuous PWM at higher torques. This provides good system efficiency at high duty cycle points when currents are relatively low, and transitions to continuous PWM methods at higher torques when current amplitude is increased and noise becomes a limiting factor for the Volt. This is only possible if the inverter provides sufficient silicon and thermal management to support the additional losses caused by continuous PWM at high current levels.

The switching frequency is the third major parameter in the motor controls which significantly impacts NVH and efficiency. With a digital control system, sample rate is a limiting factor in selecting control bandwidth. Typically (although not always), the sampling and switching frequencies are the same. A high switching frequency may be necessary to support the desired current regulator bandwidth in order to meet torque transient requirements. Also, as motor speed and fundamental frequency increase, sample rate must increase proportionately in order to maintain controllability at high speeds. These factors place a minimum acceptable limit on the switching frequency to meet control and transient response specifications. On the other hand, as motor speed is reduced towards zero, semiconductor losses near stall can cause junction temperatures to become a limiting factor. It is common to reduce switching frequency near stall to mitigate temperature rise due to longer time duration spent near the peak of the phase current. However, lower switching frequency is accompanied by high current ripple, acoustic noise and increased machine losses. Hence, it is a key trade-off in selecting the switching frequency versus motor speed characteristic.

It is important to consider the possibility of bus resonance issues on the high voltage distribution bus. Besides the traction power inverter and battery pack, there are often additional power electronic components tied to the high voltage bus, such as air conditioning compressor, accessory power DC/DC converter, etc. Each of these components will have certain reactive components on its DC input port (such as filter capacitors and possibly inductors). The large high power interconnect cables add stray inductance as well. Coupled with the input capacitance of the traction inverter, a complex electrical network exists. By design, this network has low resistance, which also implies low damping. It is important to consider the harmonic content of the current injected onto the DC bus by the power inverter, and avoid exciting critical resonant frequencies on the bus. If the electrical resonance is excited, there could be very large circulating currents which can induce noise and stress components. Harmonic currents typically center on the switching frequency and its multiples, with upper and lower sidebands at harmonics of the motor electrical frequency. In the Volt motor controls, the primary switching frequency is selected intelligently to avoid exciting the key electrical circuit resonances as much as possible.

Furthermore, the selection of switching frequency and the PWM scheme has significant impact on the machine losses. For instance, the selection of DPWM over CPWM increases the harmonics in the machine voltage. As a consequence machine iron loss increases. Moreover, the lowering of the switching frequency increases the switching ripple in the current which can also increase both machine joule and iron losses. Therefore, PWM scheme and the switching frequency in both Volt A and B machines are selected considering the tradeoffs between the overall system efficiency, NVH performance, electrical circuit resonance etc. At the end the control optimization

TEST DATA

This section will present some performance plots of the Voltec electric drive system. The drive system efficiency is plotted first. Figure 23 below plots the difference of the measured drive system efficiency performance (motor and inverter) of the Volt motor B and the simulated performance of a stranded design. For the stranded design the machine copper loss is adjusted based on the simulated phase resistance of figure 8. The measured and the simulated performance include the effect of the PWM schemes and the optimized control method as described in the earlier section. Only the area which is critical for the EREV operation, as depicted by the usage profile of figure 9, is shown here. As expected bar-wound construction shows superior performance compared to the stranded design. The bar wound test and the stranded design simulated efficiency performance is shown for a fixed stator temperature and does not include the higher temperature effect the stranded design would exhibit due to its higher resistance in the specified torque speed range.

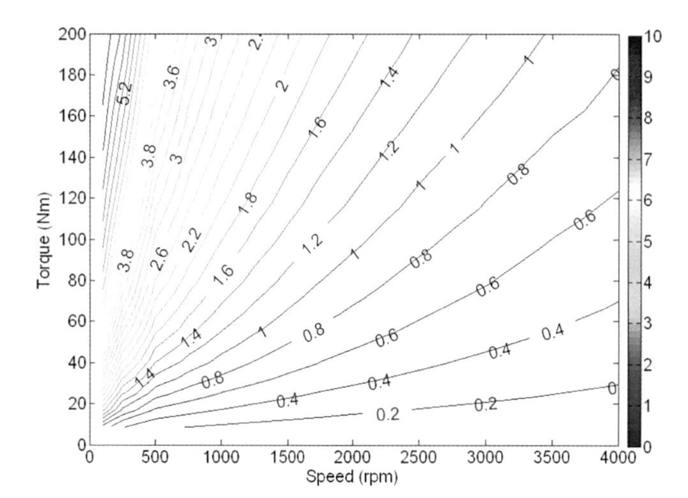

Figure 23. System efficiency percent difference of bar wound Volt motor B and simulated stranded design

The thermal behavior of the Volt motor B is presented next. First continuous torque of motor B is plotted in figure 24 and is compared against the peak torque. The high temperature test results shown in figure 24 demonstrate that the thermal performance of the Volt motor B provides a continuous operating range that completely encompasses the US06 operating region shown in figure 9. Continuous torque capability in excess of 60% of the peak capability is achieved in Volt motor B. The high level of continuous capability demonstrated by the Volt motor B is a result of the electromagnetic efficiency, the unique cooling system, and efficiency optimized torque control as explained in the earlier section. Bar wound thermal performance may be compared with the equivalent stranded design to exhibit the superiority of the bar wound machine. This is illustrated next. Figure 25 below shows the thermal performance of the Volt motor B and compares that with an equivalent stranded design for a GM internal durability duty cycle. The bar wound design has significantly better performance especially under EV mode when the machine is used extensively. The lower temperature of operation of the motor B not only improves the drive system efficiency but also extends the insulation life of the motor.

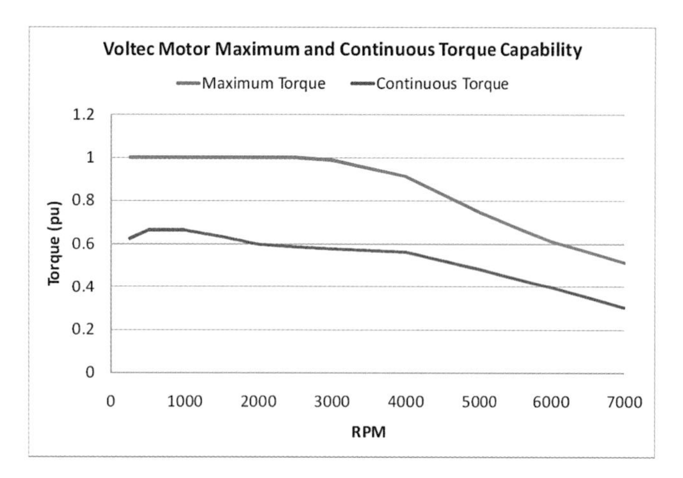

Figure 24. Volt motor B continuous torque and peak torque as a function of motor speed.

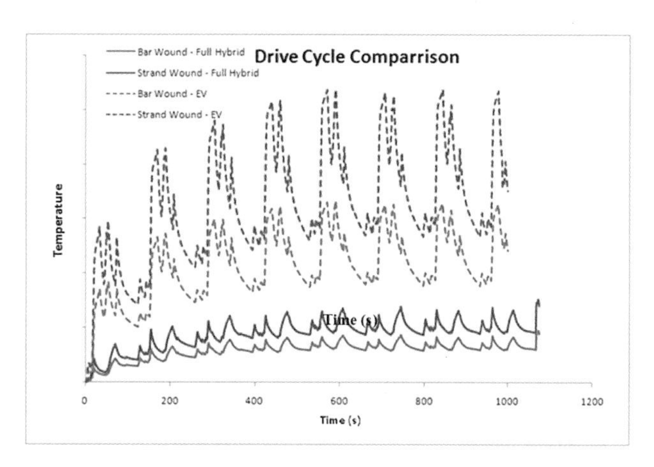

Figure 25. Chevrolet Volt motor B thermal performance as compared to an equivalent stranded design for one GM internal durability duty cycle.

Next the thermal performance of the PIM is presented. In order to validate the thermal design of the power inverter, IR camera testing was employed. Utilizing the IR camera avoids the difficulties associated with attaching thermocouples to the semiconductor junctions in a reliable and non-intrusive fashion. A special test unit was constructed to enable un-obscured viewing of the of the power semiconductor devices operating in situ during dynamometer testing. Figure 26 shows the test unit and a close up of the power module, which is unpotted and painted black to increase its emissivity. Figure 27 shows a typical measurement snapshot during operation at stall condition with the phase current vector aligned to one phase leg (a worst case condition). Post processing of the camera data is performed to identify the hotspot on each diode and IGBT device. The junction temperature versus operating condition can be plotted as seen in Figure 28. As seen in Figure 28, the junction temperature is controlled at low speed due to reduced switching frequency. As the motor speed increases, so does the motor voltage and modulation index. This leads to a shift in average current from the diode to the IGBT, resulting in lower diode temperatures. Above 3500rpm, there is a significant reduction in switching losses due to saturation of the duty cycles at high modulation indexes. This results in lower junction temperatures at high motor speeds when the motor voltage is high.

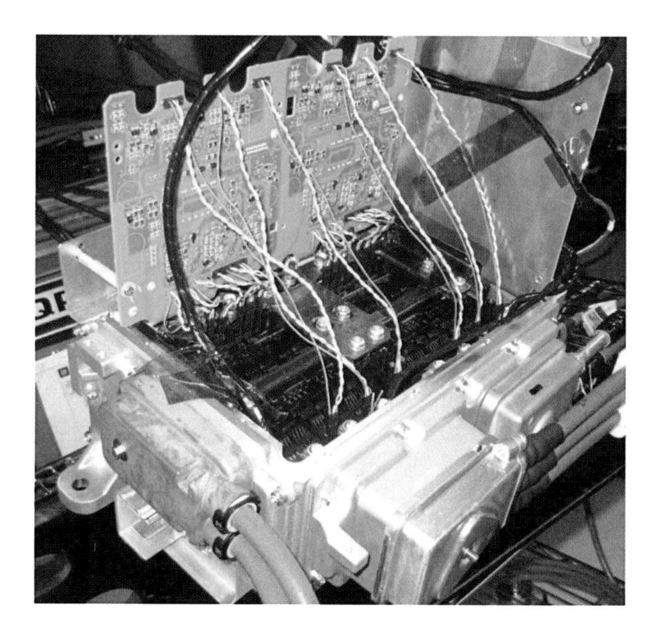

(a)

(b)

Figure 26. (a) Test setup used for power semiconductor junction temperature measurement with IR camera, and (b) close-up view of power module for one motor.

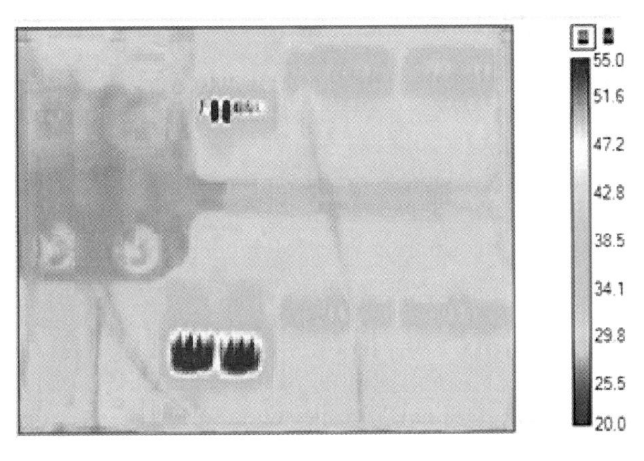

Figure 27. Typical view of IR camera during stall testing with 25C coolant. IR camera view roughly corresponds to Figure CC (b).

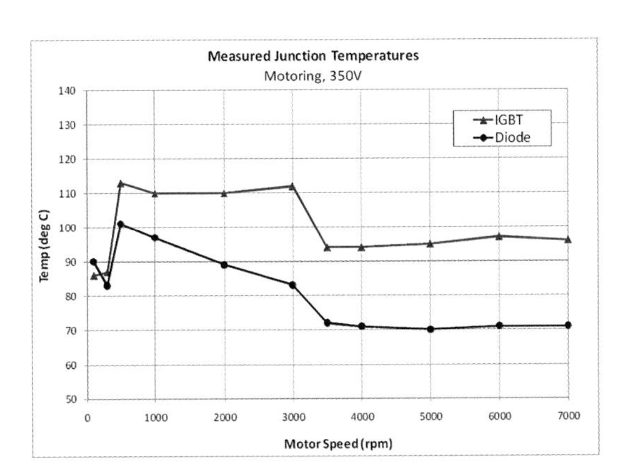

Figure 28. Measured junction temperatures versus motor speed at peak torque conditions.

Finally, noise performance of the motor A and B are presented. GM has an internal method in measuring noise behavior in the dyne setup. The frequency response function (FRF) of the test fixture is calculated first and is later subtracted from the final measurement with the machine operating on the dyno for the peak torque speed operation. This way the influence of the test fixture on the noise behavior is taken out from the measurement. The setup measures the blocked forces on the machine mount location through which noise is transmitted to the transmission casing. Figure 29 below shows the block force measurement in the motor mounting location for both motor A and B for the peak torque speed operation of the machine. This measurement is summing the top five orders for both the machines. The noise behavior in the dyne measurement (blocked force) demonstrated the low noise performance of both the machines which are later confirmed by the exceptional low noise behavior of both the machines under vehicle measurement.

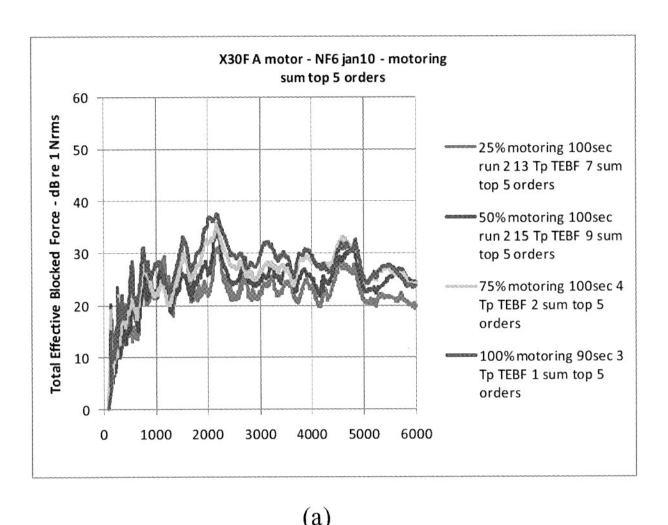

(a)

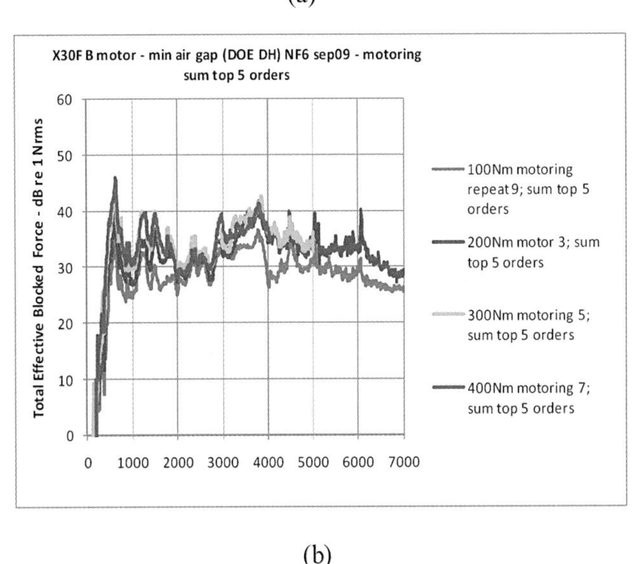

(b)

Figure 29. Blocked force measurement on the motor mount location. (a) Motor A blocked force measurement and (b) motor B blocked force measurement.

SUMMARY/CONCLUSIONS

This paper has presented the Voltec electrical drive system used in the Chevrolet Volt EREV. The construction and performance details of the traction Motor A and B are presented. Due to bar-wound construction of motor B an excellent thermal performance is exhibited. A continuous torque in the excess of 60% of the peak torque is achieved in the motor B. The bar-wound construction also improves the slot fill and reduces end-turn length. These have resulted in a reduction of in access of 30% of the dc resistance compared to an equivalent (in terms of torque and power performance) stranded design. The simulation result showed that the skin and proximity effect due to bar wound construction did not affect the performance until 8000 rpm where the bar wound resistance surpassed the stranded design resistance. However, the vehicle drive cycles for FTP urban, highway, as well as for the US06 drive cycles showed that machine significant operations all lie below 5000 rpm. As a result, bar-wound construction exhibited far superior performance compared to the stranded design for the Chevy Volt operation. Motor B rotor designed specifically with air pocket on top of the magnet has improved the spin loss by i) lowering the airgap magnet flux and ii) by lowering the harmonics of the air gap flux. This rotor has also improved saliency has compensated some loss of torque due to reduced airgap flux. A concentrated winding machine was selected for motor A mostly due to packaging issues. Special attention was given for the NVH performance of both the machines especially the concentrated winding machine. Volt EREV exhibits excellent noise performance and this can be attributed to the careful design of the Volt drive systems and the control algorithms.

This paper also presented Traction Power Inverter Module (TPIM) used for Chevy Volt. The technological superiority in selecting these power electronics components has been discussed in this paper that enabled GM to design this flat and compact structure that meets high current density and improved manufacturability requirement as required by Chevy Volt electric drive system. Design optimization, hardware and software tradeoffs have resulted in improved power electronics reliability, efficiency, performance and overall security requirements for TPIM.

A multidisciplinary approach was selected while developing the motor control algorithm to address the complexity of the electrified automotive traction system. With careful analysis and experimentation, an optimal solution was found as demonstrated in the Volt system. PWM mode and switching frequency were dynamically modified as a function of the operating condition which simultaneously satisfied requirements for efficiency, NVH, and bus resonance. The selection and verification of the final control method were done with extensive efficiency, NVH, and bus ripple testing.

REFERENCES

1. Tate, E.D., Harpster, M.O., and Savagian, P.J., "The Electrification of the Automobile: From Conventional Hybrid, to Plug-in Hybrids, to Extended-Range Electric Vehicles," *SAE Int. J. Passeng. Cars - Electron. Electr. Syst.* **1**(1):156-166, 2008, doi:10.4271/2008-01-0458.

2. Mc Cosh, D.; We Drive the World's Best Electric Car; Popular Science, January 1994.

3. Kalhammer, F. R., Koph, B. M., Swan, D. H., Roan, V. P., Walsh, M. P., Status and Prospects for Zero Emission Vehicle Technology, Report of the ARB Independent Expert Panel 2007, April 13th, 2007.

4. Nelson, D.J., Wipke, K.B., et al., "Optimizing energy management strategy and degree of hybridization for a hydrogen fuel cell SUV", *EVS-18*, Berlin, October 2001.

5. Lorenz, R.D., "Power Conversion Challenges with a Multidisciplinary Focus.", Proceedings of the IEEE Power Conversion Conference, 2002, vol. 2, pp. 347-352, Osaka, April, 2002.

6. Anwar, M., Gleason, S., Grewe, T, "Design Considerations for High-Voltage DC Bus Architecture and Wire Mechanization for Electric/ Hybrid Electric Vehicle Applications.", accepted for IEEE ECCE Annual Conference, Sept. 12-16, 2010, Atlanta, GA.

7. Islam, R., Husain, I, "Analytical model for predicting noise and vibration in permanent magnet synchronous motors", *IEE Trans. On Industry Applications*, vol. 46, pp. 2346-2354.

8. "Pulse width modulation for power converters: principles and practice," Holmes, D. Grahame, Lipo, Thomas A., IEEE Press, 2003, ISBN: 0471208140.

CONTACT INFORMATION

Khwaja Rahman can be contacted at khwaja.rahman@gm.com

ACKNOWLEDGMENTS

The authors acknowledge Michael Miller, Margaret Palardy and Daniel Berry of GM hybrid division for their help with some of the analysis results presented in this paper.

DEFINITIONS/ABBREVIATIONS

NVH
Noise, Vibration, and Harshness

HV
High Voltage

E-REV
Extended Range Electric Vehicle

PHEV
Plug-in Hybrid Electric Vehicle

ZEV
Zero Emission Vehicle

TPIM
Traction Power Inverter Module

PIM
Power Inverter Module

VSI
Voltage Source Inverter

IGBT
Insulated Gate Bipolar Transistor

CPWM
Continuous Pulse Width Modulation

DPWM
Discontinuous Pulse Width Modulation

FRF
Frequency Response Function

The Engineering Meetings Board has approved this paper for publication. It has successfully completed SAE's peer review process under the supervision of the session organizer. This process requires a minimum of three (3) reviews by industry experts.

ISSN 0148-7191

Positions and opinions advanced in this paper are those of the author(s) and not necessarily those of SAE. The author is solely responsible for the content of the paper.

SAE Customer Service:
Tel: 877-606-7323 (inside USA and Canada)
Tel: 724-776-4970 (outside USA)
Fax: 724-776-0790
Email: CustomerService@sae.org
SAE Web Address: http://www.sae.org
Printed in USA

SAE International™

The GM "Voltec" 4ET50 Multi-Mode Electric Transaxle

2011-01-0887
Published
04/12/2011

Michael A. Miller, Alan G. Holmes, Brendan M. Conlon and Peter J. Savagian
General Motors Company

doi:10.4271/2011-01-0887

ABSTRACT

The Chevrolet Volt is an electric vehicle (EV) that operates exclusively on battery power as long as useful energy is available in the battery pack under normal conditions. After the battery is depleted of available energy, extended-range (ER) driving uses fuel energy in an internal combustion engine (ICE), an on-board generator, and a large electric driving motor. This extended-range electric vehicle (EREV) utilizes electric energy in an automobile more effectively than a plug-in hybrid electric vehicle (PHEV), which characteristically blends electric and engine power together during driving. A specialized EREV powertrain, called the "Voltec", drives the Volt through its entire range of speed and acceleration with battery power alone, within the limit of battery energy, thereby displacing more fuel with electricity, emitting less CO_2, and producing less cold-start emissions than a PHEV operating in real world conditions.

The Voltec powertrain architecture provides four modes of operation, including two that are unique and maximize the Volt's efficiency and performance. The specialized electric transaxle, known as the 4ET50, enables patented operating modes both to improve electric driving range when operating as a Battery Electric Vehicle (BEV) and to reduce fuel consumption when extending range by operating with the ICE.

Historically, most EVs have used a single-speed electric transaxle with one motor and a fixed gear reduction. The single-speed gear reduction is a simple arrangement that takes advantage of the wide speed range of electric motors. While this arrangement can work well, the wide speed range of most electric motors comes at the price of a well known loss of efficiency at higher motor speeds for most types of motors. Consequently, two-speed transmissions have been proposed for BEVs that can improve both tractive effort and efficiency although with the attendant additional extra gear shift hardware and controls.

The Voltec 4ET50 multi-mode electric transaxle introduces a unique two-motor EV driving mode that allows both the driving motor and the generator to simultaneously provide tractive effort while reducing electric motor speeds and the total associated electric motor losses. This new operating mode, however, does not introduce the torque discontinuities associated with a two-speed EV drive. For ER operation, the Voltec transaxle uses the same hardware and controls that enable one-motor and two-motor EV operation to provide both the completely decoupled action of a pure series hybrid, as well as a more efficient powerflow with decoupled action for driving at light loads and high vehicle speed.

Construction of the GM Voltec 4ET50 transaxle employs significant re-use of the GM front wheel drive Two-Mode Hybrid 2MT70 transaxle with modifications to enable all-speed and full-power EV operation. A new high power driving motor, optimized generator, and modified control elements allow the two EV driving modes and the two ER driving modes to be realized in the Chevrolet Volt.

INTRODUCTION

VEHICLE TYPES AND ELECTRIC UTILITY

Hybrid Electric Vehicle (HEV)

A hybrid is defined by SAE [1] as: "A vehicle with two or more energy storage systems both of which must provide propulsion power - either together or independently." In practice, hybrid vehicles typically require a combination both sources to provide full vehicle capability. The engine is also typically the larger of the two propulsion sources, being sized to provide most of the power during high power vehicle events. The motor is typically the smaller of the two propulsion sources, being sized for transient events to maximize the amount of energy that can be captured during braking and for limited low speed EV operation.

The powertrain of an HEV incorporates significantly large electric motors, power electronics and a battery pack (or RESS - Rechargeable Energy Storage System), to make a more efficient vehicle. However, the HEV does not directly displace petroleum with electrical energy for driving the vehicle. Electric Driving Utility is defined as percentage of real world driving where propulsion power is provided by energy that comes from the electric grid. The electric utility of an HEV is 0% for all driving.

Plug-in Hybrid (PHEV)

A PHEV has been defined by SAE [1] as: "A hybrid vehicle with the ability to store and use off-board electrical energy in the RESS." These systems are, in effect, an incremental improvement over the HEV, with the addition of a large battery with greater energy storage capability, a charger, and modified controls.

Since PHEVs, commonly referred to as blended PHEVs, do not offer full electric driving capability, there are limits to EV driving based on the power capacity of the battery, electronics, interconnects, motors and associated thermal systems, as well as the kinematics of the hybrid transmission. Accordingly, there are different types of PHEV operating limits and strategies. This means a "charge depleting" mode exists where the engine comes on as the battery is discharging and both energy sources are blended to drive the vehicle. Typically, the utility of electric driving is limited in PHEVs, and the full effects of and benefits of using grid energy instead of fossil fuel can only be partially realized as analyzed and discussed in [2],[3], and[4].

Over the charge depleting range, electric energy utility, or the percentage of total energy supplied to the output from the battery during depletion, was limited to about 70% for a PHEV in real world driving [4].

Extended-Range Electric Vehicle (EREV)

SAE has not yet established a definition for an EREV. However, several automakers describe them using the definition provided in [2], "A vehicle that functions as a full-performance battery electric vehicle when energy is available from an onboard RESS and having an auxiliary energy supply that is only engaged when the RESS energy is not available."

An EREV has 100% electric energy utility over the charge depleting range. To do so, the EREV must have sufficient motor, inverter, interconnects, RESS, and thermal system power capacity. While depleting in EV mode, the engine does not assist in propulsion. Rather, an EREV relies on 100% electric energy for all wide open throttle events, high-speed freeway cruising, and extended grade conditions.

Battery Electric Vehicle (BEV)

the SAE [1] defines a BEV, or more plainly, an EV, as "A vehicle powered solely by energy stored in an electrochemical device."

The BEV is the simplest form of the electric vehicle, and has 100% electric energy utility for all driving distances and as such, like the EREV must have sufficient power capacity in the driving motor, inverter, interconnects, RESS, and thermal systems.

However, the driving distance of the BEV is severely limited to only that energy that is available in the RESS.

Table 1. Attributes of Electric Vehicles

Vehicle Type	Electric Power	Onboard Electric Storage	Grid Recharge	Electric Energy Utiltiy	Range Limit
HEV	Low to med	low	no	0%	Gasoline
PHEV	med	med	yes	~70%	Gasoline + Battery
EREV	high	high	yes	100%	Battery + Gasoline
BEV	high	highest	yes	100%	Battery

Table 1 shows a summary of HEV and BEV vehicle attributes. Full Electric Driving Capability is an attribute that defines the EREV, just like a BEV. But unlike the BEV, the vehicle's total range is not limited by the energy in the battery; rather the energy in the battery *combined* with the fuel in the tank.

EREV SYSTEM REQUIREMENTS

In order to meet the customer's expectations, the EREV system must provide excellent performance in both EV and ER modes. Table 2 lists key requirements that help to define the system.

Table 2. EREV System Requirements

Vehicle Requirement	EV Req't	ER Req't
Vehicle Acceleration	X	X
Vehicle Top Speed	X	X
Launch Acceleration	X	X
Sustained Gradeability	X	X
Fuel Economy		X
All-Electric Range	X	
Extended Range		X

Peak Tractive Effort

The requirements for vehicle acceleration, top speed, and launch acceleration together define a tractive effort curve that is required to be achieved both in EV and ER modes, as shown in Figure 1. To meet these requirements, the system should be capable of transmitting a large amount of power from either the battery or the engine during acceleration over a large range of vehicle speeds.

Continuous Tractive Effort

The requirements for Gradeability and Top Speed also define an envelope of continuous operation. Under these conditions, the system must be able to operate for long durations with adequate cooling of components. By determining the torques and powers of the electric machines under these conditions, the selection of the transmission powerflow determines both the peak and continuous sizing of the motors, inverters, and gearing.

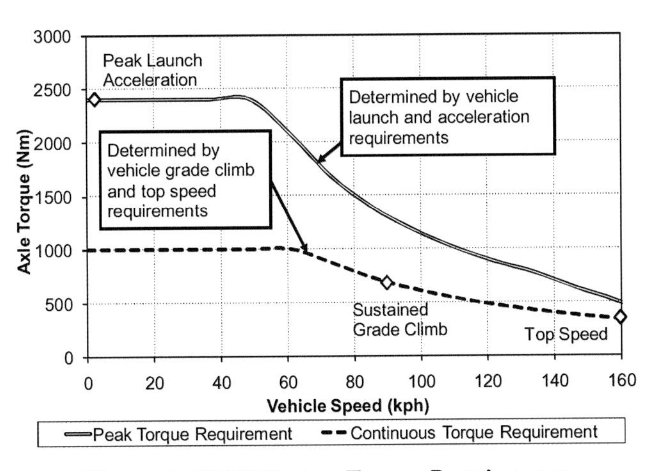

Figure 1. EREV System Torque Requirements

EV Range and ER Fuel Economy

Since range and fuel economy requirements apply under different system operating modes, tradeoffs can occur where some systems are optimized for EV range at the expense of fuel economy, or vice versa. A design goal of the system is to provide good efficiency as an EV drive, thus efficiency should be comparable to a system designed as a BEV drive unit. On the other hand, when operating in range extending mode, the system should have fuel economy comparable to the best IC engine systems available. In this situation, highway fuel economy could be expected to be particularly important to the customer, as trips which exceed the EV range are likely to involve highway driving.

EV Drive Character

An EREV has two distinct operating modes: EV and ER. When operating as an EV, the system is capable of providing quick, smooth response to driver commanded torque due to the wide constant power characteristics and inherently fast transient response capability of electric drives. It is desirable to maintain these characteristics when operating in electric drive mode and to also allow similar drive feel when in ER mode.

ELECTRIC DRIVE POWERFLOWS

One-Motor EV Drive (Single-Speed)

One-motor, single speed drives have been used on most modern BEVs. The prototypical system for a BEV uses a single motor and a fixed single-speed gear reduction connected to a differential output as was developed for the first modern BEV, the GM EV1[5].

Lever diagrams are useful, simple representations of driveline kinematic arrangements, where torque and rotating speed relationships are represented with forces and displacements using a schematic linear linkage and free-body diagram scheme [6]. They provide a linear analog to the rotational reference system of a powertrain that many find easier to comprehend. The kinematic arrangement, or *powerflow*, of the single speed EV drive is shown in Figure 2 in the form of a lever diagram.

A typical single-speed One-Motor EV drive combined (inverter × motor × gear reduction) efficiency map is shown in Figure 3 in the axle torque - vehicle speed domain. This figure uses a 7.0:1 overall gear ratio from the motor speed to the wheel speed. The efficiency was defined as traction power delivered to the axle divided by the DC electric power input as shown in Equation 1.

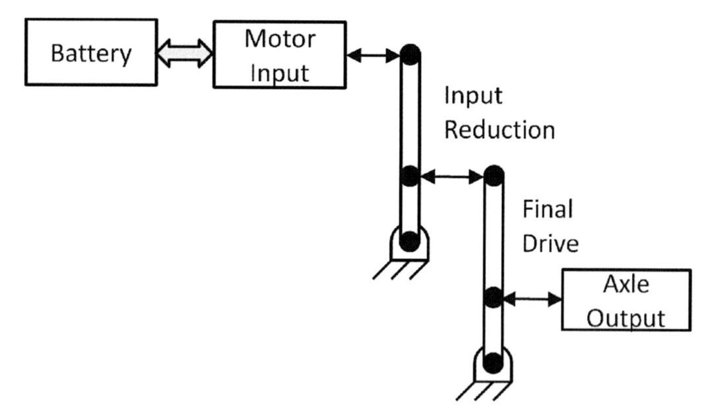

Figure 2. One-Motor EV Drive (Single-Speed) Powerflow

$$\eta_{EV\ drive\ type} = \frac{P_{traction,\ net\ output}}{P_{electrical,\ net\ input}} \quad (1)$$

Efficiency was calculated for the entire tractive effort envelope using correlated, modeled component data and the same motor controls. Each mechanical and electrical component was separately modeled, including each torque transmitting and spinning element.

The level road load line of a reference BEV, assumed to be similar to the Chevrolet Volt, was overlaid on the efficiency. From the figure, a BEV with a single-speed drive experiences the most efficient point of operation at approximately 75 kph (47 mph) at loads that are much higher than the road load.

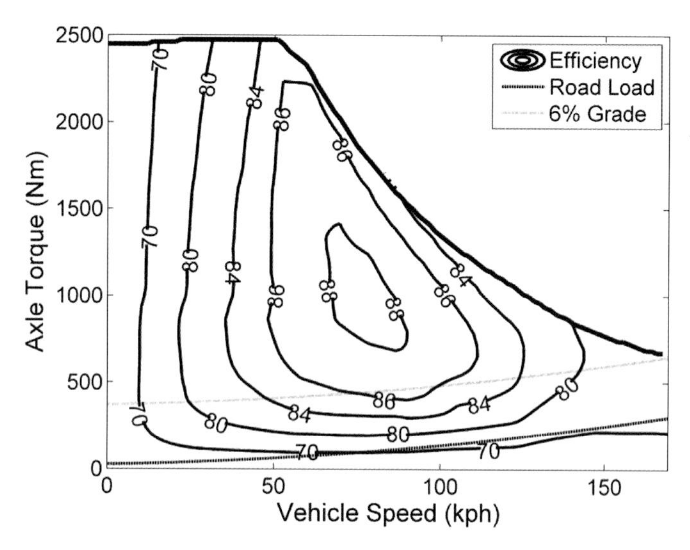

Figure 3. One-Motor EV Drive (Single-Speed) Efficiency Map

Figure 4 shows the same efficiency map with operating points added that represent time durations of the US urban, US highway, and higher intensity USO6 Schedules. This figure shows that many of the US urban and highway schedule driving points are at speeds and loads below the most efficient operating point, and only the highway and US06 schedules spend a significant portion of time at speeds higher than the most efficient operating point.

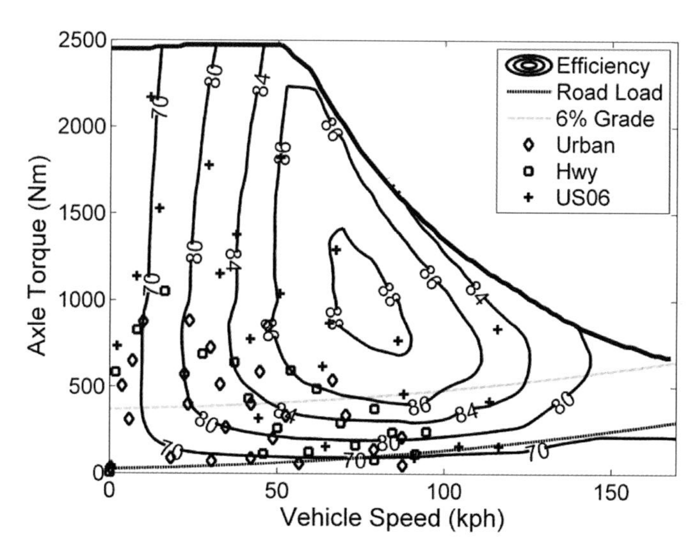

Figure 4. One-Motor EV Drive (Single-Speed) Efficiency Map with Operating Points

$$\eta_{EV\ drive,\ sched.} = \frac{E_{traction} + E_{regen\ capture}}{E_{electrical} + E_{braking}} \quad (2)$$

Schedule-averaged efficiencies of a single-speed drive in the reference BEV on the three schedules were calculated according to Equation 2. This calculation is similar to Equation 1, but in addition to including the conversion from input (electrical) to output (traction) energy for propulsion, the energy capture efficiency from mechanical (braking) to electrical (regen capture) energy during regenerative braking was also included. As a sign convention, all energies in Equation 2 are positive for a typical drive schedule. This equation integrates the power over the entire operating envelope of the referenced schedule. For the single-speed EV drive, the following efficiencies were found:

$$\eta\ EV_{Single\text{-}speed,\ Urban} = 77.2\ \%$$

$$\eta\ EV_{Single\text{-}speed,\ Hwy} = 75.5\ \%$$

$$\eta\ EV_{Single\text{-}speed,\ US06} = 80.9\ \%$$

It is interesting to note that while component peak efficiencies can be in the mid or even high 90 percents, the reality of their application yield much less. Relevant measures of an EV drive must account for all the losses in the driveline over a defined and useful driving envelope, like the reference vehicle test schedules. In this case, actual EV drive efficiency percentages are in the upper 70s to low 80s.

Two-Speed EV Drive

Two-speed drives have been proposed for BEVs [7], offering a possible reduction of overall motor size necessary to meet a given tractive effort requirement and improved overall efficiency. There are many ways that a two-speed transmission can be mechanized. Figure 5 shows a typical implementation, along with the clutch state table that enables the active modes.

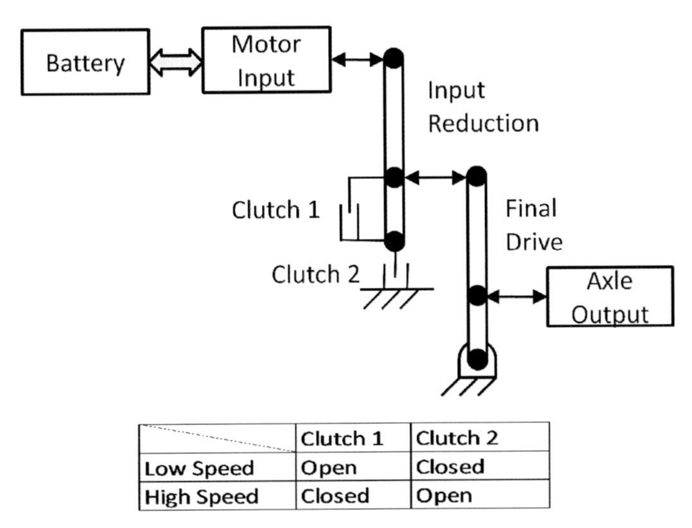

	Clutch 1	Clutch 2
Low Speed	Open	Closed
High Speed	Closed	Open

Figure 5. Two-Speed EV Drive Powerflow

The example single-speed electric drive can be modified to add a second gear, at 2.16:1 for purposes of improving EV drive efficiency. Mechanical spinning loss reductions in second gear were modeled, but no additional energy losses were used to represent the gear shift mechanism. Figure 6 shows the efficiency of adding a second gear, again with time duration points of the US Urban, US Highway, and US06 Schedules.

Efficiencies above 100 kph improve compared to the single-speed One-Motor EV drive, illustrating that the second gear would be used often on the highway. The two-speed schedule-averaged efficiencies are as follows:

$$\eta\ EV_{\text{Two-speed, Urban}} = 79.3\ \%$$

$$\eta\ EV_{\text{Two-speed, Hwy}} = 80.4\ \%$$

$$\eta\ EV_{\text{Two-speed, US06}} = 84.1\ \%$$

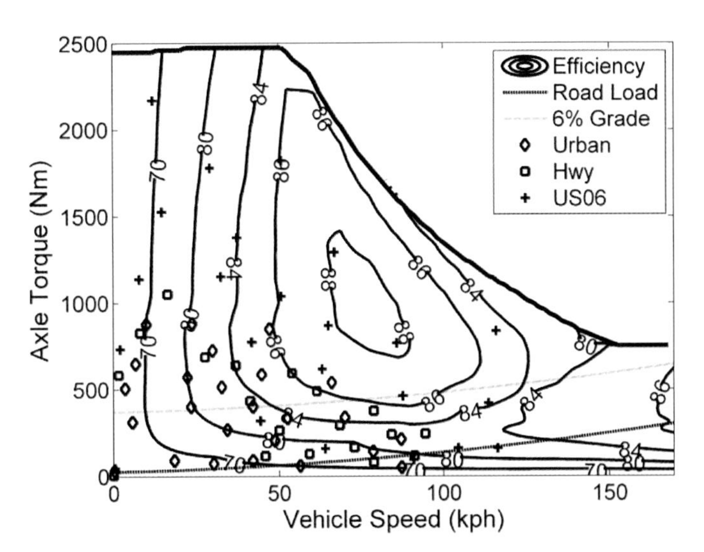

Figure 6. Two-Speed EV Drive Efficiency Map with Operating Points

Two-speed drives are rarely used in practice, in part due to the expense of the ratio changing mechanism and the controllability and driveability of the gear shifts.

Two-Motor EV Drive

In the case where more than one electric motor can be made available to drive the vehicle, a proprietary EV drive using a combination of two motors is possible [7]. In this arrangement, the output torque of the electric drive must be reacted by both input motors through a differential at a fixed ratio through a planetary gearset. The output speed is a linear combination of the speeds of both motors.

With this two-motor arrangement, motor speeds can be adjusted continuously, for greatest tractive effort or for greatest overall efficiency. Since power is speed × torque, the power split between the two motors used to drive the vehicle is controlled by varying the motor speeds.

There are many ways that a Two-Motor EV drive could be mechanized, Figure 7 shows a typical implementation. In this diagram, Motor A has a 3:1 gear reduction to the output axle and Motor B has a 7:1 gear reduction.

To further elaborate on the continuously adjustable motor speeds, take a simple Two-Motor EV drive with a 6:1 gear reduction from each motor to the output, one motor could be held at rest while the other runs at a 6:1 speed reduction. Alternatively, both motors could run at 3:1, or a third split could be 4:1 and 2:1. This speed reduction split can be controlled to continuously vary the ratio in time, although each motor must support torque proportional to its gear reduction, in this case equally at a 6:1 ratio. For this example, the output axle speed is equal to the sum of Motor A and Motor B divided by 6. The diagram in Figure 8 shows the motor speed combinations at various output speeds when the motors have equivalent reduction gearing to the output. Since both motors react the full output torque, the power provided by both driving motors is proportional to the motor speeds. Hence, the power-split ratio is the same as the speed-split ratio.

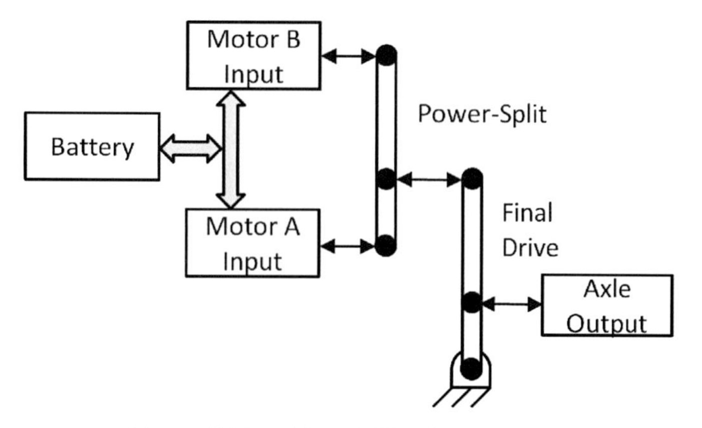

Figure 7. Two-Motor EV Drive Powerflow

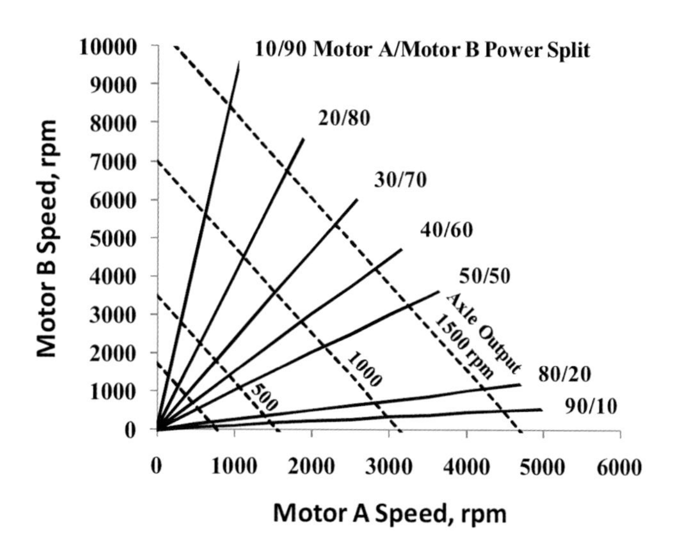

Figure 8. Two-Motor EV Speed Ratio (also Power-Split Ratio) as a function of Motor Speeds

To predict the relevant Two-Motor EV drive efficiency, a second motor and inverter was added to the One-Motor EV Drive. When the second motor was used, additional inverter losses were incurred, albeit with overall power consumption on each somewhat reduced based on the motor operating points. The mechanism and torque ratios illustrated in Figure 7 were used in evaluating efficiency, consequently impacting the losses for the torque transmitting and spinning elements.

The Two-Motor EV drive is a compelling option when considered within the EREV application. An obvious arrangement for the EREV is the Series Electric Drive, where there is an electric machine that is used as a generator when the ICE provides range extension. During electric driving, which may be for two thirds or more of all driving in the real world [2], this motor would remain un-used in a series hybrid arrangement. However, with a two-motor configuration, the generator could be enlisted as a second driving motor to help reduce overall driving losses in EV operation.

When operating the Two-Motor EV drive, the most efficient combination of motor input speeds can be selected to meet the output speed and torque. The efficiency map for only the Two-Motor EV drive is shown in Figure 9.

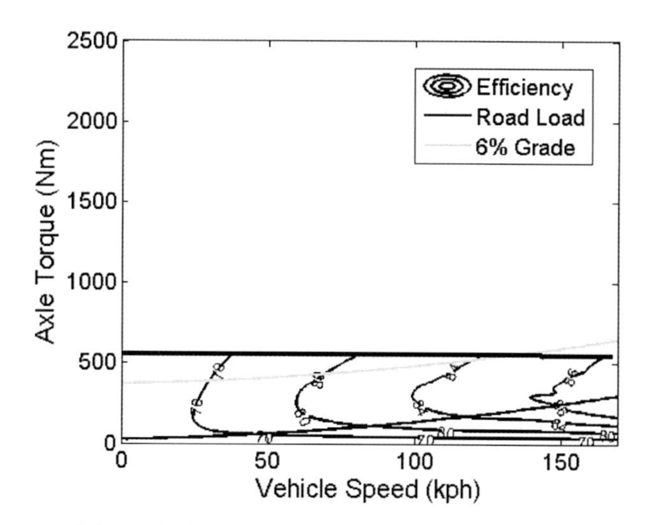

Figure 9. Two-Motor EV Drive Efficiency Map

Using Two-Motor EV to displace less efficient single-speed One-Motor EV operation, schedule-averaged efficiencies with Two-Motor EV drive enabled are somewhat better than the One-Motor EV drive alone, and comparable to the two-speed EV. As was the case for the two-speed EV, Two-Motor EV enables better efficiency at higher vehicle speeds. This benefit can be seen in Table 3.

Table 3. Summary of Schedule Averaged Efficiencies for Various Electric Drives

	Urban	Highway	US06
EV, Single-Speed	77.2%	75.5%	80.9%
EV, Two-Speed	79.3%	80.4%	84.1%
EV, Two-Motor	79.6%	81.1%	84.5%

EXTENDED-RANGE POWERFLOWS

Series One-Motor Extended-Range (with Single-Speed EV output)

An obvious choice for extended-range, or charge sustaining, operation in an EREV is the series arrangement. In this arrangement, the engine is connected directly to an electric generator. Series drive has no direct mechanical connection from the engine to the output drive. The output of the series drive consists of the electric drive motor with gear reduction to the output, which is identical schematically to a One-Motor EV drive as shown in Figure 2. Figure 10 shows the lever diagram for the Series One-Motor ER powerflow.

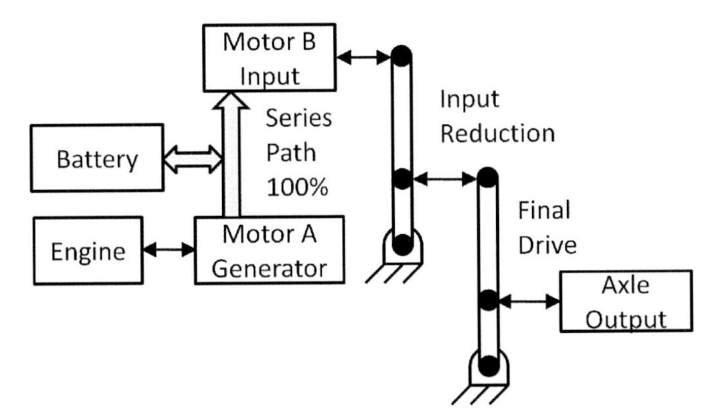

Figure 10. Series One-Motor Extended-Range Powerflow

The benefits of this arrangement are simplicity and complete decoupling of output speed and load from ICE speed and load. Of course, in practice, ICE speed and load must be limited by NVH and thermal considerations. The drawbacks of the series arrangement with single-speed output are two-fold:

- Because of the single-speed One-Motor EV drive, efficiency declines at higher driving speeds,
- 100% of the driving power must go through the Series path [from the ICE shaft (mechanical) to the generator(electrical) to the Motor (electrical) to the output axle (mechanical) power], and suffer the associated conversion losses.

The efficiency of an extended-range powerflow having an engine decoupled from the output can be calculated by measuring the output mechanical traction power delivered to the drive axles road divided by the input mechanical power from the engine (assume no power provided from the battery), for a given engine load line. The engine load line assumed is one that gives the best efficiency for a given power level. The engine reference load line used for the efficiency calculations is shown in Figure 11.

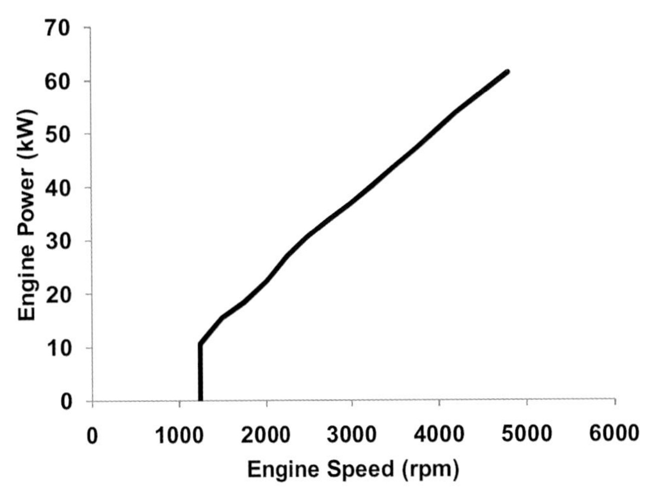

Figure 11. Reference ICE Load Line

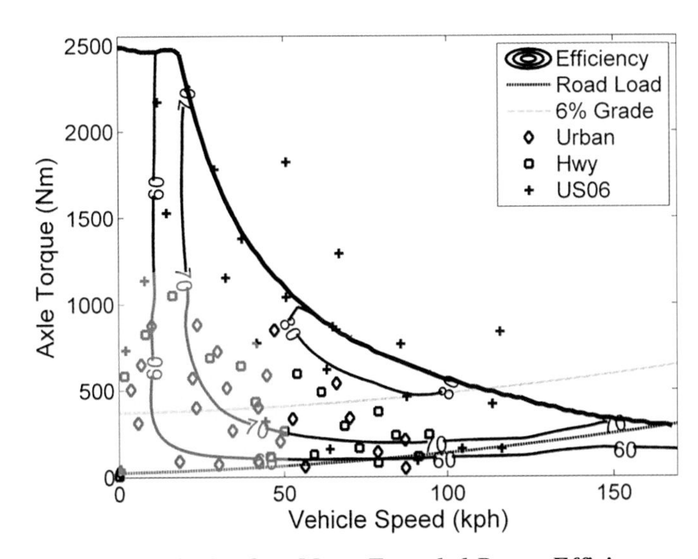

Figure 12. Series One-Motor Extended-Range Efficiency Map

Figure 12 shows the extended-range efficiency losses of the series path, along with the road load line and schedule operating points for the same example vehicle. The schedule operating points above the peak axle torque curve require battery power to supplement capability beyond what the engine can generate.

During extended-range operation, the ICE may be turned off at low power levels and when the vehicle is at rest, as indicated by the shaded area of Figure 12.

Combined Two-Motor Extended-Range (Output Split) Powerflow

There are several ways that power-split architectures can be arranged. The characteristics of the fundamental ways to split shaft power are developed analytically in [9]. There are three fundamental types of splitting mechanical power:

1. Input power-split
2. Output power-split
3. Compound power-split

The input power-split and compound power-split powerflows have been developed thus far for application and are in production HEVs [10, 11]. The output power-split has not, until now, been used for production vehicles.

Figure 13 shows an example output split powerflow. Like the series powerflow, this powerflow also has no direct connection between the engine and the output, providing freedom to control engine speed and load independent of output speed and load.

The primary benefit of the output split powerflow is improved efficiency due to the reduction in series path losses. With an output split, less of the total power must go through series path conversion, as some of the engine shaft power indirectly reacts axle output torque. The amount of power that does require series path conversion is a function of the driving speed and the engine speed, and is, in general, between 0 and 70%.

The drawback of the output split is that, like the Two-Motor EV drive, the total tractive effort of the powerflow is limited to the weaker of either side of the split, in this case Motor B, or the sum of Motor A and the ICE.

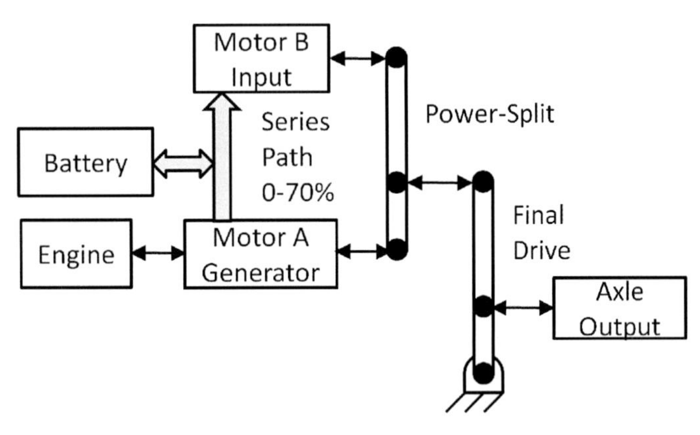

Figure 13. Combined Two-Motor Extended-Range Powerflow

A comparison of the two range-extending powerflows can be made by evaluating the efficiency of the Combined Two-Motor powerflow under the same conditions as the Series One-Motor powerflow. Using the same reference engine load line as shown in Figure 11, we can generate the efficiency map shown in Figure 14.

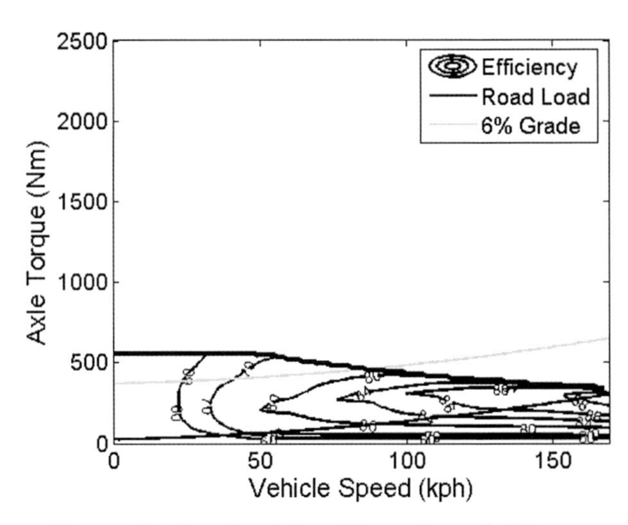

Figure 14. Combined Two-Motor Extended-Range Efficiency Map

It is useful to consider the efficiency along the road load line and 6% grade lines when comparing Series vs. Combined extended-range modes. Figure 15 illustrates that the Series One-Motor ER offered efficiencies in the 55 to 65 percent range along the road load line above 50 kph. Over a similar region, the Combined Two-Motor ER offered efficiencies between 70 and 88 percent, highlighting the efficiency advantage provided by an output split powerflow at light loads.

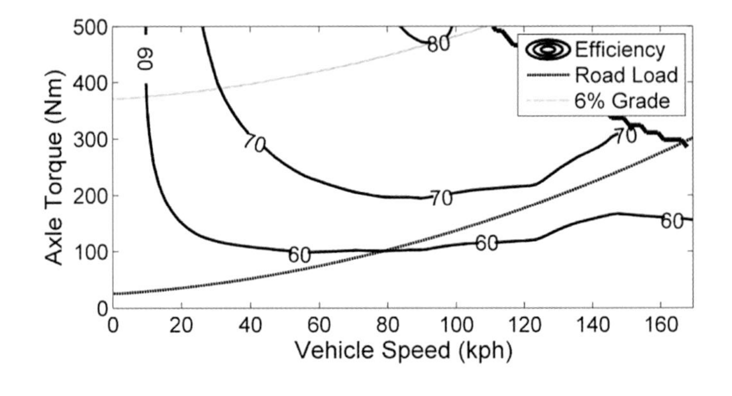

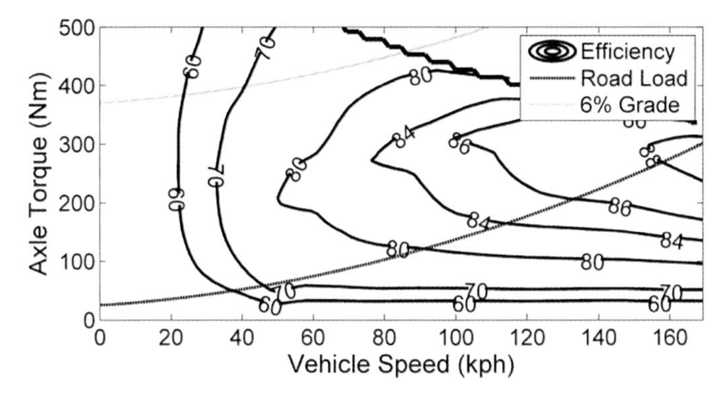

Figure 15. Comparison of Series One-Motor (Upper) vs. Combined Two-Motor (Lower) Efficiency at Highway Cruising Conditions

THE VOLTEC 4ET50 POWERFLOW

As discussed, Two-Motor EV drive and Combined Two-Motor ER operation offered by the Voltec powerflow provided advantages over the more conventional EV drives and Series operation. One counter-intuitive drawback of these two-motor driving modes was the reduced tractive effort capability.

The Voltec Electric Transaxle is a Multi-Mode transaxle that provides a charge depleting, full-performance EV drive with a range-extending feature when the ICE is turned on. Battery charge and discharge, is of course possible during extended-range operation, but that is not developed in this paper.

In EV operation, the Voltec Electric Transaxle automatically reconfigures its powerflow between two modes:

- ***One-Motor EV:*** The Single-Speed EV Drive powerflow, which provides more tractive effort at lower driving speeds.
- ***Two-Motor EV:*** The Output Power-Split EV Drive powerflow, which has greater efficiency than One-Motor EV at higher speeds and lower loads.

In extended-range operation, the Voltec Electric Transaxle automatically reconfigures its powerflow between four modes, the two EV modes above and:

- ***Series One-Motor ER:*** The Series extended-range powerflow that provides more tractive effort at lower driving speeds.
- ***Combined Two-Motor ER:*** The Output Power-Split extended-range powerflow that has greater efficiency than Series at higher speeds and lighter loads.

The proprietary Voltec Powerflow [12] uses two rotating clutches and one stationary clutch to change the operating modes, as shown in Figure 16. The clutch state table shows how the basic arrangement is reconfigured. The principal benefit of the Multi-Mode Voltec electric transaxle when compared to a simpler single-speed EV drive with series range extender is improved vehicle efficiency in both EV and ER modes of operation.

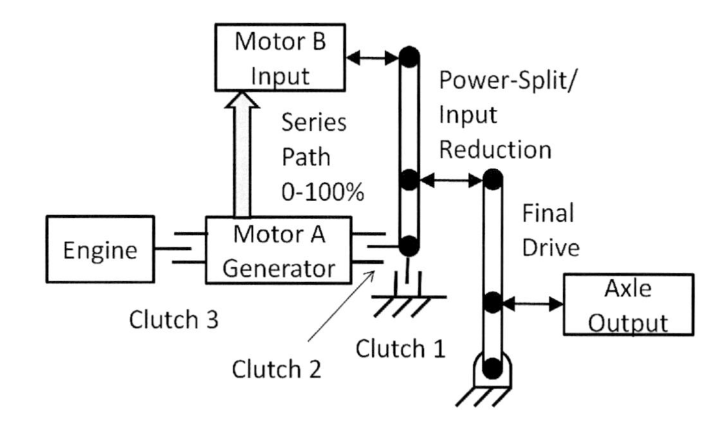

	Clutch 1	Clutch 2	Clutch 3
One-Motor EV	Closed	Open	Open
Two-Motor EV	Open	Closed	Open
Series One-Motor ER	Closed	Open	Closed
Combined Two-Motor ER	Open	Closed	Closed

Figure 16. The Voltec Electric Transaxle Powerflow

In a vehicle typified by the Chevrolet Volt, the additional Two-Motor EV and Combined Two-Motor ER modes improved EV range at highway speeds by 1 to 4% and fuel economy at highway speeds by 10 to 15%. Table 4 and Table 5 list the percentage of time each mode was used over the specified drive schedule for EV and ER driving, respectively. These tables reiterate the efficiency improvements that both Two-Motor EV and Combined Two-Motor ER offer at higher vehicle speeds.

Table 4. Percentage of Time Spent in Each Mode by Drive Schedule during Electric Operation

	Urban	Highway	US06
1-Motor EV	93%	49%	58%
2-Motor EV	7%	51%	42%

Figures 17 and 18 show the approximate operating envelope for the Voltec propulsion system in the Chevrolet Volt. In tractive effort space, the figures show which regions each of the four modes are typically utilized. Modes are automatically selected based on tractive effort, efficiency and NHV considerations.

The combined efficiency maps of the two EV and two extended-range modes are shown in Figures 19 and 20, respectively.

Table 5. Percentage of Time Spent in Each Mode by Drive Schedule during Extended-Range Operation

	Urban	Highway	US06
1-Motor EV	71%	11%	21%
2-Motor EV	3%	25%	5%
Series 1-Motor ER	20%	6%	16%
Combined 2-Motor ER	6%	58%	58%

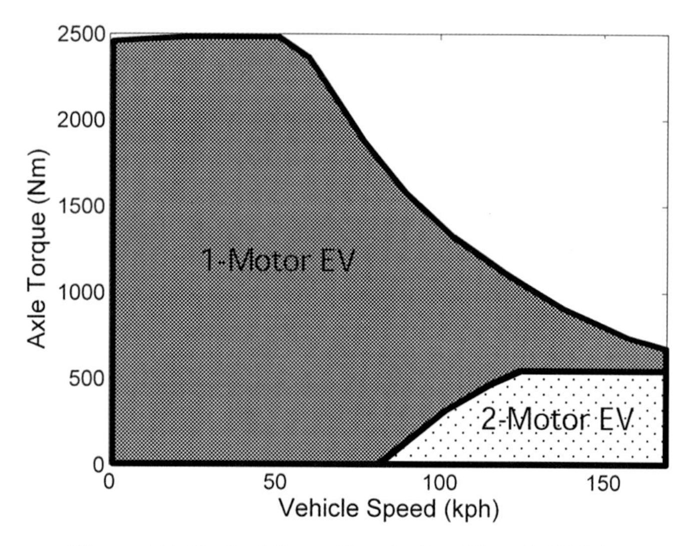

Figure 17. Typical Operation during Electric Driving

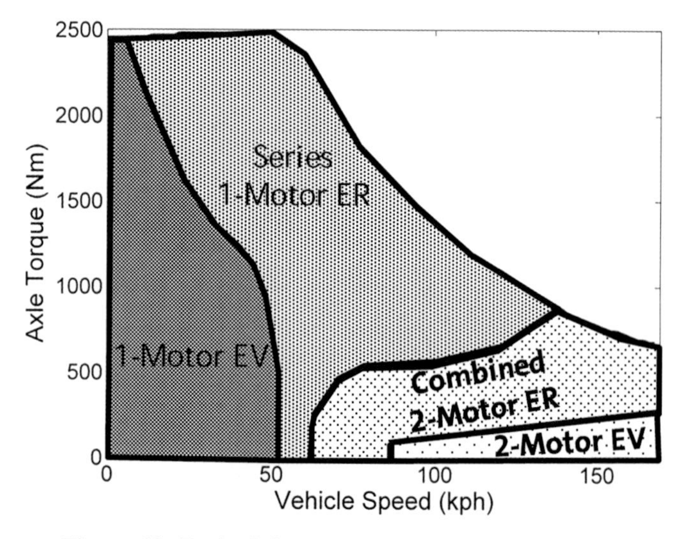

Figure 18. Typical Operation during Extended-Range Driving

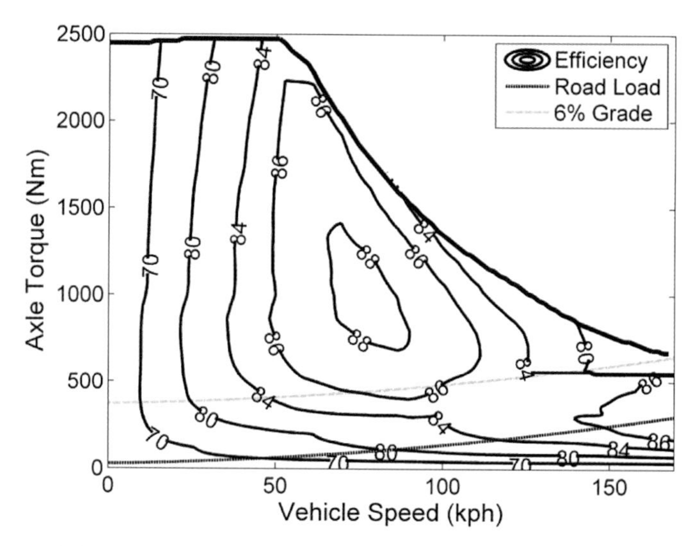

Figure 19. Composite Electric Driving Efficiency Map

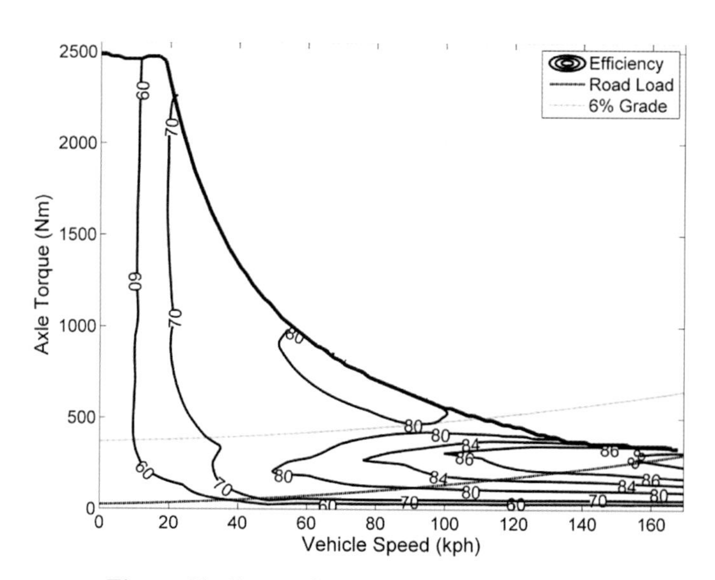

Figure 20. Composite Extended-Range Driving Efficiency Map

THE EXECUTION OF THE VOLTEC 4ET50 ELECTRIC TRANSAXLE

Electric and hybrid vehicles, with a few exceptions, have not yet reached levels of production that are considered "high-volume" in the automotive industry. The tooling and engineering costs to produce robust, reliable, durable transaxles or drive units for EVs or HEVs are still very significant parts of the vehicle cost. A smart strategy of common components was necessary to provide an efficient, cost-effective propulsion system.

Over the past several years, GM developed a drive unit for front-wheel drive HEVs, the 2-Mode 2MT70 [13]. GM had also developed ways to use the 2MT70 effectively for PHEVs. During conception of the Voltec EREV drive unit, both a clean-sheet electric transaxle design and an electric drive unit based on maximizing reuse of the tooling and components from the 2MT70 were considered. GM was able to develop a modified arrangement of the 2MT70 that minimized new components while providing a compelling combination of advantages in energy efficiency. The evolution of the Voltec 4ET50 from the 2-Mode 2ET70 is depicted with lever diagrams in Figure 21.

The 2MT70 drive unit had been optimized to propel a mid-size SUV with the capability to tow a mid-size trailer with high fuel economy and aggressive acceleration, primarily by using a powerful 3.6L V6 engine rather than electric motors. Its components were carefully examined for use in a four-passenger EREV with primarily electric propulsion. Since an EREV must deliver full performance on electric power alone, the electric size of the transaxle for the smaller Volt vehicle and engine was actually a good fit for many 2MT70 parts originally tasked for a larger, but less electric, HEV.

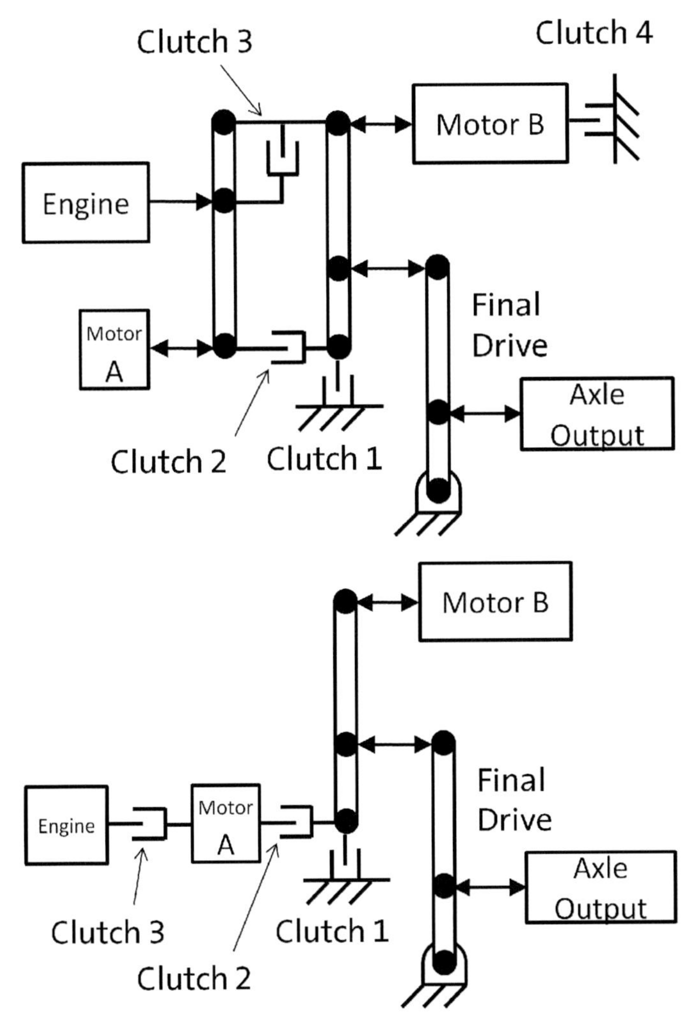

Figure 21. Comparative Lever Diagrams of 2-Mode 2MT70 (upper) and the Voltec 4ET50 (lower)

The Voltec drive unit was able to reuse a very large number of components from the 2MT70, but certain differences were necessary, as depicted in Figure 22. First, the gearing system was simplified by reducing a gearset. Second, the simplified gearing allowed for the removal of one clutch which had been used in the 2MT70 to directly couple the engine to the output to enable fixed-gear operation at high vehicle speeds, a feature that was not needed or desired for the Voltec. Third, the electric motor used for primary propulsion was made larger and optimized for continuous EV operation. Forth, the electric pump in the Voltec is the primary source of lubrication and cooling within the drive unit, so pump motor was designed to operate at high voltage, rather than from 12V as an auxiliary pump (not shown). A few other changes in the drive unit included the final drive, engine damper springs, extension housing, and the addition of a valve body spacer plate to accommodate the removal of the clutch.

As mentioned previously, the gearing system in the Voltec drive unit needed three clutches, a single set of planetary gears, a transfer gear, and a final drive system, of which all but the final drive were common with the 2MT70 drive unit. The single set of planetary gears had three gear elements, or nodes, that, in conjunction with the clutches, were capable of combining the output of the primary propulsion motor and the secondary propulsion motor (typically a generator) to power the vehicle. Use of the two motors allowed both Two-Motor EV operation with the engine off and Combined Two-Motor ER operation with the engine on. The final drive ratio was reduced to allow electric propulsion up to 160 km/hr (100 mph).

The three clutching elements contained in the Voltec drive unit include: a brake for the ring gear of the single planetary gear set, a clutch between the generator and the ring gear, and a clutch between the engine and the generator. The first two elements were common in function and structure with those in the 2MT70. The third clutch, between the engine and generator, was also common in structure but very different in function. The 2MT70 version of this clutch was used as part of a direct connection between the engine and the wheels for the 2MT70's 1st gear and 3rd gear.

The primary propulsion motor in the Voltec drive unit has distributed bar-windings, which improved thermal performance for long-duration and continuous loads when compared with the concentrated windings of the comparable motor in the 2MT70. Bar-windings were impractical for the 2MT70, as designed for the mid-size SUV, because of space constraints within the engine compartment if coupled with a large 4-cylinder engine. The engine used with the Voltec drive unit is a compact 1.4L engine, which allowed the necessary space for the larger and higher rated motor.

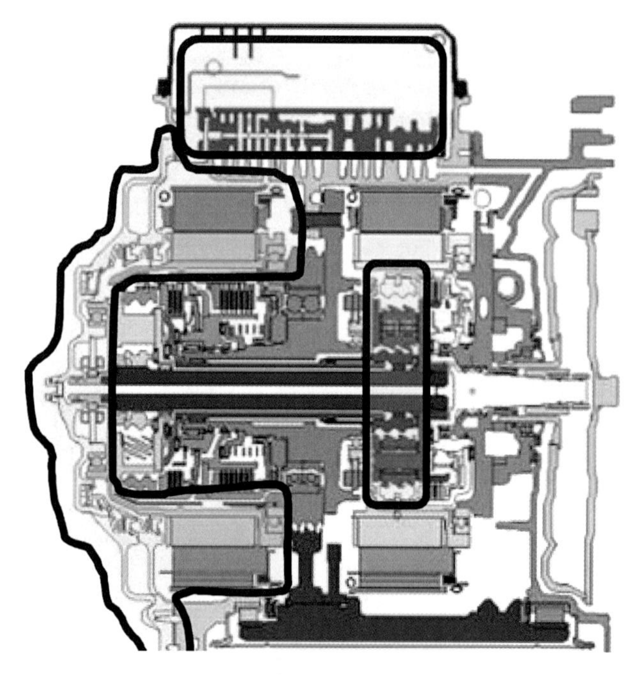

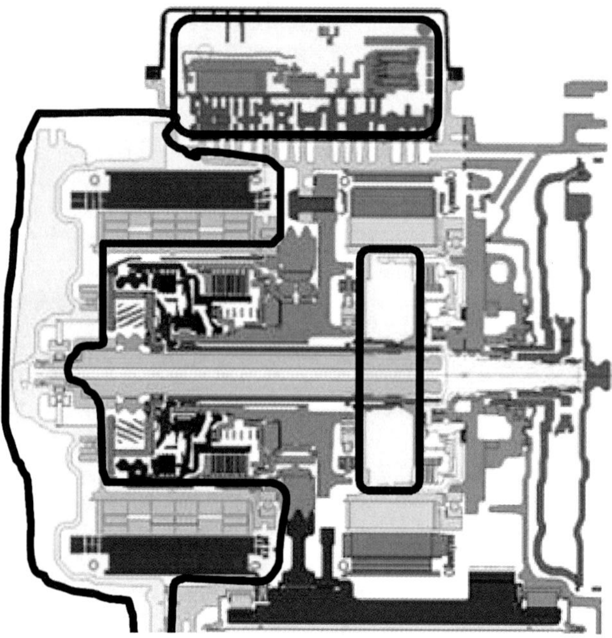

Figure 22. Cross Section of the 2-Mode 2MT70 (upper) and Voltec 4ET50 (lower) with Key Unique Parts Highlighted

Figure 23 shows the oil cooling schemed section of that motor. The bar wound motor has 30% lower DC winding resistance than a comparable conventional wire wound motor. Additionally, the winding head surface area exposed to the transmission oil used for cooling the motor is 60% higher. This means that the motor has exceptional continuous power duty, which is essential for a full electric performance necessary in an EREV having good dynamic performance.

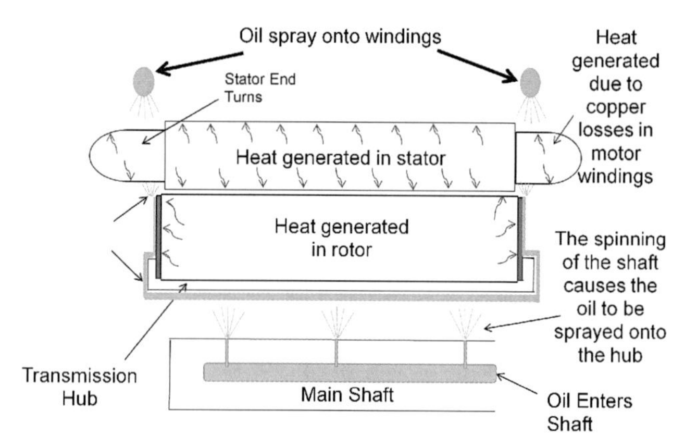

Figure 23. Voltec 4ET50 Drive Motor Section

CONCLUSIONS

1). Plug-in EREVs have 100% electric energy utility, exactly like battery electric vehicles for the prescribed initial charge depleting operation. The 100% utility under all load conditions is a defining characteristic of EREVs that is higher than the estimated 70% electric utility associated with blended PHEVs. Accordingly, the EREV displaces more fuel and creates the least CO_2, and regulated emissions.

2). EREV propulsion is especially challenging, since it requires full electric operation in charge depleting operation, as well as efficient operation during extended-range, or charge sustaining, operation.

3). The GM "Voltec" 4ET50 Electric transaxle has four modes that enable the Chevrolet Volt and other EREVs the combination of great electric-only range and tractive effort during charge depleting (EV) operation, as well as great fuel economy and tractive effort during extended-range operation (engine on). The four modes are:

- ◦ ***One-Motor EV:*** Single-speed, single motor EV drive effective for low speed, high load driving
- ◦ ***Two-Motor EV:*** EV drive effective for high speed, light load driving with propulsion split between two motors that have a continuously variable speed ratio
- ◦ ***Series One-Motor ER:*** Series drive that adds extended-range capability to One-Motor EV
- ◦ ***Combined Two-Motor ER:*** Output power-split drive that adds extended-range capability to Two-Motor EV

4). Utilization of Two-Motor EV drive resulted in a 1 to 4% improvement in EV range at highway speeds over a single-speed configuration.

5). Utilization of Combined Two-Motor ER resulted in a 10 to 15% improvement in fuel economy at highway speeds over a Series-only configuration.

6). The GM Voltec 4ET50 Electric Transaxle was derived from the GM 2MT70 2-Mode hybrid system, with significant modifications for EREV functionality, while achieving a 76% part number reuse on the transaxle.

REFERENCES

1. SAE International Surface Vehicle Information Report, "Hybrid Electric Vehicle (HEV) & Electric Vehicle (EV) Terminology," SAE Standard J1715, Rev. Feb. 2008.

2. Tate, E., Harpster, M.O., and Savagian, P.J., "The Electrification of the Automobile: From Conventional Hybrid, to Plug-in Hybrids, to Extended-Range Electric Vehicles," *SAE Int. J. Passeng. Cars - Electron. Electr. Syst.* **1**(1):156-166, 2008, doi:10.4271/2008-01-0458.

3. Rousseau, A et al.; Impact of Real World Drive Cycles On PHEV Fuel Efficiency and Cost for Different Powertrain and Battery Characteristics, Advanced Automotive Battery Conference, Long Beach, CA, June 2009.

4. Tate, E.D. and Savagian, P.J., "The CO_2 Benefits of Electrification E-REVs, PHEVs and Charging Scenarios," SAE Technical Paper 2009-01-1311, 2009, doi: 10.4271/2009-01-1311.

5. Mc Cosh, D.; We Drive the World's Best Electric Car; Popular Science, January 1994.

6. Benford, H.L. and Leising, M.B., "The Lever Analogy: A New Tool in Transmission Analysis," SAE Technical Paper 810102, 1981, doi:10.4271/810102.

7. Resele, P.E. and Bitsche, O., "Advanced Fully Automatic Two-Speed Transmission for Electric Automobiles," SAE Technical Paper 951885, 1995, doi:10.4271/951885.

8. Savagian, P; Differential Motor Drive, USPTO Patent 5,310,387, May 1994.

9. Conlon, B., "Comparative Analysis of Single and Combined Hybrid Electrically Variable Transmission Operating Modes," SAE Technical Paper 2005-01-1162, 2005, doi:10.4271/2005-01-1162.

10. Abe, S., Sasaki, S., Matsui, H., Kubo, K., Development of Hybrid System for Mass Productive Passenger Car, (JSAE Proc. 9739543), 1997.

11. Grewe, T.M., Conlon, B.M., and Holmes, A.G, "Defining the General Motors 2-Mode Hybrid Transmission," SAE Technical Paper 2007-01-0273, 2007, doi:10.4271/2007-01-0273.

12. Conlon, B., Savagian, P., Holmes, A., Harpster, M.; Output Split Electrically-Variable Transmission with Electric Propulsion Using One or Two Motors, USPTO Patent, 7,867,124, January, 2011.

13. Hendrikson, J., Holmes, A., and Freiman, D., "General Motors Front Wheel Drive Two-Mode Hybrid Transmission," SAE Technical Paper 2009-01-0508, 2009, doi:10.4271/2009-01-0508.

CONTACT

Michael A. Miller can be contacted at michael.6.miller@gm.com.

DEFINITIONS, ACRONYMS, ABBREVIATIONS

BEV
Battery Electric Vehicle

EREV
Extended-Range Electric Vehicle

ER
Extended-Range

EV
Electric Vehicle

HEV
Hybrid Electric Vehicle

ICE
Internal Combustion Engine

PHEV
Plug-in Hybrid Electric Vehicle

The Engineering Meetings Board has approved this paper for publication. It has successfully completed SAE's peer review process under the supervision of the session organizer. This process requires a minimum of three (3) reviews by industry experts.

ISSN 0148-7191

SAE Customer Service:
Tel: 877-606-7323 (inside USA and Canada)
Tel: 724-776-4970 (outside USA)
Fax: 724-776-0790
Email: CustomerService@sae.org
SAE Web Address: http://www.sae.org
Printed in USA

High Voltage Power Allocation Management of Hybrid/Electric Vehicles

2011-01-1022
Published
04/12/2011

James D. Marus
General Motors Company

doi:10.4271/2011-01-1022

ABSTRACT

As the automotive industry moves toward producing more advanced hybrid/electric vehicles, high voltage Rechargeable Energy Storage Systems (RESS) are now being implemented as the main power source of the vehicle, replacing the need for the traditional Internal Combustion Engine (ICE) altogether or just during certain parts of a drive cycle. With this type of architecture, it is becoming a necessity to equip these vehicles with devices that can draw their power from the high voltage (HV) RESS. These HV devices are not only used to support the propulsion of the vehicle but to perform other necessary vehicle functions as well.

With demands of high voltage power from multiple systems ranging from RESS thermal conditioning, cabin thermal conditioning, RESS charging, and vehicle propulsion, power demands can exceed the available power of the vehicle. This creates a perplexing problem of how to manage the high voltage power consumption between the different vehicle systems. This document describes how a controls method to manage the HV loads of a vehicle can be implemented to optimize the use of available power and ensure that power consumption remains within the operating bounds of the RESS. This document will also go into detail regarding when these types of vehicles may get into power constraining conditions and how a control system would manage power allocation in each specific condition.

INTRODUCTION

A hybrid/electric vehicle encounters power constraining conditions when the demand for HV power exceeds the available power output of the vehicles power source. When multiple HV devices are attempting to consume power, it is not an acceptable solution to simply shutdown non-propulsion HV systems during power constraining conditions. A smart trade off of power must be implemented to ensure the most optimum allocation of available power is achieved.

Each power constraining condition may be different, so a dynamic power distribution strategy must be used. Each HV system on the vehicle will have specific parameters that need to be monitored to determine the importance of full power operation relative to a specific power limited condition. These parameters will be used to optimize power allocation during power constraining conditions.

GENERAL CONTROLS ARCHITECTURE

Every HV device on the vehicle will report desired power consumption to a central algorithm, referred to as the Central Power Allocation Manager (CPAM), that will manage the trade-offs of allocating HV power to each individual device. The CPAM will monitor parameters relative to each HV devices' sub-system function in order to intelligently weigh the benefits of power limiting one system over another. The vehicles HV energy source will report the available HV power that may be distributed to support operation of all HV systems on the vehicle. The CPAM will then have all the information needed to allocate power in the most optimum manner when the vehicle is in power limited situations. Figure 1 is a representation of a controls architecture template that encompasses this layout, consisting of three HV devices and one Energy Source. Note that the number of Energy Sources and HV devices may vary from different vehicle architectures but the fundamental design remains the same.

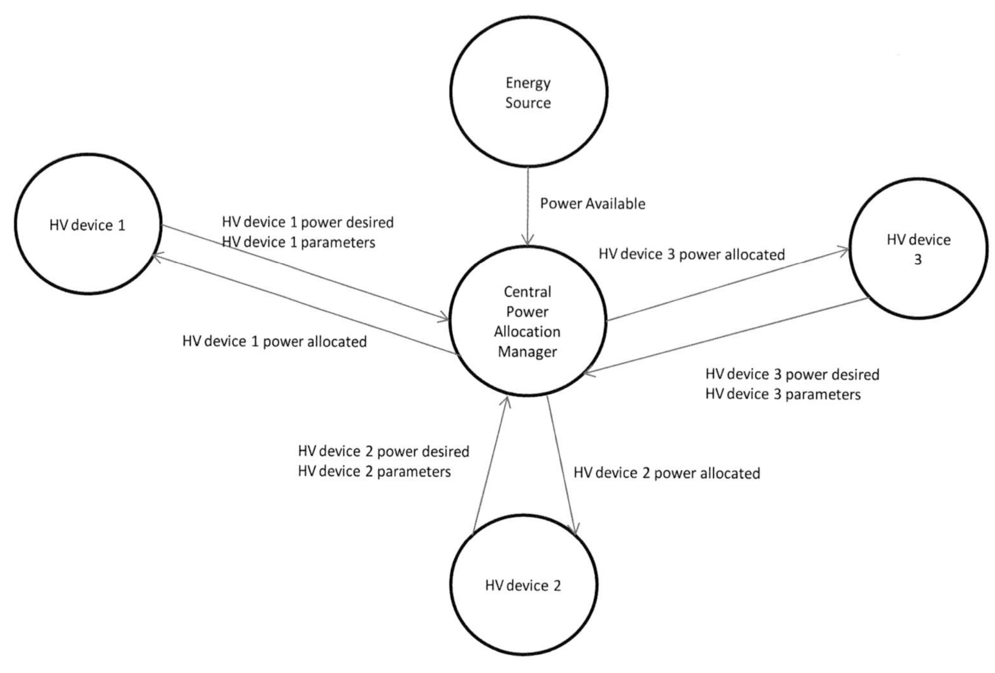

Figure 1.

CONTROLS INTEGRATION

This section will discuss how the above controls architecture is integrated into an electric vehicle and how the controls will operate under specific power limited conditions. For the purpose of this integration example we will be assuming the electric vehicle will consist of the following HV components:

- Energy Source:
    - HV RESS
- HV Consumers:
    - Electric Motors
        - Primary function is to provide propulsion of the vehicle
    - Air Conditioning (AC) Compressor
        - Primary function is to condition the vehicle's cabin
    - Cabin Heater
        - Primary function is to heat the vehicle's cabin
    - RESS Conditioning Compressor
        - Primary function is to condition the vehicle's RESS
    - RESS Heater
        - Primary function is to condition the vehicle's RESS

Selecting Relative Parameters for HV Devices

The first step in integrating a HV device into the CPAM design is to select all relative input parameters that would be useful for the CPAM to monitor for optimizing power allocation. Parameters that should be monitored must be relative to the devices primary function. For example, the RESS Heater and RESS Conditioning Compressor's primary function is to condition the vehicle's RESS to an ideal operating temperature. In this case the RESS temperature is a relative parameter that should be monitored by the CPAM. Figure 2 displays the CPAM design for this specific example with all the HV devices and relative parameters integrated into the controls design template that was shown in figure 1.

Power Limiting Conditions

The customers driving style and drive terrain will be big factors as to when the electric motors will require large amounts of power, possibly exceeding the power output capabilities of the RESS. An aggressive driver may perform many wide open throttle (WOT) maneuvers or push the vehicle to its maximum speed capabilities forcing the vehicle into a power limited condition. Even a normal driver that encounters any excessively steep grade climbs may find it difficult to maintain vehicle speed with the power that is available. When the RESS is reaching a low state of charge (SOC) or the vehicle is in extremely cold climates, the

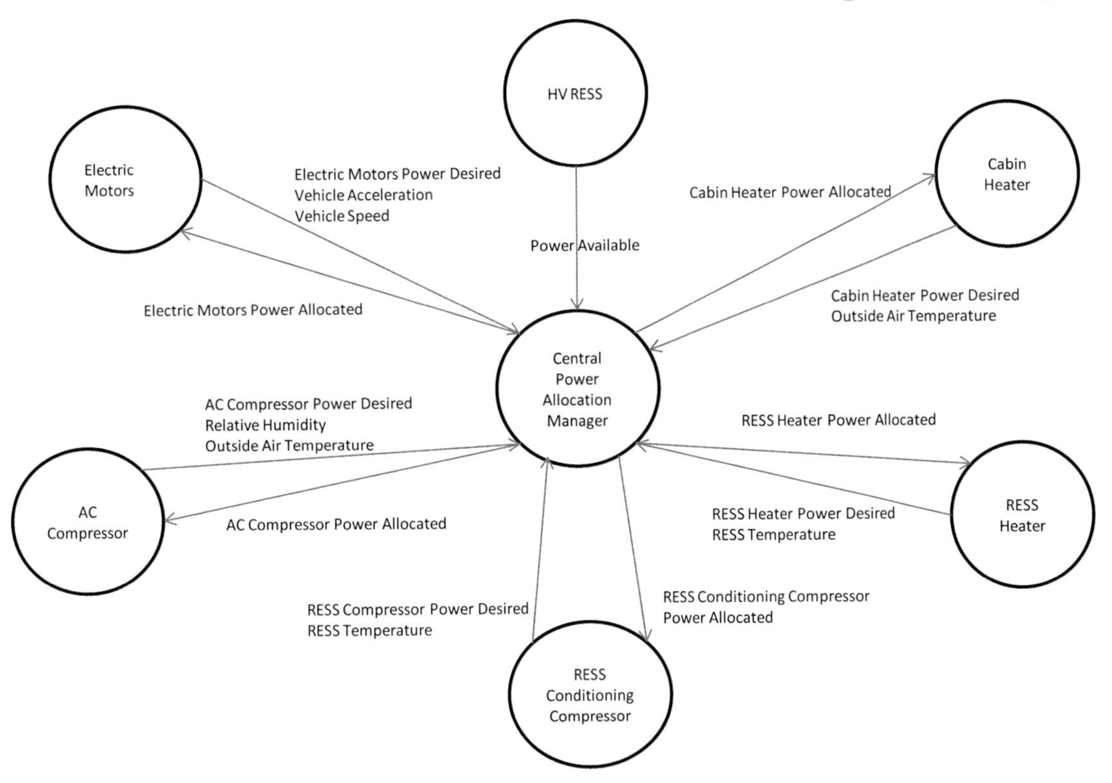

Figure 2.

maximum output power of the RESS may be limited, inducing power constraining conditions. Each power limited event should be placed into one of two groups, short term or long term. A short term power limited event will be in the magnitude of a few seconds while a long term event may last several minutes. The CPAM algorithm must be dynamic enough to allocate power differently for both long term and short term events as the priority of each HV device will differ in these situations.

Hot Ambient Conditions

In hot climate conditions certain HV devices will need to operate in order to meet specific vehicle needs. These hot climates usually require the operation of the RESS Conditioning Compressor to maintain the RESS at its ideal temperature and the operation of the AC Compressor to keep the vehicle's cabin comfortably conditioned. When the vehicle is in a power limited event in hot ambient a perplexing condition is created in which the RESS is now unable to meet the power demands of all three HV devices.

In this specific scenario the CPAM will be responsible for allocating power to the following HV consumers; the electric motors, RESS Conditioning Compressor, and the AC Compressor. The CPAM algorithm will take the power requests from all HV devices and optimize how much power each device should be allocated based on the relative parameters of each device. The algorithm will be dynamically calibrated to allocate power differently as the parameters change throughout the drive cycle. Below are some fundamental rules that the CPAM power allocation strategy may be designed around for this particular vehicle and situation.

1. As the vehicle speed or acceleration performance decreases, more power should be allocated to the electric motors.

2. The AC Compressor should only be allocated enough power to support a reasonable cabin comfort level. As the outside air temperature and humidity increase, the power allocated to the AC compressor should increase as well.

3. The RESS conditioning compressor should only be allocated enough power to ensure the RESS temperature does not exceed an extreme temperature limit causing significant harm or damage to the battery pack reducing its overall life. This should be at a temperature above the ideal RESS temperature set point.

Cold Ambient Conditions

Much like in hot climates, the CPAM will be forced to allocate power between three HV devices in cold climates as well. The operation of the RESS Heater and Cabin Heater to maintain ideal RESS temperatures and properly heat the cabin will be required. Below are some fundamental rules that the CPAM should follow for this particular vehicle and situation.

1. As the vehicle speed or acceleration performance decreases, more power should be allocated to the electric motors.

2. The Cabin Heater should only be allocated enough power to adequately meet any safety/windshield de-icing requirements and support a reasonable cabin comfort level. As the outside air temperature decreases, the power allocated to the Cabin Heater should increase.

3. The RESS Heater should only be allocated enough power to ensure the RESS temperature does not decrease to a point that would de-rate the power output of the RESS, inducing a more severe power limited event.

Charging Events

Many hybrid/electric vehicles will also be equipped with an on board charging device capable of connecting to an external AC power source with the main purpose of charging the RESS. When this process is occurring, there is a limited amount of HV and LV power that the on board charger is able to supply to the vehicle. Any operation of a RESS thermal conditioning device throughout the charge process will in turn leave less power available to support charging the RESS, thus increasing the overall charge time required. Below are some fundamental rules that the CPAM power allocation strategy may be designed around for allocating the charger power available during a plug in charging event.

1. If the RESS SOC is extremely low, less power should be allocated to support RESS thermal conditioning devices. As the SOC increases, more power should be allocated to support the operation of RESS thermal conditioning.

2. If the RESS temperature decreases to a point at which it is too cold to be charged without a high potential for battery cell plating to occur, then all available charger power should be allocated to support RESS thermal conditioning.

3. If the RESS temperature increases to a point at which it is too hot to reach a full charge termination voltage (due to dynamic voltage limits of the RESS decreasing at high temperatures), then all available charger power should be allocated to support RESS thermal conditioning.

4. The power allocated to any RESS thermal device should be limited to the output power of the on board charger to ensure the SOC does not decrease during a plug in charge event.

SUMMARY

In summary, HV power allocation management is crucial to support proper system functionality when the vehicle is in power limited conditions. Integrating the CPAM controls design into a vehicle will ensure power is allocated in the most optimum manner under all vehicle conditions.

CONTACT INFORMATION

James D. Marus
james.marus@gm.com
General Motors Company
Milford Proving Grounds
M/C 483-316-250
3300 General Motors Road
Milford MI, 48380

DEFINITIONS/ABBREVIATIONS

RESS
Rechargeable Energy Storage System

HV
High Voltage

LV
Low Voltage

SOC
State of Charge

CPAM
Central Power Allocation Manager

AC
Alternating Current

ICE
Internal Combustion Engine

AC Compressor
Air Conditioning Compressor

WOT
Wide Open Throttle

The Engineering Meetings Board has approved this paper for publication. It has successfully completed SAE's peer review process under the supervision of the session organizer. This process requires a minimum of three (3) reviews by industry experts.

ISSN 0148-7191

SAE Customer Service:
Tel: 877-606-7323 (inside USA and Canada)
Tel: 724-776-4970 (outside USA)
Fax: 724-776-0790
Email: CustomerService@sae.org
SAE Web Address: http://www.sae.org
Printed in USA

SAE International™

CHAPTER FIVE:

Codifying the Car

Model-based design was a key enabler to Volt's rapid development.

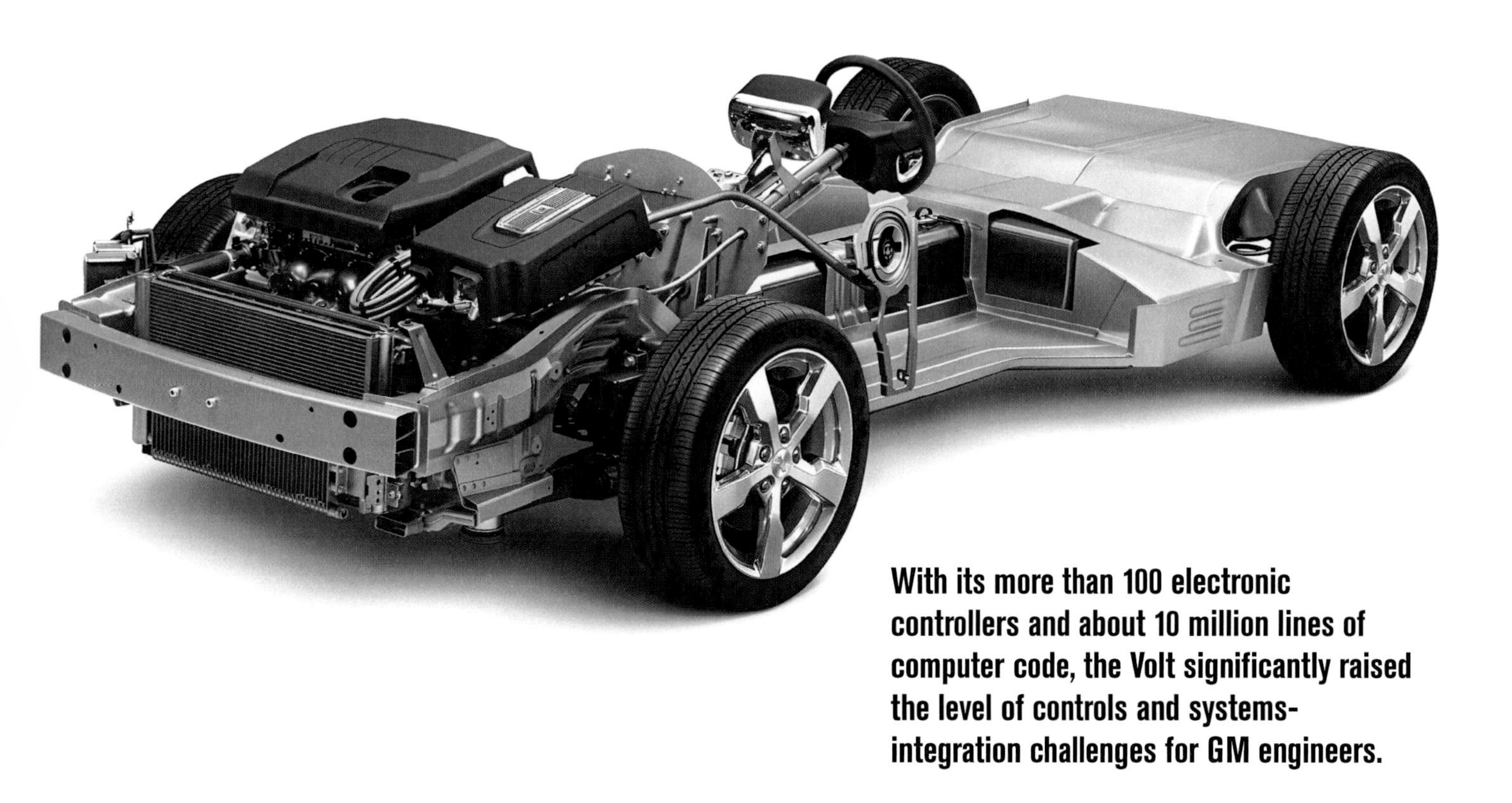

With its more than 100 electronic controllers and about 10 million lines of computer code, the Volt significantly raised the level of controls and systems-integration challenges for GM engineers.

Codifying the Car

Greg Hubbard, Senior Manager of Hybrid and Electric Drive Controls, said GM's expertise in model-based design allowed Volt engineers to demonstrate early mules "far earlier in a program than had ever been done before."

Key enablers in Volt's rapid development were model-based design, automatic code generation, and knowledge gained from the 2-mode hybrid program.

With GM's announcement in early 2007 that Volt would be a production program another piece of news broke that was almost as unique as the car itself. Volt's development activities would be more open and accessible to the media than any product development program in GM history, the company's engineering leadership proclaimed.

This appeared to be a risky strategy because the car's sophisticated energy-storage system was being developed on the critical path. A battery had to be invented. However, as the program took flight an overall air of confidence emerged during discussions with the engineers, scientists, and technicians. As more time passed, more confidence predominated. Only after Volt's launch did GM engineers divulge one of the key reasons for their confidence which, they assert, was real.

"Our high level of confidence definitely came from having experience in model-based controls, which enabled us to describe how we wanted the propulsion system to behave," explained Greg Hubbard, GM's Senior Manager, Hybrid and Electric Drive Controls. "It gave us the ability to test those algorithms so when the first hardware arrived, we could demonstrate the state of development from an early stage."

He said those demonstrations came "far earlier in a program than had ever been done before." This had significant effect on time (and cost) savings. The more you get the model right, the more you prevent late fixes in the vehicle. Time spent debugging an algorithm in the development stages generally subtracts from time available for tuning and calibrating at the vehicle level.

In the software world, a generally accepted rule of thumb assumes there will be on the order of 0.1 to 1 residual software defects per 1000 lines of code written (counted after stripping comments and blank lines). Thus, if a control system has one million lines of code, it will have approximately 1000 residual defects. And as noted by Dr. Barry Boehm in his 1981 book, *Software Engineering Economics*, the cost of fixing a defect rises in each subsequent phase of a project.

Boehm's research found that it costs up to 100 times more to fix a problem during operation than it would have cost to correct it during the requirement definition phase. Thus given the exponential software growth content in Volt, the GM engineers chose a path that minimized risk by leveraging previous experience.

Applying 2-Mode teachings

Even before GM launched the all-new 2-Mode hybrid system for trucks in 2007, engineers assigned to the program were already beginning work on the Volt's electrified drive system. Hubbard recalled that the team was "making natural extensions of what we'd learned on 2-Mode in terms of controlling the motor, the power electronics, and the drive unit."

The Volt significantly raised the level of controls and systems-integration challenges for GM engineers. The car uses more than 100 electronic controllers, with about 10 million lines of computer code, to shunt power rapidly and seamlessly among the car's battery pack, power inverter, traction motor, combustion engine/generator, and other subsystems.

By comparison, Boeing's new 787 Dreamliner relies on a mere eight million lines of code, and the F-35 Joint Strike Fighter is relatively code-deficient, with "just" 5.7 million lines controlling the sophisticated aircraft's systems. (As an historical reference, the average GM automobile in 1990 used one million lines of code.)

"While the Volt definitely has a different vehicle operating philosophy, it also has aspects that are true for any advanced propulsion system—diagnostics, safety systems, and some performance attributes, for example," Hubbard said. "We had modules of essentially MATLAB and Simulink files that we were able to extend for the Volt usage, based on our 2-Mode work."

GM has been using MATLAB, Simulink, and other tools supplied by The MathWorks for nearly fifteen years, with heavy early emphasis on the powertrain and research areas, noted Wensi Jin, MathWorks' Automotive Industry Manager. Model-based design allows engineers to capture the system dynamics and control algorithms, including diagnostics, in a modeling environment while the hardware is being developed. Then they refine the models of the systems being controlled—including the battery, combustion engine, and traction motor, in the case of Volt— to the point where they're happy with it, and it's time to move over to the hardware.

Having the models provides engineers with the confidence to rigorously test the systems being developed before moving on and trying to debug algorithms on the car, which is time consuming, costly, and difficult at best, Jin explained. GM engineers use the MATLAB tool chain as the basis for teamwork—a common language that enables efficient requirements definition, algorithm design and test, and calibration.

In the Volt program, a development path based on a traditional prototype design process was not feasible, given the vehicle's all-new battery technology being developed in parallel with the vehicle. The team couldn't afford to wait for new hardware. That's where the benefits of their 2-Mode learning became obvious. The challenge in the program (as with all hybrid development) was to get many controls to work together seamlessly, and to manage a complex system design in less time.

Wensi Jin, Automotive Industry Manager at The MathWorks, said models give engineers the confidence to rigorously test the systems being developed before moving on and trying to debug algorithms on the car.

According to Hubbard, the clutches inside the Voltec 4ET50 multi-mode electric transaxle allowed the engineers to use and reuse some of the 2-Mode functionality for turning the clutches on and off, and changing between drive modes.

"With GM methods on how to architect these models, and using the Mathworks tools to do so, we were able to use the models from one program to another," he explained. The control system was verified using closed-loop simulations. dSPACE Rapid Controls Prototyping (RCP) and hardware-in-the-loop (HIL) tools were used for ECU software development, validation and verification, for technology development of powertrain, vehicle dynamics, body electronics, safety, and finally integration testing.

Autocoding benefits

Automatic code generation using The MathWorks' Real-Time Workshop Embedded Coder was vital to meeting Volt's aggressive program timing, Hubbard said. It played an essential role in enabling the code generation from model on down. The Real-Time Workshop Embedded Coder product generates C and C++ code from Simulink and Stateflow models that engineers say has the clarity and efficiency of hand-written code.

"It's one of the strong advantages of the Mathworks tools," Hubbard explained. "We have a single

Codifying the Car

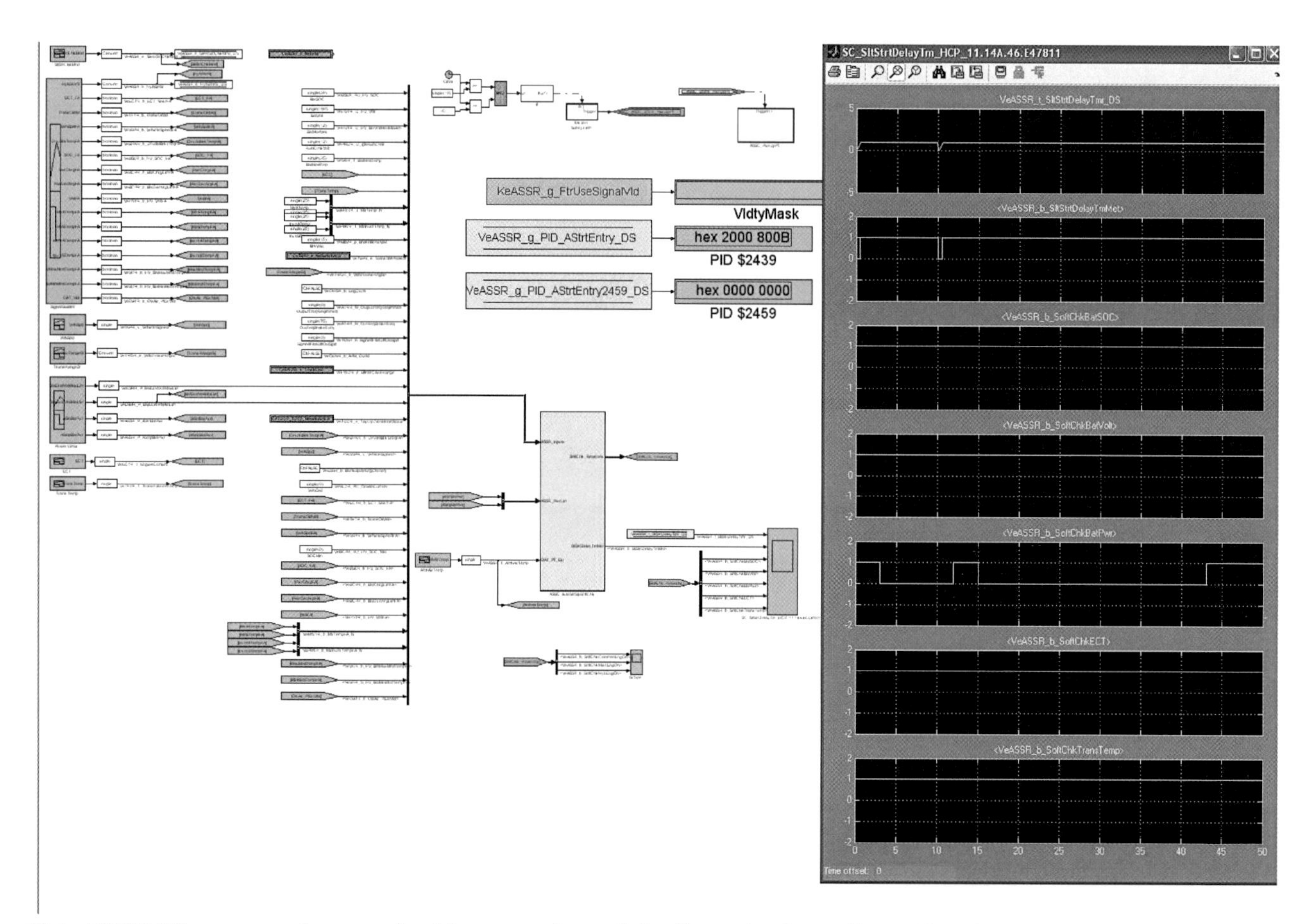

This MATLAB screen shows a test harness for a "ring" – a test configuration for an algorithm. This particular algorithm deals with "on" and "off" commands for the Volt's combustion-engine. Blocks on the left include the test scenarios. The test results (strip chart) on the right shows how those signals change as the test vector proceeds.

source for how a particular function should behave. So if we have a function for how we determine how to start the engine once the battery charge has depleted and we've decided to turn on the generator, we have Simulink models that define how we want those functions to work."

He said the Volt engineers describe the functionality in the Simulink tools, and then can generate code from those models—and that code is used directly in the computers in the car.

"So by having a single source it means the designers of the algorithms, the testers and calibrators of those algorithms and software, are all using the same 'gold source,' I'll call it," he said. "It functions one way. We don't introduce problems with translation errors in going from, say, a Microsoft Word spec and someone writing C code based on that. Having multiple interpretations of a function, that's been one of the keys in being able to move more quickly. It's a single source in how the functions should operate."

Nearly 100% of the software for many of Volt's modules was generated automatically, Karla Wallace, GM's Senior Manager of Powertrain Control

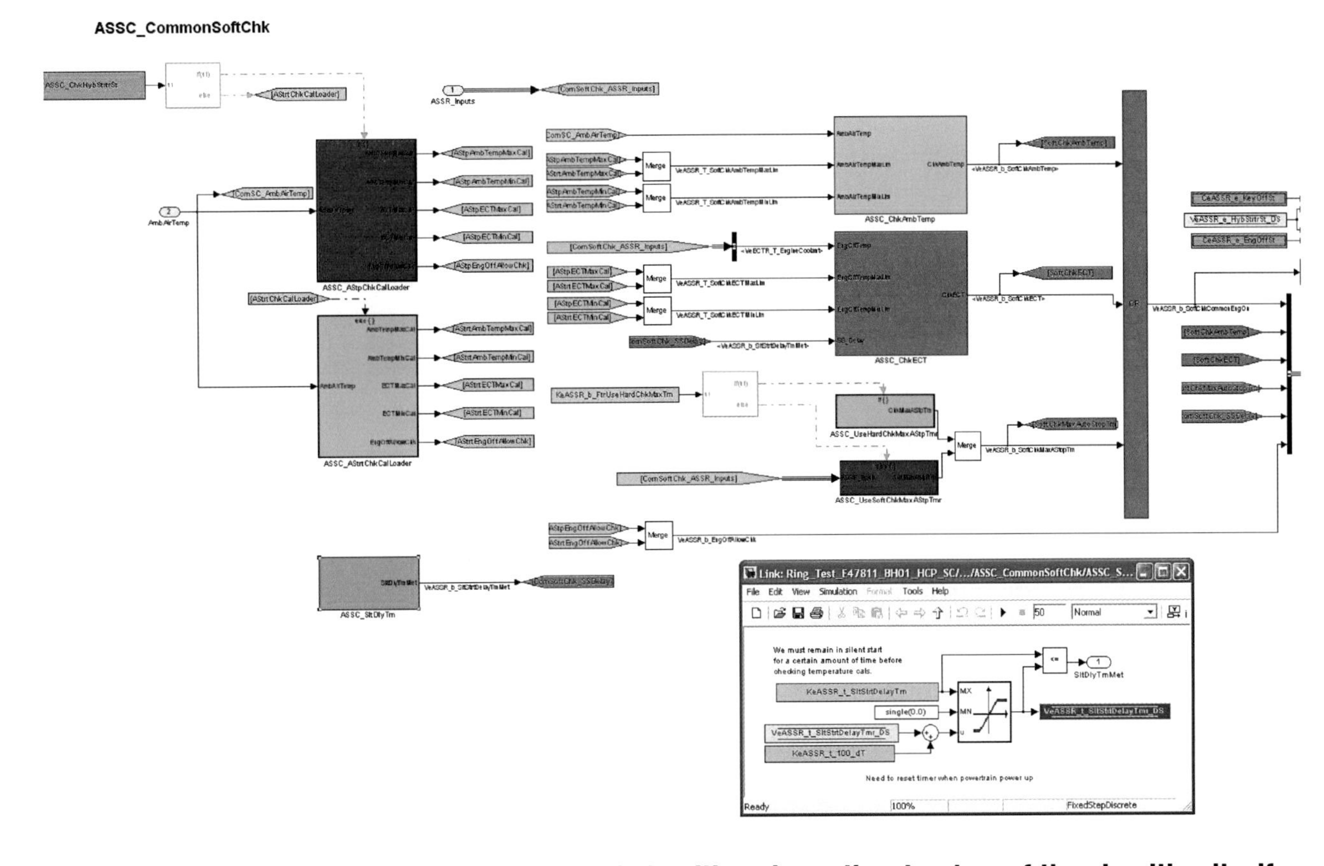

A "deeper down" level of the generator-control algorithm shows the structure of the algorithm itself showing reasons why the Volt's generator would need to be turned on or off.

Electronics Engineer, noted during the 2010 SAE Convergence Conference on Automotive Electronics. Hubbard estimated that using automated code generation instead of hand coding gives GM a 30-35% efficiency gain.

Hubbard recalled that GM completed its first prototype using code-generating tools in 1997-2000. He said during that period engineers were only able to comfortably bring one function to be autocoded. With Volt, approximately 10,000 functions were accomplished using models and autocode.

GM engineers started talking publicly about generating control code from the MATLAB and Simulink models—and how it was on par with handwritten code for production programs—in 2005.

"You develop the control system, then model it 100% in full detail. And you also model what's being controlled," noted The MathWorks' Jin. "That's why GM started looking at generating code directly, rather than spend a lot of manpower rewriting code line by line, following what's in the model."

Wallace told the Convergence audience that her company is moving toward more of a virtual desktop environment, with tools including those from MathWorks that make it appear to the engineers that they're in a vehicle running tools used for calibration.

"Engineers can also use simulation in a batch mode, running many simulations and getting data back in an integrated format," she said.

Wallace noted that GM research engineers have been running batch simulations for powertrain designs on high-performance computers, not desktops. After the move to desktops, those large

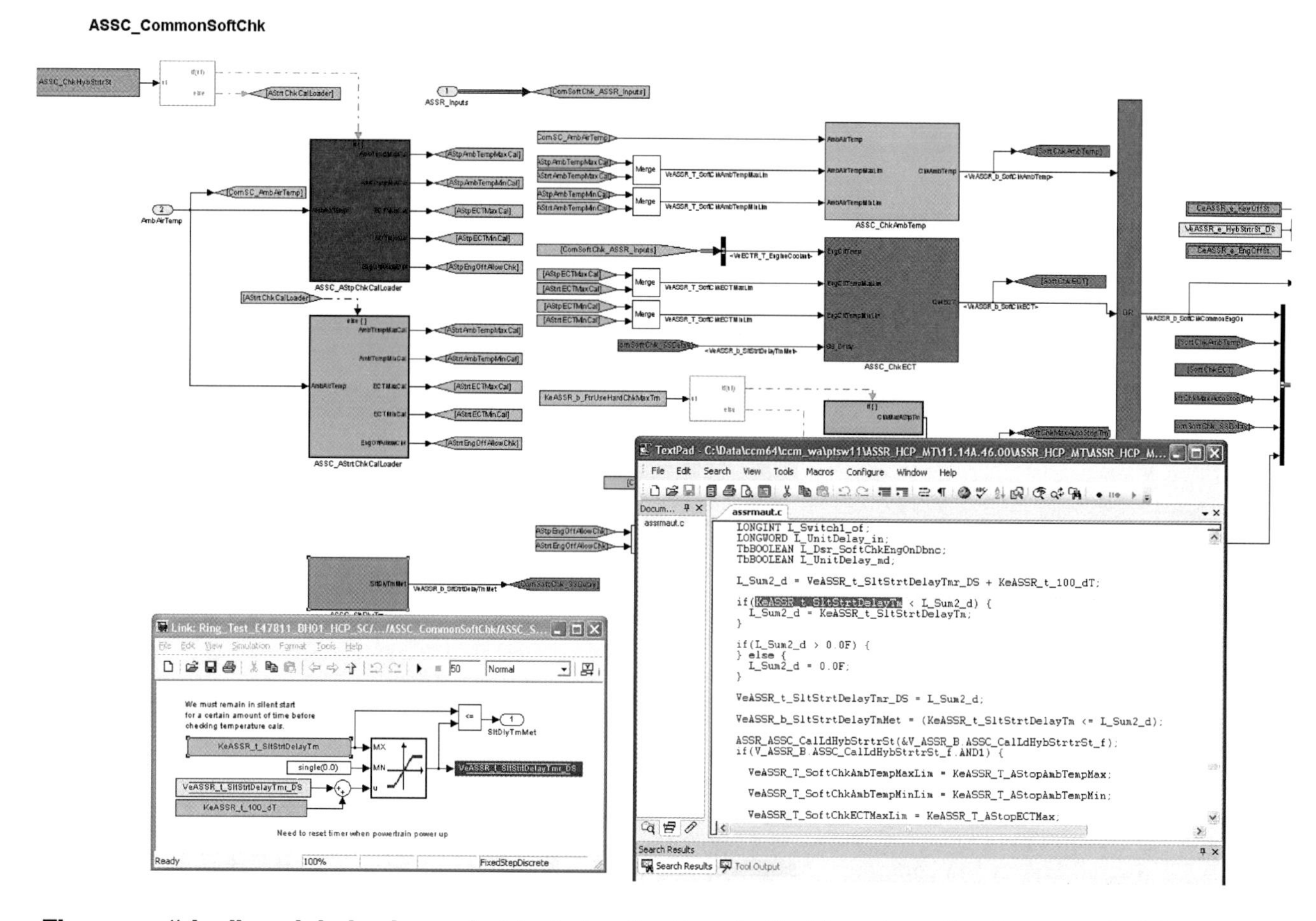

The same "ring" module is shown, but includes the C code that is automatically generated from the Matlab/Simulink model. This C code is then compiled and linked with other automatically generated code and handwritten code and downloaded to the Volt's propulsion ECU.

systems will be used after the research team has a high degree of confidence in a design. These grid-computing systems run thousands of test runs, setting the stage for real-world prototypes that can test critical functions and verify the simulations.

The Volt program is in the vanguard of these trends. What did GM's software engineering group learn from it?

"We'll continue to build on past lessons learned—EV-1, 2-Mode, and now Volt," Hubbard said. "Our lessons learned in smoothing engine stopping and starting are being applied to the new e-Assist system (which enters production in MY2012 on Buicks).

"We're using the same models to control these different systems. We're getting generational learning much faster on each iteration," he asserted. "It benefits us to use these single sources that people can understand; it allows us to go faster."

CHAPTER SIX:

Sweating the body details

Bob Boniface's design team significantly revised the original concept car's exterior form to give the production Volt more EV range.

Extensive wind-tunnel work gave Volt a shape that's slicker than it looks. But engineers aren't happy with the curb weight.

Advanced powertrain engineers and eco-enthusiasts argue convincingly that Volt's technology crown jewel is its electrified propulsion system. And they're right. But the car's overall efficiency, and success in the marketplace, also hinges on its aerodynamics, styling, package efficiency, and occupant protection.

The body form and construction count as much for electrified vehicles as it does for conventionally powered ones, perhaps even more. Witness Toyota's Prius, whose overall wedge shape and tall greenhouse make it far from a handsome car. But

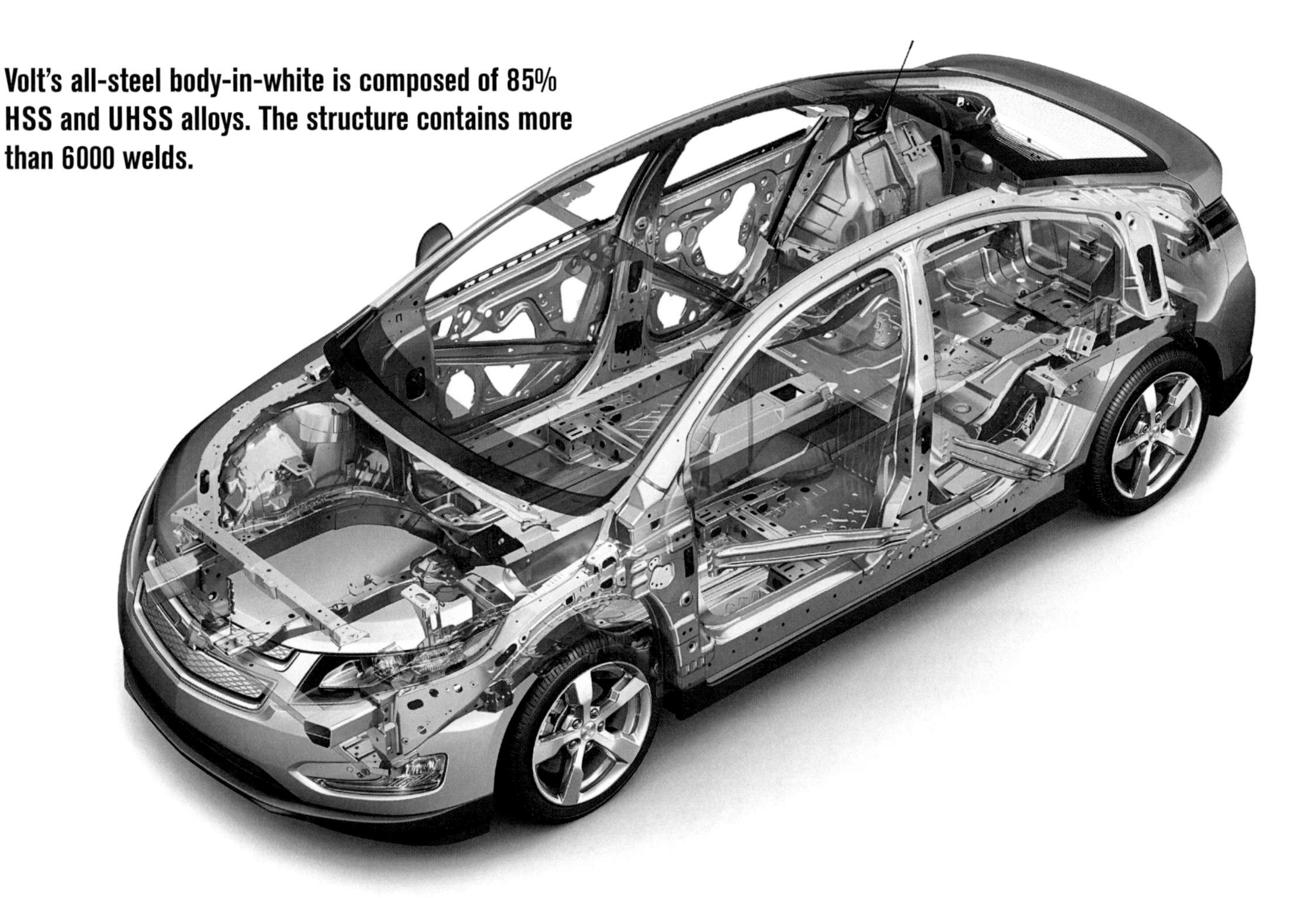

Volt's all-steel body-in-white is composed of 85% HSS and UHSS alloys. The structure contains more than 6000 welds.

that look struck a major chord with hybrid early adopters and cemented it to the point that Honda paid homage with the current-generation Insight.

The "Prius look," often satirized by editorial cartoonists, became the face of the hybrid movement.

But when General Motors began developing Volt as a concept car for the 2007 Detroit auto show, Vice President of Global Design Ed Welburn's team had no intention of following the crowd. Nor did they simply stuff the E-REV powertrain into a mildly re-styled existing body.

"The challenge to the designers wasn't to design the most beautiful car imaginable and accept the compromises you have to make to do so," explained former GM Vice Chairman Bob Lutz on GM's Fastlane blog. "It was, make no compromise to fuel efficiency and electric range, and then do the most beautiful design possible around those aerodynamic dictates."

The task of directing Volt's four-seat body design was given to Bob Boniface, a graduate of Detroit's College for Creative Studies who had worked in Chrysler's styling studio before joining GM in 2003. Boniface had worked on the Corvette and Camaro design teams prior to being assigned to a new group dedicated to hybrid, EV, and fuel-cell vehicle design based at GM's Advanced Design Center.

Their sporty Volt concept car, with its extremely squat greenhouse, tall beltline, and big-shouldered stance, echoed the new Camaro and helped make Volt an immediate sensation. But soon after, in early 2007, reality struck when GM put the concept car into the wind tunnel.

"Our early tunnel work showed we had some work to do to improve the car's aerodynamics," Welburn told *AEI*. He said everyone involved recognized the car's drag had to be reduced significantly

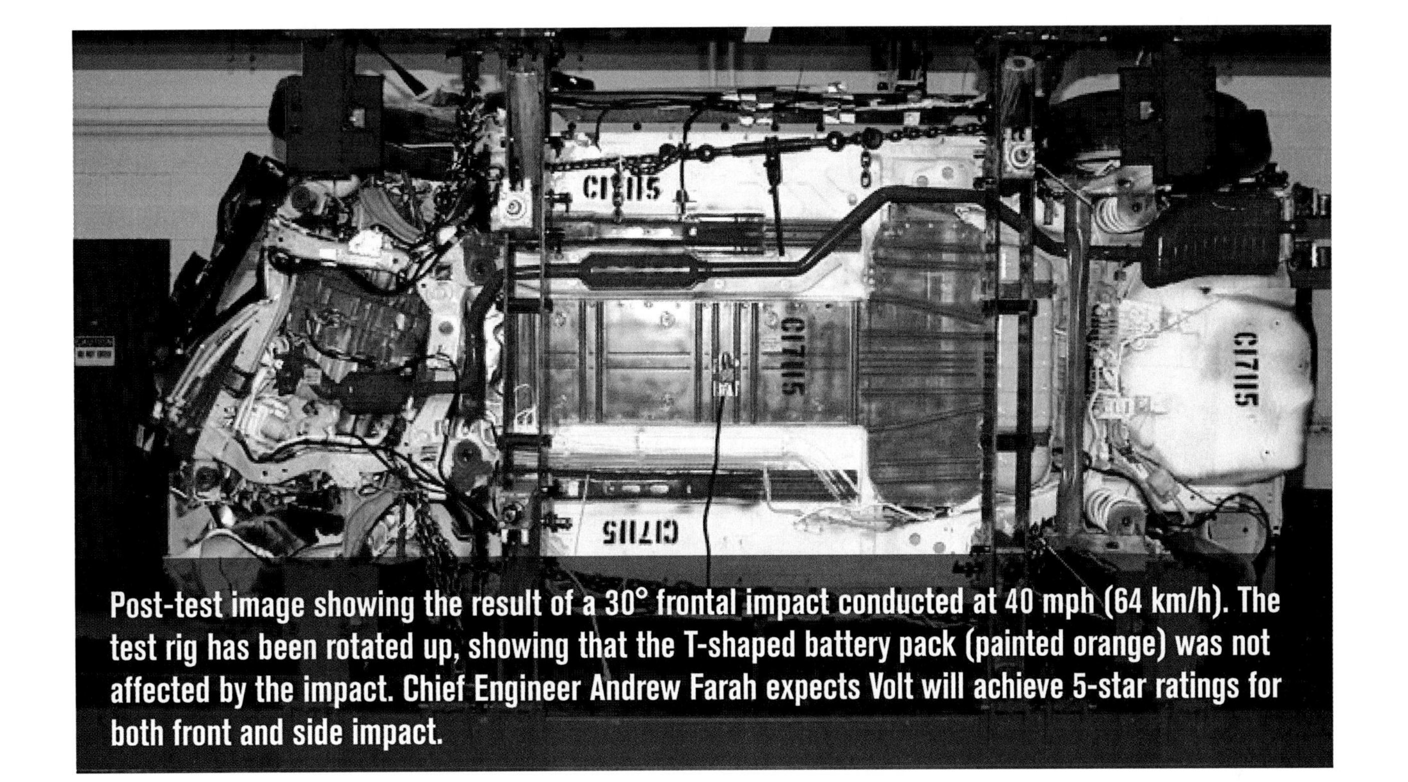

Post-test image showing the result of a 30° frontal impact conducted at 40 mph (64 km/h). The test rig has been rotated up, showing that the T-shaped battery pack (painted orange) was not affected by the impact. Chief Engineer Andrew Farah expects Volt will achieve 5-star ratings for both front and side impact.

for it to deliver 40-mi (64-km) battery range. An E-Flex Systems Design Studio with Boniface at its helm was established to prioritize Volt's body development while developing styling proposals for future electrified vehicles.

"When Volt was approved for production, I decided we needed a dedicated mix of designers and engineers from the show-car team, working together with people from vehicle aerodynamics and the production side, and they needed their own creative space," Welburn said. The E-Flex design group grew to 50, and more than a year of aerodynamic development followed.

The evolved Volt was revealed in September 2008 during GM's centennial celebration in Detroit. This was essentially the production body form, the result of approximately 500 h of wind-tunnel development, according to Boniface. At the GM 100 event, he walked around the improved Volt detailing the changes to *AEI* and other media. At first glance the car looked more conventional, with a taller greenhouse, and less Camaro.

But the real aero gains were found at the corners. The leading edges of the front fenders were rounded and refined to create consistent laminar airflow down the body sides. The car's front fascia was made flush and the frontal air intake was now handled by a horizontal opening below the grille rather than by the grille itself, which became ornamental. The rear corners were resculpted, their edges sharpened.

Boniface said much attention was given to the rear spoiler, rocker panels, and the "speed" (rake) of the A- and C-pillars to minimize turbulence over the roof, and reduce drag overall.

"We sweated the details in so many areas," he recalled, "and gained many counts of aero in the end."

Boniface eventually revealed that the production Volt's coefficient of drag (Cd) would be 0.28. This was significantly less "slippery" than the EV1's groundbreaking 0.19 Cd, which was achieved with the help of partially faired rear wheelhouses. The Volt team wanted to avoid fender skirts.

Volt's new lower-drag shape achieved a lower Cd number than that of GM's reigning aero king, the Corvette (0.29 Cd). It also beat Honda's Insight (0.32 Cd) and Toyota's Prius.

"We tested both a 2010 Prius with 17-in wheels as well as the new Insight," he noted. The Prius came in at 0.30 Cd, and that number was also verified in the tunnels of both Ford and Chrysler." Toyota officially claimed 0.25 Cd for Prius.

Curb-weight compromises

As a concept vehicle, Volt suggested GM would deploy significant lightweight-material content if a production version was approved. The extended-

An interior **trendsetter**

Paul Haelterman

Volt is in the vanguard of vehicle cabins that are stylish and also energy-efficient, says automotive interiors expert Paul Haelterman, Managing Director of Global Automotive Consulting at IHS Automotive.

Q: What interests you most about Volt from an interior-sourcing perspective?

Haelterman: That Lear Corp. is the second-largest supplier on Volt, based on dollar value. The battery cell supplier LG Chem is the top content supplier. Lear supplies the Volt seats but they also have the charging hardware sets. They picked up the electronic technology when they purchased United Technologies Automotive a while back.

Volt's seating configuration is unique—four buckets. You get a manual height adjuster for driver and passenger, and a leather option. That's about it.

Q: Volt's interior team chose to use the car's seats to heat and cool the occupants directly, rather than conditioning the cabin air which gobbles electricity. Will we see more of this strategy?

Haelterman: Yes. Reducing energy consumption of key components such as the HVAC and other electronic devices is a smart strategy that's gaining a lot of attention. So are EPS [electric power steering] and electric coolant pumps. Everybody needs to reduce parasitics across the vehicle. The problem with EVs is there's no heat engine—or in the case of stop-start hybrids when the IC engine shuts off at a stop you no longer have thermal capability.

Q: With more EVs coming into the market, do you expect more use of seats as primary heaters?

Haelterman: Yes. It's the easiest, lowest-tech solution. With seat heaters you're not talking that much expense. Currently the technology costs around $10 per seat. If you install them as standard equipment, it makes everybody's life easier.

Q: When will electric HVAC become more viable?

Haelterman: As the number of EV and high-voltage hybrid programs increase. In many cases the OEMs are jumping to this seat-based cabin heating solution as a temporary step until the really good electric HVAC solutions are ready.

Q: When will electric HVAC hit the market in volume?

Haelterman: I believe it will be the middle of the decade before electric HVAC starts to show up in any appreciable numbers, maybe 1-2% of the HVAC systems market. And it will be the end of this decade before we see real volume—perhaps 10-15% penetration by 2020.

range powertrain was designed to be plugged in for charging, thus mandating a relatively large and heavy battery pack. Mass-reduction countermeasures on the concept were led by various thermoplastic body elements developed by the former GE Plastics (now part of Saudi-owned SABIC Innovations).

"We were able to take mass out of the Volt in order to optimize its overall efficiency," commented Jon Lauckner, who brainstormed the Volt concept and was GM's Vice President of Global Program Management at its debut. Lauckner strategized that high composites content, some of it using recyclate, would also help boost the car's eco-cred. GM consulted GreenOrder, an independent auditor on all things environmental, regarding the benefits of using certain materials.

When the concept rolled out in Detroit wearing Noryl GTX front fenders, veteran autowriters recalled GM's past quality issues due to use of composite fenders on various models. Volt's door panels were molded in Xenoy iQ resin (a collaboration of GE Plastics and Azdel), and its roof panel and decklid made with Lexan GLX and coated in Exatec.

These and composites applications, including plastics window glazing, offered a 30-50% mass reduction per part, according to Amanda Roble, Executive Director of the former GE Plastics' Automotive business. Even the car's low-voltage wire harness was made from GE's nonhalogenated plastics, which Roble estimated was worth a 25% weight reduction compared to traditional wire.

But the dream of diverse exterior-body panel materials evaporated almost immediately when Volt was green-lighted for production, according to Vehicle Line Executive Tony Posawatz. "Incorporating a lot of new materials ideas into concept vehicles is part of the reason we do concepts," he said.

The decision to base Volt on the global Delta II's 100% steel architecture offered scale, cost savings, and proven assembly quality, Posawatz noted. The body is commendably stiff in torsion, at 24-25 Hz. That's due to extensive finite-element modeling, a commitment to high-strength and ultra-high-strength steel alloys (comprising 85% of the overall structure, according to Posawatz), more than 6000 welds, and using the battery pack as a semistructural member.

Chief Engineer Andrew Farah and other engineers on the Volt team are on record as not being happy about Volt's 3781 lb (1715 kg) curb weight. This is perhaps the car's major engineering compromise due to the need to carry a 16 kW·h battery, use the relatively large iron-block 1.4-L ICE, and get the car into production on an extremely aggressive timetable with high build quality.

"Looking ahead, we can optimize many of the technologies as aspects of the car that are somewhat suboptimized on Volt, due to the program timing and our focus on getting the battery absolutely right," Farah explained.

Examples of GM's plans for future E-REVs include the Cadillac Converj coupe, which was mothballed during the automaker's financial crisis prior to bankruptcy, and the Chevrolet MPV5, an E-REV-powered crossover-type people mover concept shown at the 2010 Beijing Auto Show.

The six-passenger MPV5's design and aero details are based on the production Volt, and showcase the architectural stretch of the Delta II-derived platform. MPV5's 108.6-in (2758-mm) wheelbase is 0.6 in (15 mm) longer than Volt's. Its body measures 180.5 in (4585 mm) overall, 7 in (178 mm) longer than Volt. It's also 2.9 in (74 mm) wider and 7.1 in (180 mm) taller.

MPV5 was created by GM's North America Crossover Exterior Design team, noncoincidentally headed by Bob Boniface, with input from GM's Holden studio. The concept is a good indicator of how GM aims to leverage Volt learnings to spin many new E-REV body styles and vehicles.

The centerpieces of Volt's cockpit are the multiconfigurable instrument cluster and center stack, the latter styled to resemble an iPod or similar portable device.

The Volt MPV5 is the most recent Volt spinoff. The Delta II-based concept was revealed at the 2010 Beijing auto show and points to GM's intent to proliferate EREV iterations.

Aerodynamic Development of the 2011 Chevrolet Volt

2011-01-0168
Published
04/12/2011

Nina Tortosa and Kenneth Karbon
General Motors Company

doi:10.4271/2011-01-0168

ABSTRACT

This paper presents some of the challenges and successful outcomes in developing the aerodynamic characteristics of the Chevrolet Volt, an electric vehicle with an extended-range capability. While the Volt's propulsion system doesn't directly affect its shape efficiency, it does make aerodynamics much more important than in traditional vehicles. Aerodynamic performance is the second largest contributor to electric range, behind vehicle mass. Therefore, it was critical to reduce aerodynamic drag as much as possible while maintaining the key styling cues from the original concept car. This presented a number of challenges during the development, such as evaluating drag due to underbody features, balancing aerodynamics with wind noise and cooling flow, and interfacing with other engineering requirements. These issues were resolved by spending hundreds of hours in the wind tunnel and running numerous Computational Fluid Dynamics (CFD) analyses. The end result is a unique electric vehicle that, at start of production, has the lowest coefficient of drag, C_D, of any Chevrolet sedan.

INTRODUCTION

The Chevrolet Volt (Fig. 1) is an electric vehicle with extended-range capability powered by the Voltec electric propulsion system (Fig. 2). It consists of a 16 kWh lithium-ion battery, an electric drive unit, a 1.4-liter internal combustion engine generator, and a vehicle charge port. Able to re-charge the battery from a standard wall outlet, many drivers will consume no gasoline and utilize only battery power that provides the first 25-50 miles of range. Afterwards, the engine generator can deliver electric power for 300 more miles if needed. Extending the life of the battery to achieve the initial electric range and making the Volt as fuel efficient as possible for the total extended mileage were key goals during its development. Aerodynamics played a crucial role in that quest. Engineers spent over 500 hours in the wind tunnel and thousands of cpu hours on the computer to refine the air flow around the vehicle.

Figure 1. The 2011 Chevrolet Volt.

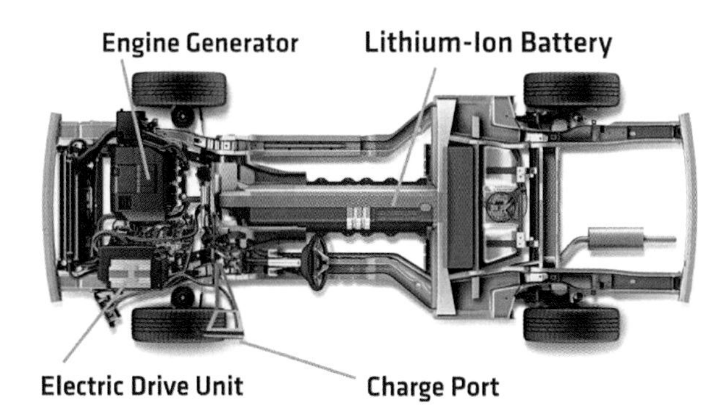

Figure 2. The Voltec electric propulsion system.

VOLT AERODYNAMICS

Drag Requirements for Fuel Economy

Aerodynamic requirements on any given vehicle are determined by top speed, handling, and fuel economy needs. For efficiency cars, aerodynamic drag is to be minimized for vehicle energy conservation. Once the electric and extended range objectives were determined on the Volt, the required drag coefficient was backed out. On conventional vehicles, aerodynamic drag is usually the third largest energy loss, behind powertrain friction and vehicle mass. On the Volt system, losses from the electric drive unit are less significant, leaving mass and aero drag as the main fuel economy enablers. However, in an electric vehicle where much of the weight is fixed due to battery capacity, efficiency is much more sensitive to percentage changes in C_D, rather than mass. This fact put aero development to the forefront of the Volt engineering process.

Figure 3 shows aerodynamic drag performance of a number of production vehicles in the North American small car market since the 2001 model year. GM and competitor data were consistently gathered in the General Motors Aerodynamic Laboratory (GMAL). Clearly, minimizing drag area (C_DA) is key to all hybrid and electric cars in this segment. A saleable Volt was tested per the new SAE Recommended Best Practice J2881, "Measurement of Aerodynamic Performance for Mass-Produced Cars and Light-Duty Trucks" [1], and achieved a C_D of 0.28 in GMAL.

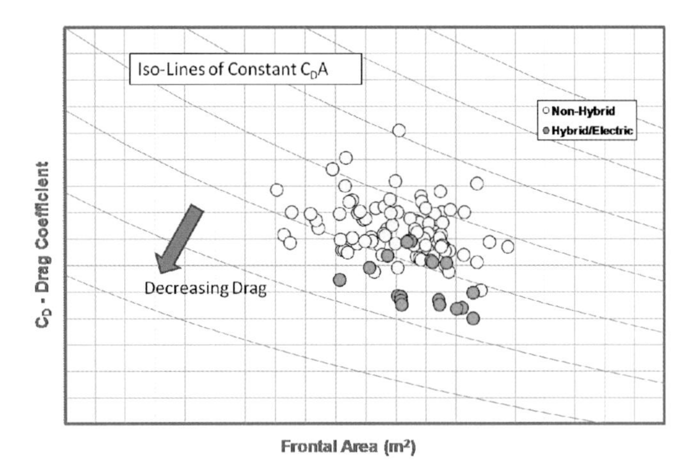

Figure 3. Aerodynamic Drag of North American Small Car Market, 2001-2011

Development Process

The development process began shortly after the Volt concept car debuted at the North American International Auto Show in January 2007 (Figure 4). Engineering efforts started in earnest as the concept design was proportioned to fit on to GM's small car platform. This provided the first aerodynamic evaluation as shown in the scale model of Figure 5.

Figure 4. Volt concept vehicle

Figure 5. First reduced scale wind tunnel test.

GM's aero methodology relies on several complementary tools. Drag reduction is developed mainly in the wind tunnel starting with reduced scale clay models and then full scale clay models, where design studies can be quickly sculpted and tested to a high degree of accuracy. Computational Fluid Dynamics (CFD) is the primary method to analyze and optimize front end airflow, with its faithful representation of nderhood and underbody details. CFD evaluation of drag is lso used when physical parts cannot be obtained or abricated in the clay models. Lastly, functional prototypes, alled IVER vehicles, allow for final fine tuning and onfirmation of the aerodynamic performance. Each tool's trength is utilized to the fullest, and each play an important ole in the process.

Jsing all the available tools was necessary to achieve the /olt's aggressive C_D requirement. The time line in Figure 6 hows the history of drag coefficient and some of the ignificant milestones in the development process. Also hown are Vehicle Technical Specifications (VTS) and the Glide Path. From beginning to end, vehicle drag was reduced by 150 counts.

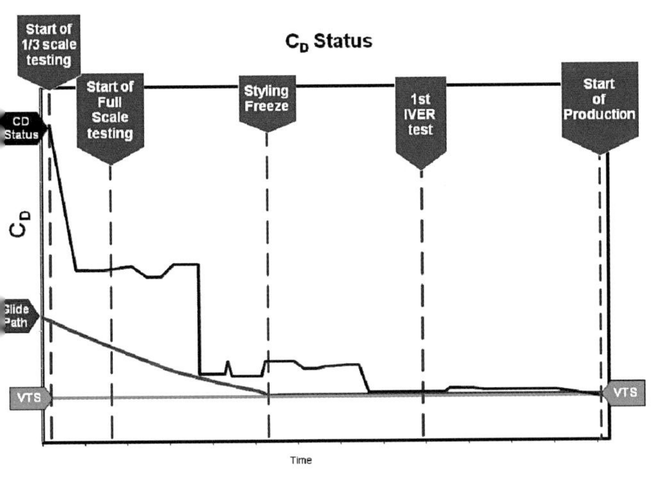

Figure 6. Timeline of the Volt's Aerodynamic Development

Reduced Scale Testing

The majority of the exterior surface development was done on a 1:3 clay model in the wind tunnel, and over 500 hours of testing were spent improving the basic shape of the vehicle. Particular attention was paid to shaping the front and rear fascias while maintaining styling cues. Reduced scale testing was ideal for early aerodynamic development of the Volt because less clay needed to be sculpted for a given geometry change. It was also this early 1:3 scale testing that showed the pedestal door mounted mirror design had lower drag than the more traditional patch mounted design.

GM's approach to vehicle aerodynamics development mandates simulation of the airflow through the engine compartment and underbody to account for any interaction with the exterior surface flow. This methodology is applied even in reduced scale, where grilles, radiators, and chassis components are represented while evaluating the styling surface drag. Figure 7 shows the model installed on the GMAL reduced-scale balance. The clay construction is enhanced with rapid prototype components. Testing of the Volt also showed that both underbody panels and an airdam would be needed to achieve aggressive C_D targets, with follow up work planned for the full size model.

Figure 7. Reduced scale (1:3) model installed on the reduced scale balance of the GM Aerodynamics Lab (GMAL). Stereo lithography grille, mirror, and wheels enhanced the clay model fidelity.

Full Scale Testing

Once the aerodynamic drag of the 1:3 scale model was significantly reduced, development moved into the full scale testing phase. To check for consistency, the last surface used on the 1:3 model was milled on to the full scale clay and tested in the wind tunnel as the first data point. The C_D results between reduced and full scale models were in agreement, paving the way for further tuning of the exterior surface and component details, including A-pillars, mirrors, underbody panels, and the airdam.

Figures 8 and 9 illustrate some of the exterior aerodynamic drag features of the Volt. Nose curvature allows for smooth airflow around the front corners, hood, and wheel openings. A set-back airdam is also optimized for the front end to reduce underbody drag and to maintain a low stagnation point. At the back of the vehicle, sharp vertical edges and a refined decklid spoiler achieve clean separation and minimize drag within a short rear overhang design.

One of the biggest challenges in developing the Volt was the outside rear view mirror. The styled mirror had to meet global market vision requirements, aero performance, and wind noise performance, all within the design theme of the car. Aerodynamic coefficients and sound pressure level are both measured and developed in GMAL. The swept, door-mounted design initially had a low C_D increment, but its aero-acoustics was not acceptable. Significant wind tunnel optimization resulted in a quiet mirror that maintains a low drag contribution.

Figure 8. Front view of the full scale clay model installed on the full scale balance of the GM Aerodynamics Lab (GMAL)

Figure 9. Rear view of the full scale clay model

Computational Fluid Dynamics (CFD)

Throughout the Volt aerodynamic process, non-intrusive visualization and in-depth analysis from CFD explained complex airflow behavior and complemented the hardware development in GMAL. The review of simulation results during wind tunnel tests was a great communication tool to help designers understand why modifications to the clay surfaces led to changes in drag coefficient. Over 1200 unique designs and operating points were computed in CFD to develop the aerodynamic drag and front end airflow.

Engineers applied flow simulation to rank early studio themes and to assess the aero performance of styling features. Figure 10 shows isosurfaces of P_{total} =0, colored by turbulent kinetic energy. These regions of low energy airflow highlight the major drag regions on the car and were useful in side by side comparisons of different vehicle shapes. The surface restricted particle traces in Figure 11 shows how the body sides and wheel arches were streamlined during the course of development.

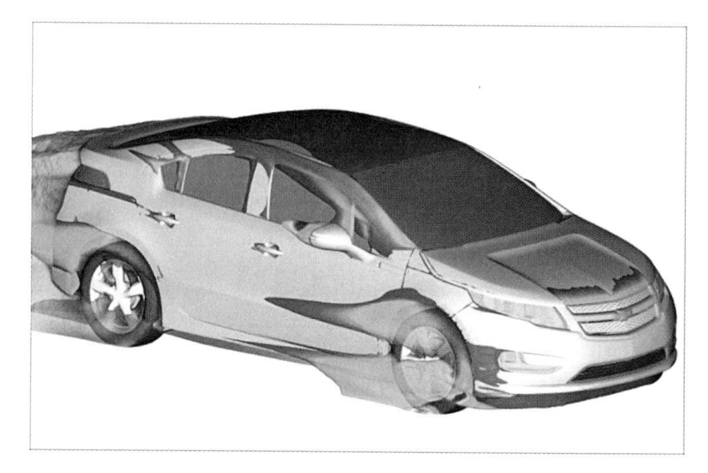

Figure 10. Isosurfaces of P_{total} = 0, colored by turbulent kinetic energy

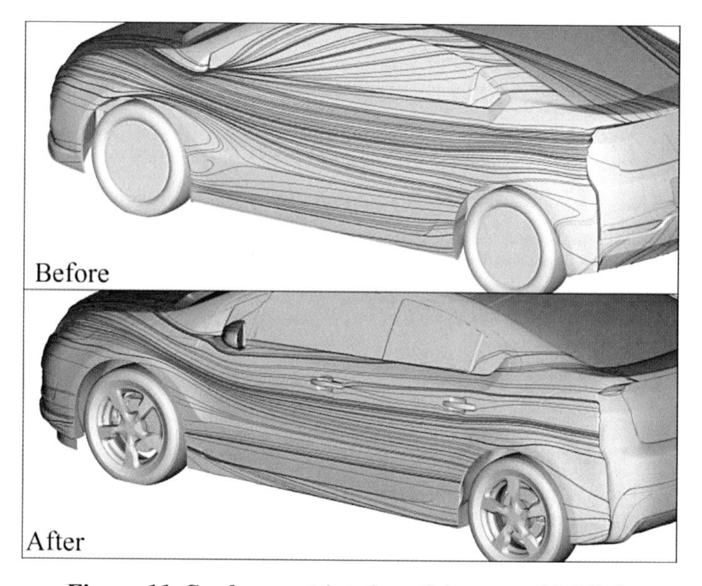

Figure 11. Surface restricted particle traces highlight body side streamlining

In particular, the flexible nature of the virtual computer model was critical to designing the underhood layout and front end airflow (FEAF) performance. The grille openings, condenser/radiator/fan module (CRFM) orientation, and sealing were developed in detail with simulation. Figure 12 shows the heat exchangers for the five unique cooling loops in the Volt. CFD testing optimized the amount and direction of airflow to these heat exchangers as well as the engine air induction in order to meet high temperature driving conditions. The efficient air delivery system allowed designers to close much of the grille opening, thereby reducing the parasitic cooling drag through the engine bay.

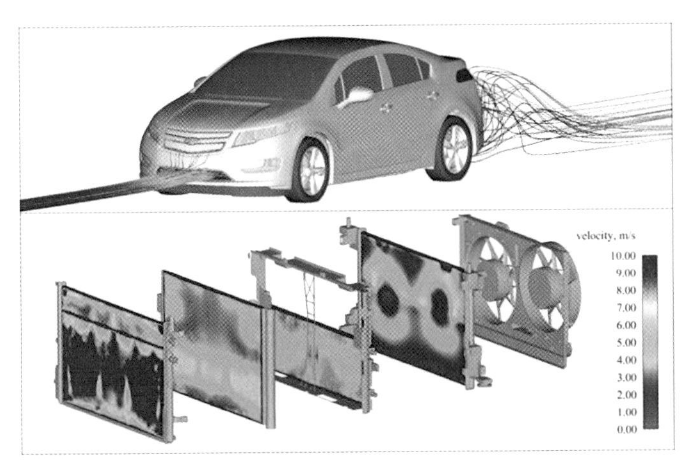

Figure 12. (Top) The majority of front end airflow enters through an optimized lower grille opening. (Bottom) An exploded view shows the velocity profile on the five heat exchanger cooling package.

The high-definition CFD model helped design the underbody splash panels ("duckbills") and wheel liners. Pressure and velocity results from the computations showed the airflow interaction under the hood, how the panels contribute to aerodynamic drag, and the areas of improvement (Figure 13). Further simulation focused on a venting strategy to reduce the drag while still allowing the splash panels to provide water protection to the Volt's electrical components.

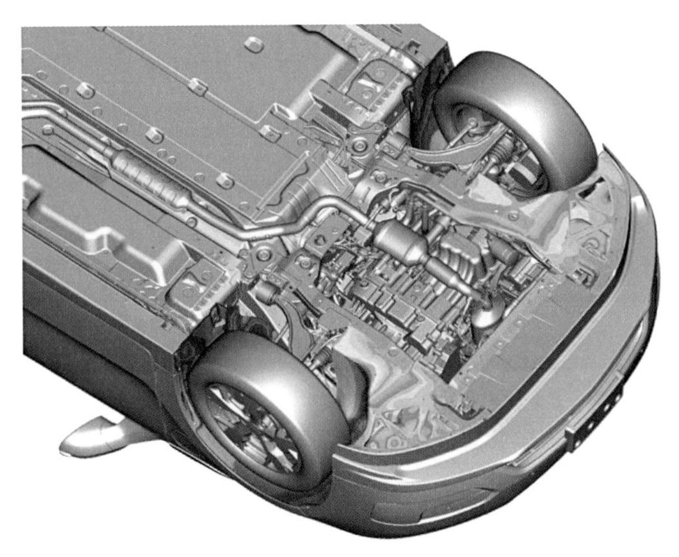

Figure 13. High definition virtual models helped guide low-drag, vented splash panels (duckbills). Pressure contours are shown.

Integration Vehicles

The first sheet metal prototypes are known as the integration vehicles or IVERs (Figure 14). It is here that all the modeling and simulation is proven out on functioning hardware. These vehicles had the final exterior surface and complete underhood components needed to validate the C_D status established with the clay models. Excellent correlation was achieved between the final full scale clay and integration vehicles, indicating that all key aerodynamic features were captured throughout the entire development process.

The integration vehicles allowed for final adjustments to the duckbills and confirmed the drag contribution of the styled wheels. Most importantly, interior sound pressure levels were measured in the fully-dressed cabin. Wind noise performance of the mirrors, wipers, and A-pillar were validated.

Figure 14. Sheet metal "Integration Vehicle"

Summary of Key Aerodynamic Features

Many aerodynamic enablers were identified during the course of wind tunnel testing and CFD analysis. Some of them affected styling and are clearly visible to the customer, while others were more engineering oriented and not easily seen.

• **Closed Front Grille** - Due to lower cooling flow requirements and efficient baffling to the electric propulsion system, it was not necessary to maintain a traditional wide open front end. Much of the upper grilles were blocked, resulting in 0.010 C_D improvement.

• **5mm Kick-up on Rear Spoiler** - To help trip the separation on the rear deck-lid, a kick-up was added resulting in about 0.005 C_D reduction.

• **Rear Sharp Trailing Edges** - As with the spoiler, to help control the flow separation at the rear sides of the vehicle, sharp trailing edges helped reduce C_D by about 0.010.

• **Mirrors** - Door mounted mirrors, as opposed to the more traditional patch mounted mirrors, helped cut another 0.008 off the C_D value.

• **Airdam** - An enabler for both aerodynamics and cooling flow, the airdam shields underbody components from the airstream and improves flow through the several heat exchangers. This part alone reduced the C_D by 0.025.

• **Underbody Panels** - Working with the airdam, the underbody panels also shield components from airflow. In total, the 4 underbody panels reduced the C_D by 0.013.

• **Lowered Ride Height** - The ride heights were lowered to the minimum allowable by good engineering judgment and corporate standards, resulting in 0.010 C_D improvement.

• **Duckbills** - Properly vented 'duckbills', a splash panel forward of the front wheel house liner and aft of the airdam, helped vent the engine compartment, reducing C_D by 0.004.

• **Body Sides and Wheel Openings** - Minimizing the body side features within the styling theme also helped reduce drag significantly.

Conclusions: Lowest Drag Chevrolet Sedan

As the Volt hits the showroom, it is one of the most aerodynamic vehicles on the road and has the lowest drag of any Chevrolet sedan. The Volt's drag coefficient of 0.28 was measured per the SAE Recommended Practice J2881 at the GM Aerodynamics Laboratory. Designers and engineers worked hand-in-hand to deliver the necessary styling and aero performance to set the Volt apart from conventional cars. Hundreds of wind tunnel hours and thousands of computer cpu-hours were used in the development process. From the front grille openings to the rear trailing edge, all components exposed to airflow were studied. From start to finish, the Volt C_D was reduced by 33 %, providing an additional seven miles of battery electric range and 50 miles of extended range.

REFERENCES

1. SAE International Surface Vehicle Recommended Practice, "Measurement of Aerodynamic Performance for Mass-Produced Cars and Light-Duty Trucks," SAE Standard J2881, Issued June 2010.

CONTACT INFORMATION

Lead Aero Development Engineer
nina.tortosa@gm.com

Lead Aero CFD Engineer
kenneth.j.karbon@gm.com

ACKNOWLEDGMENTS

We would like to acknowledge the following individuals for their support during development in achieving very aggressive C_D requirements.

Greg Fadler, Aero Group Manager

Max Schenkel, GM Technical Fellow for Aerodynamics

Young Kim, Volt Lead Exterior Designer

Robert Boniface, EREV Design Director

DEFINITIONS/ABBREVIATIONS

C_D
Coefficient of Drag

CFD
Computational Fluid Dynamics

GMAL
General Motors Aerodynamics Laboratory

The Engineering Meetings Board has approved this paper for publication. It has successfully completed SAE's peer review process under the supervision of the session organizer. This process requires a minimum of three (3) reviews by industry experts.

ISSN 0148-7191

Positions and opinions advanced in this paper are those of the author(s) and not necessarily those of SAE. The author is solely responsible for the content of the paper.

SAE Customer Service:
Tel: 877-606-7323 (inside USA and Canada)
Tel: 724-776-4970 (outside USA)
Fax: 724-776-0790
Email: CustomerService@sae.org
SAE Web Address: http://www.sae.org
Printed in USA

CHAPTER SEVEN:

A chassis that Cruzes

To speed development and minimize cost, Volt shares key underpinnings with its high-volume cousin.

Volt's underbody structure (concept shown) has some unique features. Its deep center tunnel accommodates the large, T-shaped battery, which helps the body achieve a claimed 24-25 Hz bending frequency. The structure contains a high content of high- and ultra-high-strength steel alloys.

A chassis that Cruzes

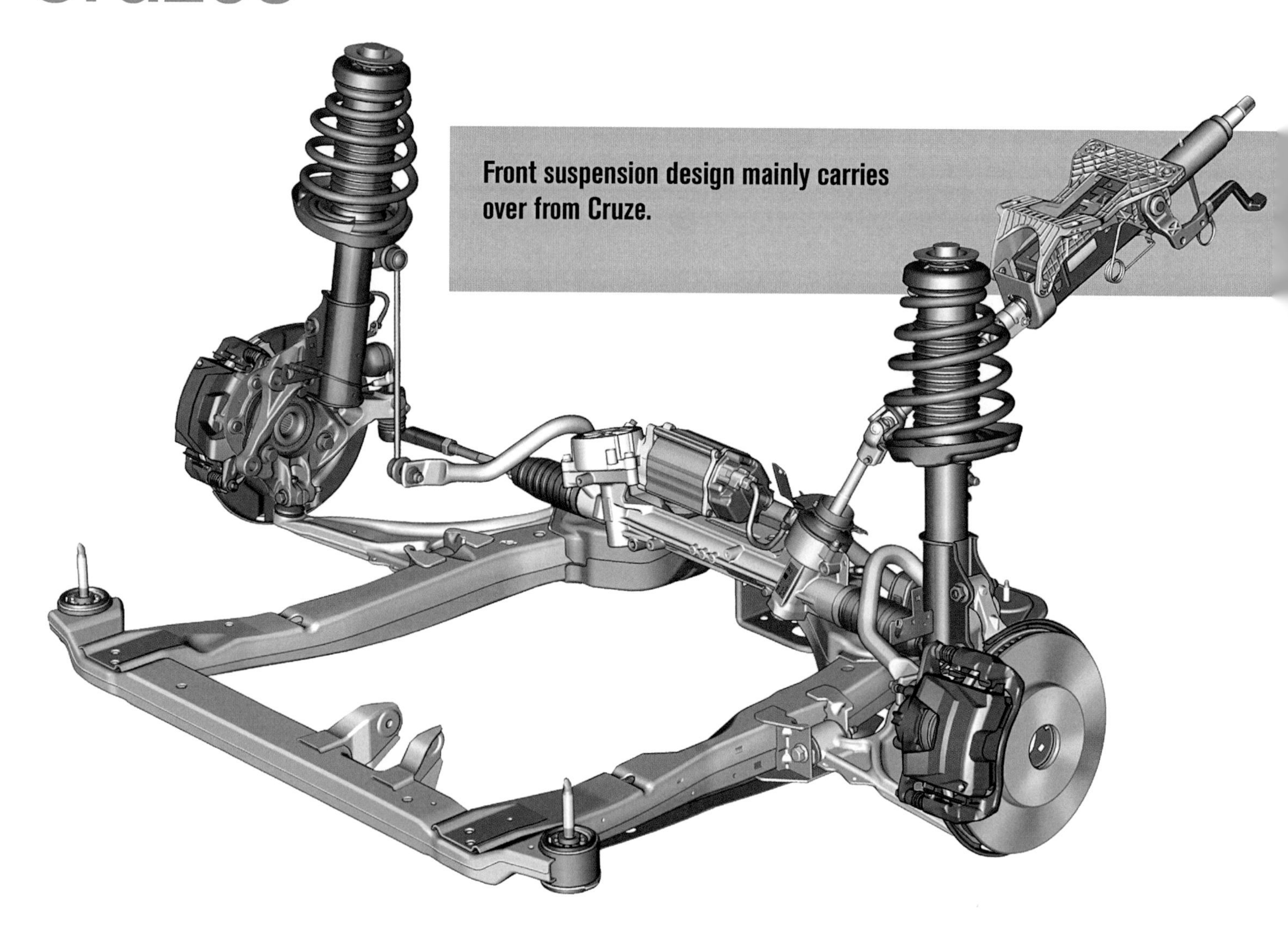

Front suspension design mainly carries over from Cruze.

Bringing an all-new advanced battery and electrified powertrain to market in less than four years required a significant investment in engineering resources for General Motors, including hiring hundreds of electrical engineers and battery technologists. So when it came to deciding on the car's body structure and chassis systems, Volt's strategic planners wisely chose not to give the vehicle an all-new chassis and suspension system to go with the high-tech drivetrain.

"We had a lot of stuff on the critical path with this new powertrain. Leveraging the Delta II architecture for the body and suspension made sense in terms of keeping the program on target," former GM Vice Chairman Bob Lutz told *AEI* in early 2009.

That doesn't mean Volt is merely a Chevrolet Cruze in disguise. The two vehicles share some of their bills-of-material, but Volt's underbody stampings and assembly are significantly different. In particular, the geometry of its deep center tunnel is designed to accommodate the 5.5-ft-long (1.7-m-long) T-shaped battery pack and its attendant electrical and coolant-pipe connections.

Various underbody members are dedicated to supporting the 375-lb (170-kg) battery pack.

GM engineers also note that Volt's underbody contains greater content of high-strength and ultra-high-strength steel than its conventional cousin.

Volt and Cruze are built in different assembly

plants, but they could easily be processed on the same line without any tooling changes. Key dimensions are very close—they share a 105.7-in (2685-mm) wheelbase—but Volt's front and rear tracks (61.2 in/1554 mm and 62.1 in/1577 mm, respectively) are slightly wider and thus more muscular-looking than Cruze's.

Volt's front MacPherson-strut-type suspension hardware and geometries are virtually identical to those of Cruze. Same goes for their rack-mounted electric power steering system with ZF steering gear. It's a dual-pinion system (one pinion is used for steering, the other to add assist) with variable assist.

A combined electric motor and sensing unit monitors steering angle, and delivers appropriate assist to the steering gear in all scenarios. The system draws its power from a 12-V battery in the rear of the vehicle.

Volt's semi-independent rear suspension geometry is based on the standard Delta II with subtle unique changes to compensate for Volt's different center of gravity, heavier curb weight (3781 lb/1715 kg vs. Cruze's 2900 lb/1315 kg), and resulting different handling characteristics.

The compound-crank torsion beam features a double-walled, U-shaped profile at the rear. Volt's system adds a variable-section cross-car beam, to which the control arms are attached with a patented "magnetic-arc" welding process.

Front and rear hydraulic ride bushings help eliminate road harshness.

Volt's electrohydraulic regenerative braking system is, of

Helping GM put the brakes on Volt

General Motors tapped TRW's experience in developing electrohydraulic brake-control technologies for the 2006 T900 hybrid trucks and the Saturn Vue Hybrid and Vue Fuel Cell Vehicle for the Volt program. Dan Milot, Chief Engineer for TRW's North American Advanced Control Systems group, explains.

Q: What specifically did TRW contribute to the 2011 Volt program?

Milot: TRW supplies the braking systems actuation and slip controller, the ABS/ESC functionality, to the vehicle. This also encompasses the regenerative braking. It required us to begin working with GM very early in the program.

Q: What did your team learn from this program?

Milot: Volt's type of powertrain is different from the GM Two Mode hybrid. The Two Mode has a transmission that's mated with the two electric motors, but the ICE is the primary engine. Volt is primarily an electric-motor-driven vehicle. Its electric drive mode produces scenarios where the vehicle NVH is significantly different than any other car.

With Volt's configuration, the effects of wheel inertia from the traction motor are different than in a vehicle with a conventional powertrain. That includes during ABS and traction control activation, and even the interface with the powertrain for developing torque. We did a lot of work around that to ensure our ABS, ESC, and even the regen blending will quickly generate the actual braking torque for us. Then we determined the friction braking response time to make it all seamless to the driver.

Q: What was GM's brief to TRW for tuning the 'feel' of Volt's braking?

Milot: Putting 'drag torque' into the regen system is meant to replicate the deceleration feel of a manual transmission. Compared with some electric vehicles that have fairly aggressive regen 'feel' due to a high calibration, Volt's braking is tuned to feel more like a conventional vehicle. Lifting off the throttle is not intended to be the primary means of braking; that's for the brake pedal.

The GM engineers provided their own level of what we call 'base brake' software, which we integrated into our control module. That had to interface with the powertrain that's doing the actual brake blending. There was a lot of work to achieve this. We had the powertrain guys and the braking guys working closely together every day to make sure it is a seamless experience for the driver.

Q: As an engineer, what did you learn from working with GM on Volt?

Milot: A big learning was the car's NVH, or lack of it. With an EV when you come to a stop, it's dead silent. If your braking system has to turn a motor on, or if anything comes on, you can hear it.

You have to be careful how you control your motors and do your mounting, and pay attention to many other details in these vehicles.

We spent a lot of time with GM, including directly with their NVH team, to find solutions to make the system quieter and quieter.

A chassis that Cruzes

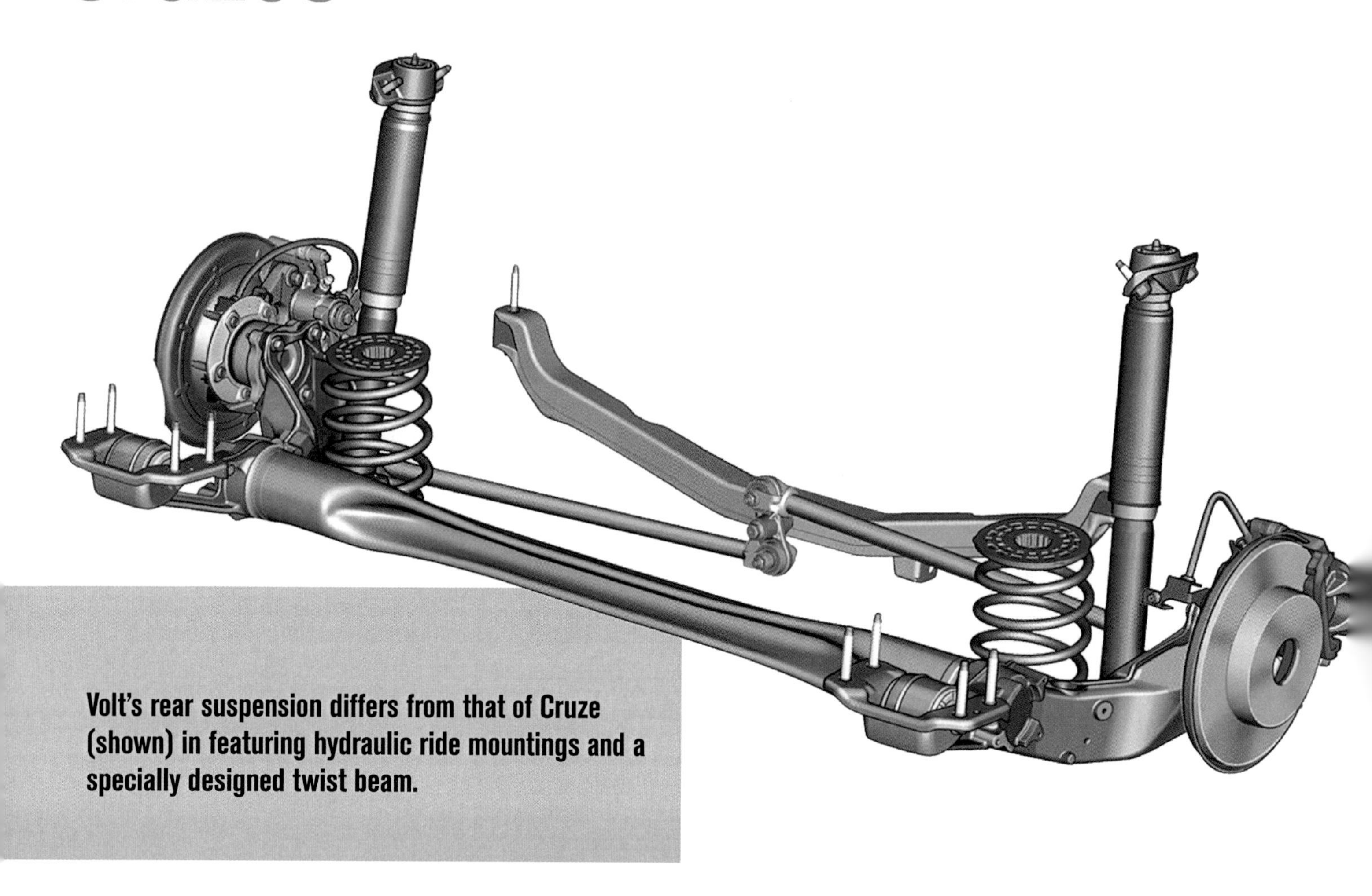

Volt's rear suspension differs from that of Cruze (shown) in featuring hydraulic ride mountings and a specially designed twist beam.

course, different from Cruze's conventional hydraulic brakes. It is calibrated to capture energy up to 0.2 g for transfer back to the battery. The friction braking system features large rotors with a GM-patented coating known as FNC that is impregnated into the rotors. According to Volt Chief Engineer Andrew Farah, the new coating protects against corrosion and promotes longer life.

The brake system's careful blending algorithms provide 100% regen, 100% friction braking, or any combination of the two. (See sidebar).

Rolling assurance

Farah's vehicle-dynamics team aimed to ensure Volt did not sacrifice rolling-resistance efficiencies for good handling performance. From the earliest days of the program, GM worked closely with Goodyear to develop a tire that helped save propulsion energy without sacrificing ride quality and grip.

The resulting Goodyear Fuel Max tires are standard on Volt, as well as the Cruze Eco model. GM tests showed they help Volt achieve 1 additional mile of range on electric-only mode, said Farah.

The main challenge in developing tires for EVs is balancing ride quality, grip, and low-noise operation, noted Chad Melvin, Goodyear's Project Manager for the Volt and Cruze programs. His company's engineers refer to balancing tread wear, traction, and rolling resistance as "the performance triangle." Melvin recalled the tire's development path.

"At the same time GM was developing the Volt, we were in development of our Assurance Fuel Max line. The tread pattern for the Volt came concurrently with our replacement Fuel Max tire, so it seemed like a perfect fit."

He said engineers in Goodyear's Replacement Tire group developed the pattern in collaboration with GM, the two teams working in what he described as "excellent synergy."

"We were able to use the low-rolling-resistance technology that we were developing for the replacement market for the Volt as well. There were some caveats, because the Volt tires were developed specifically for it. But, visually, the tread patterns of the Volt and Cruze Eco tires are visually the same."

During the tire's development, Goodyear used its advanced modeling techniques to develop a tread cavity and mold shape that would allow both fun-to-drive and reduced rolling resistance qualities. Melvin's team worked closely with GM, whose engineering team provided kinematics and compliance data to go with Goodyear's rolling-resistance modeling work.

"The parallel development allowed us to model how the Volt was going to handle on the road," Melvin said. Optimization of that process created the mold shape.

The new tires also feature the latest iteration of Goodyear's advanced compound technology, which incorporates what Melvin calls "functional polymers." The Volt tire has a "dual-cap" tread—it has two compounds.

"This allowed us to provide a bit more traction than our replacement line of Fuel Max tires, which has a functional polymer as well," Melvin noted. "The Volt is just a slight tweak of that, because GM wanted to get a sportier feel with better stopping distance as well. We were able to achieve that through the modeling and compounding."

He said the Assurance tire's low-rolling resistance is achieved through design and not by using higher pressures than conventional tires.

In terms of 'hurrah' moments in his team's effort, Melvin cited creating the mold shape using advanced tooling, then finding the tire achieved the level of performance that the tire maker's model had predicted.

"That's always a good feeling," he said, adding that the advent of electrified vehicles has Goodyear dialing up its game in designing quieter-running tires that don't sacrifice safety and fun-to-drive qualities.

"We didn't go into this program thinking we're going to design an ultimate low-rolling-resistance tire for GM and the Volt," Melvin noted. "We also needed to make the vehicle fun to drive, with good traction, and best-in-class stopping distance.

"Achieving the overall balance was a challenge, and the results are something that will impact our future developments for these types of vehicles," he said.

Goodyear's new Assurance Fuel Max tire was designed for Volt and the Cruze Eco. GM tests show the tires give the car an extra mile range in EV mode.

Co-Development of Chevy Volt Tire Properties to Balance Performance and Electric Vehicle Range

2011-01-0096
Published
04/12/2011

Dean Degazio
General Motors Company

David Hubbell
The Goodyear Tire and Rubber Co.

Mark Popilek
General Motors Company

Nick Hill, Zachary Chappell, Rachel Graves and Tom Szelag
The Goodyear Tire and Rubber Co.

doi:10.4271/2011-01-0096

ABSTRACT

As an innovative electric vehicle with some new approaches to energy usage and vehicle performance balance, the Chevy Volt required a special relationship between the OEM and tire supplier community. This paper details this relationship and how advanced tools and technology were leveraged between OEM and supplier to achieve tire component and overall vehicle performance results.

INTRODUCTION

The Chevy Volt extended range electric vehicle requires tires that can provide exceptionally low rolling resistance to meet electric-only operating range goals while maintaining good ride, handling, and braking performance. These goals required tires that were beyond any typical tires that were available. GM worked with Goodyear from a very early stage on the program through production to achieve a tire construction that met the needed performance balance. This was done through advanced technology and integration with the unique characteristics of the electric vehicle.

This paper will detail (1) the initial requirements development process and early tire analytical/physical evaluations by suppliers to show capability to meet requirements and understand tradeoffs, (2) the new technology employed to separate and balance different areas of performance in compounding, construction, and tread and (3) the performance achieved and final development between vehicle and tire supplier in delivering the production tire.

BODY

Extended Range Electric Vehicle (EREV) Early Performance Definition and the Need for Special Tire Performance

From its inception and in the earliest stages of defining an "extended range electric vehicle" (EREV) at General Motors, the vehicle that would become the 2011 Chevy Volt, had some clear defining characteristics:

- The vehicle would have an electric-only range of approximately 40 miles to address the fact that ~80% of U.S. drivers drive 40 miles or less each day
- The vehicle would plug-in charge to replenish the full charge, ~40 mile range, overnight
- The vehicle would require a large onboard Li-Ion battery pack that would add considerable mass both itself and through its supporting thermal systems
- The vehicle would be based on an existing compact vehicle architecture, including suspension systems, and would have

to carry the extra weight and related packaging challenges of the electric propulsion and energy storage systems

• The car should operate and behave as a conventional vehicle with minimal tradeoffs for electric operation to the driving operation, driver workload and feel, and have North American all-season performance

The General Motors advanced vehicle dynamics team was tasked with comprehending the aspects of the extended-range electric vehicle that made it different from conventional vehicles and understanding both (1) the gaps to competitive performance and determining competitive targets in the compact segment and (2) strategies to address the adverse impacts of the electric vehicle on required performance.

In assessing competitive and customer needs for extended range electric vehicles, advanced vehicle dynamics engineers acknowledged the gap that had emerged between the performance and "fun to drive" qualities of current hybrid vehicles (and their compromises made for fuel economy) and the customer perception of top European and other recognized leaders in "fun to drive" vehicle dynamics. As with all new vehicles several years from production, the challenge of meeting and winning against other competitive OEMs also becomes not a battle to match current cars but to match increasing performance in next-generation vehicles and assuming incremental improvement in each product cycle by all competitors (see Fig. 1). The additional challenge with the EREV was how to identify targets and vehicle attributes that would deliver globally recognized thresholds of "fun to drive" and competitive vehicle dynamics acknowledging that an extended-range electric vehicle would inevitably have propulsion and energy storage requirements that would offset driving dynamic performance attributes. Offsetting innovations or new approaches would be required to get back to competitive performance levels.

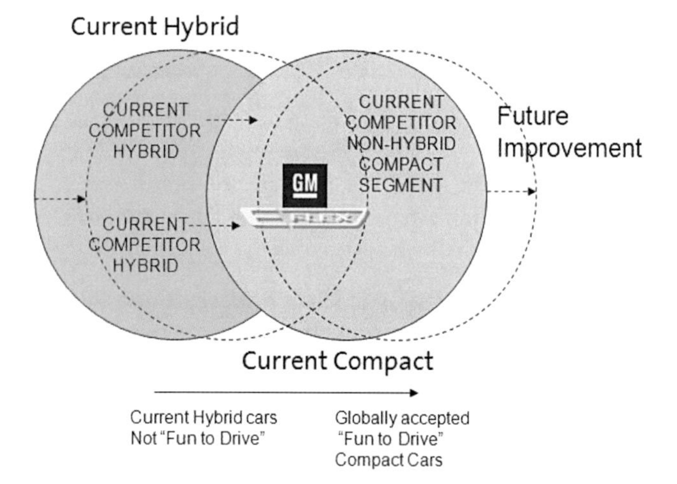

Fig. 1. Current vs. future vehicle performance

Very early, the program took the approach that the EREV should have vehicle dynamics performance targets and a target competitive vehicle set that expanded outside of current hybrids. "Fun to drive" characteristics were acknowledged in the highest level program definition documents as critical areas of development-prioritized just below the defining electric vehicle and propulsion characteristics. It was recognized by the engineering and program leadership that delivering a capable and "fun to drive" car was not "expected" given the current level of hybrid performance in the market but, rather, "fun to drive" driving dynamics would be a significant "surprise" positive factor for an electric vehicle customer if delivered at sufficient levels. With this focus on "fun to drive" attributes, a set of performance targets aligned with a non-Hybrid set of benchmark competitors was identified to drive vehicle development and system designs.

"Fun to Drive" Enablers and Electric Vehicle Realities - Designing the First Virtual EREV Chassis

After establishing a set of non-Hybrid benchmarks for overall ride, handling, braking, and steering feel performance influencing buyer behavior, a set of overall desired vehicle targets emerged. At this point, suspension, steering, braking, and other subsystem and component requirements could be effectively communicated and digested by chassis engineering and other design groups. To begin to comprehend building a real car and how the electric propulsion system would influence the vehicle dynamics behavior, building a virtual model of the vehicle was the next logical step (see Figs. 2, 3, and 4).

- Model based on Global Compact Arch HB5 parameters
- WB: 2685 mm; 2P: 1768 kg (59.9% front)
- McPherson Front Suspension
- Twist Axle Rear Suspension
- Geometry provided by GME
- C-Factor: 58.1 Overall Steering Ratio: 16
- Tire: P215/50R17 with F&M from MTA Database @ 240kpa

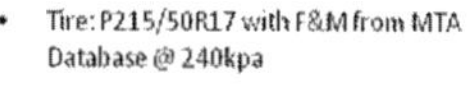

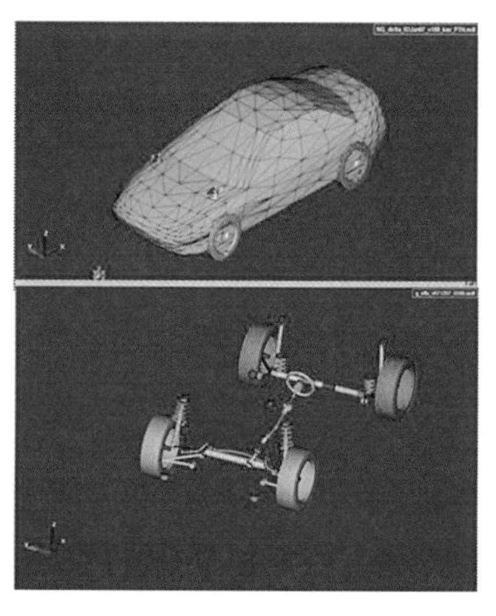

Fig. 2. Early EREV ADAMS GM Vehicle Dynamics simulation model

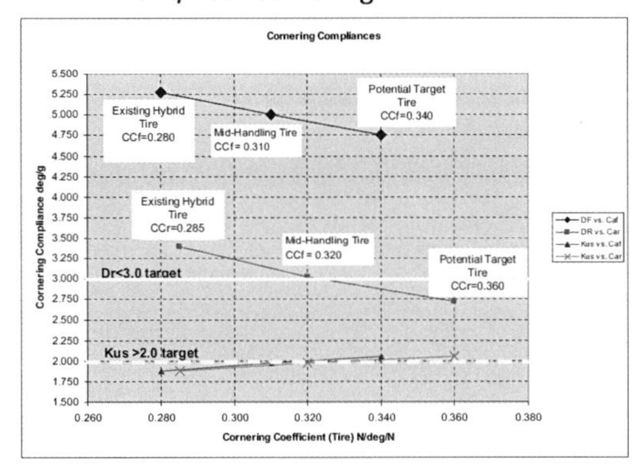

Fig. 3. Early EREV ADAMS GM Vehicle Dynamics simulation model results showing Cornering Compliance (understeer-related handling vehicle property) vs. required Cornering Coefficient of the tire

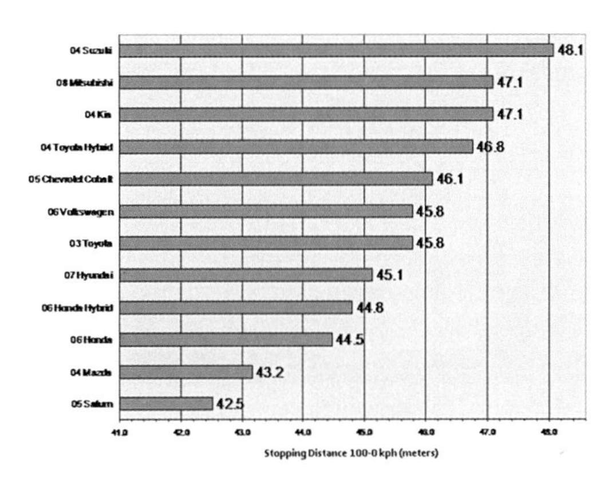

Fig. 4. Late-2007, EREV benchmark braking dry stopping distance 100-0 kph performance on Consumer Reports type test

To understand and develop requirements and the designs further, an analytical model of the EREV was created. Development of the chassis model was greatly accelerated by using the existing kinematic models developed by General Motors Europe on the parent architecture. The EREV program also leveraged some of the physical hardware, suspension, and prototype tire data from the parent architecture that was ahead of the EREV in development. The parent architecture also comprehended many vehicle combinations including heavier variants, so modifications to accommodate the EREV energy storage mass in some cases already had existing concepts that could be borrowed.

Vehicle Dynamics Targets vs. the Chassis Components to Achieve Them

Having a carryover-based, global vehicle architecture chassis design allowed the team to quickly model the new EREV chassis and lumped vehicle parameters. Using these models, the initial performance assessments at new EREV vehicle masses and packaging requirements immediately showed gaps to "fun to drive" levels of performance. This was expected but was immediately daunting.

In some cases, the steering system, braking system, and other carryover systems provided a great baseline set of design and tuning bandwidth to deliver performance-this leveraged the strength of the global platform. However, the following suspension assumptions driven by both reuse and mass considerations revealed some significant challenges:

- Reuse of front strut geometry
- Reuse of rear torsion beam suspension and constrained handling attributes
- Mass increases involved with EREV-specific propulsion/ energy storage immediately saturated most design bandwidth with existing suspension component sets

Given the need to reuse the rear, torsion beam suspension without a Watts link or any other suspension re-design or lateral stiffness enhancements due to mass and packaging, it became immediately clear that the overall vehicle performance would be strongly impacted by the tire.

Given this reality, tire properties for Vehicle Dynamics-further constrained by rolling resistance requirements to meet energy and electric range goals for the program-became an immediate focus. While seeking to design the rest of the vehicle component set to deliver performance and not overburden tires, it became clear that program scope, reuse, and mass reduction would just not permit additional suspension enhancement. So defining a tire balance and obtaining a supplier to deliver the needed tire balance was critical to a "fun to drive" overall vehicle.

Compounding this logic, the team understood that the parent architecture had design variants for Europe vs. other regions of the world using the same rear torsion-beam suspension system but with summer-only vs. all-season tires that provided substantially different handling and braking traction attributes. The ability to select more performance-oriented tires at the expense of all-season or fuel economy allowed the simple suspension system to still meet regional competitive metrics. The EREV would need to do something similar while in an all-season tire market-with the additional initial elements of (1) the added load reserve requirements from the propulsion and energy storage additional mass and (2) the need for "hybrid-type" rolling resistance performance.

All of these factors, revealed through both past experience and objectively through simulation of the EREV early designs, highlighted the harsh reality that tires with balanced performance would be required to deliver the goals of the program. The tires required could be specified via simulation but did not physically exist in any known current tire.

OEM-Required Performance - Traditional GM Tire Requirements Balancing and the Need for Advanced Technology

General Motors tire requirements development is a balancing process between several groups. Performance engineering, chassis engineering, and the tire and wheel systems group would generate a statement of requirements based on all-season tires already in existence and a "robust" enough balance of requirements to permit sourcing to proceed (1) with all global sources having likely capability to produce tires on conventional tools and (2) tire attributes might be "stretched" in 1 or 2 areas to preference handling over ride or rolling resistance but not stretched in too many areas at once. For an all-season tire, this process would generally result in requirements close to the average for a typical tire in production today or "stretched" within the distribution of known tires from all suppliers. This gives programs realistic expectations and a high probability of success in achieving the needed goals for fuel economy and overall performance.

However, on paper, and at an early stage, it was clear the ideal tire for the EREV required what amounted to "outlier" improvements in performance balance between rolling resistance, cornering coefficient, and braking dry traction while still providing threshold snow and wet properties for overall vehicle all-season capability and standard tread wear and durability.

The disconnect between a typical tire and what was required was extremely clear (see Figs. 5 and 6). Technology would be required that, if available, only existed in the Research and Development areas of the supply base, and not their mainstream production engineering groups. What the EREV tire required were the following seemingly intractable goals:

• Innovation in tire construction and design to generate a strikingly aggressive balance of rolling resistance, handling, and braking traction properties

• Development of physical tires demonstrating this performance much earlier than a normal program to allow a better-than-average confidence level leaving the advanced development stage that tire balance was achievable-and resulting estimates of electric vehicle range and vehicle dynamics performance were believable

• Create a relationship with a tire supplier willing to take risks on a production timeline and with significant existing technology to leverage-and bring them onboard before and during the sourcing process

What emerged was an invitation to the current tire suppliers to participate in a competitive tire process starting with simulation and proceeding with physical prototype tire builds in two separate early tire submissions to demonstrate capabilities on early vehicles. The OEM would provide chassis attributes for the new EREV vehicle to drive tire supplier predictive models, as virtually developed tires were seen as the best, or perhaps the only way, to generate needed performance-balanced constructions rapidly. The OEM would also build a physical vehicle representative of EREV chassis properties communicated in the early virtual math-based models and actually have vehicle dynamics experts drive the prototype tires at the GM Desert Proving Ground and provide feedback.

The goal of this project was threefold:

• Gain access to tire technologies and how they could help provide a new level of vehicle performance in the EREV and compact vehicles

• Develop a collaboration in simulation and in early vehicle development with a supplier that could pay dividends throughout the EREV and future programs

• Incentivize and communicate the OEM's interest in leveraging new technologies on new vehicle programs through a different sourcing process and different set of purchasing assumptions

It was through this early EREV advanced tire program that at first several tire suppliers and, then, ultimately, just Goodyear emerged as the OEM's partner on this advanced tire project.

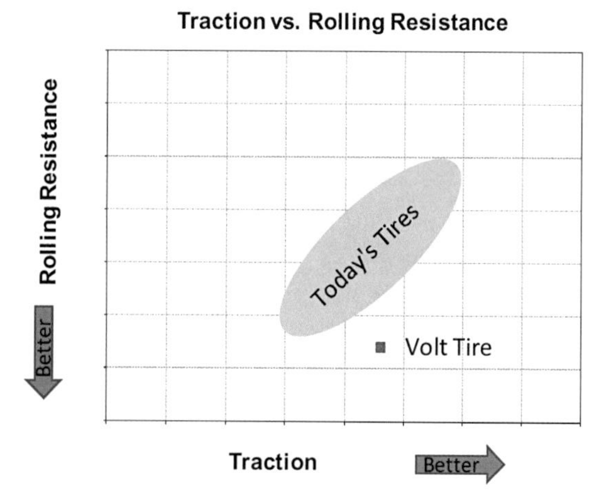

Fig. 5. EREV tire targets for Traction vs. Rolling Resistance properties vs. measured distribution of existing tires

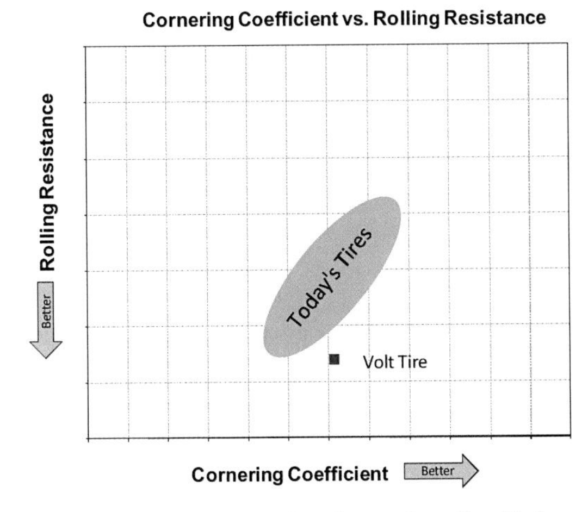

Fig. 6. EREV tire targets for Cornering Coefficient vs. Rolling Resistance properties vs. measured distribution of existing tires

The Virtual Tire Process

The modern radial passenger car tire is a highly optimized composite structure. In attempting to further increase any one performance parameter such as tread wear, it is often found that there are trade-offs in other performance parameters. The management of these performance trade-offs is paramount to having success in a tire development program. Goodyear has spent considerable effort in developing predictive models, not only for pure tire performance items such as rolling resistance or traction, but also in cases of tire-vehicle interaction such as subjective handling or NVH.

Tire development for OEM applications has customarily been a "build and test" heuristic process. The initial submission is based upon the tire designer's best calculation as to what will work on the platform in question and the initial goals and specifications provided by the OEM. The OEM will then test and provide feedback on this first submission to the tire designer for the next submission. The process is then repeated through several iterations until the OEM receives the performance that they are looking for. This iterative approach works well for established platforms and applications, but can struggle with new concepts. It is also cost and labor intensive and there is a long cycle time between submissions (see Fig 7).

The incorporation of tire and vehicle predictive models into the design process can represent a significant opportunity for reduced cycle time and costs as well as minimizing the potential performance trade-offs. By constructing and evaluating virtual submissions at the very start of a project, as well as between builds, initial designs will be closer to the programs targets from the start, and more time will be available for small refinements (see Fig 8). This more flexible approach also enables the tire designer to anticipate challenges that might otherwise not be foreseen prior to first submission (see Fig 9).

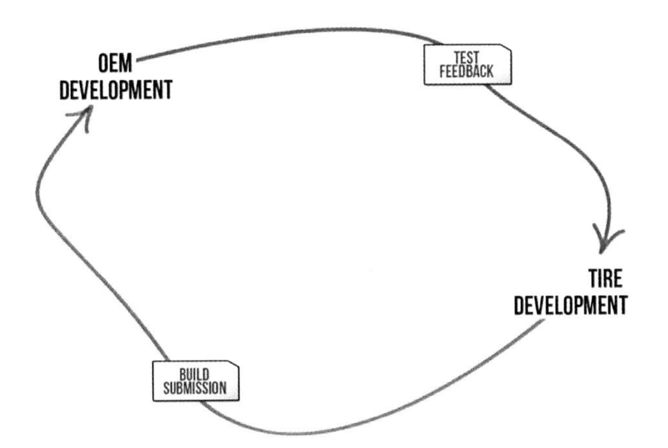

Fig. 7. Traditional process

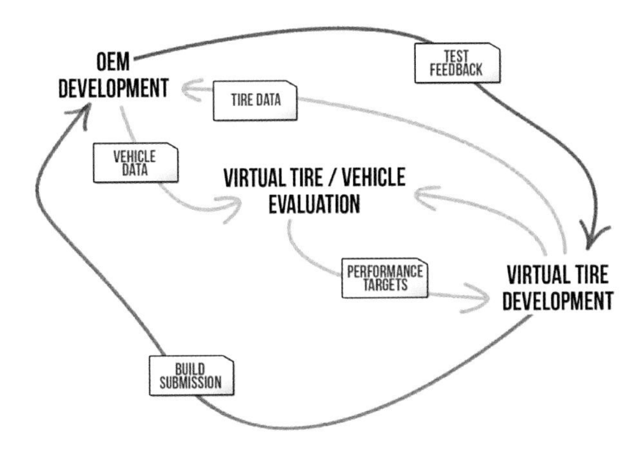

Fig. 8. Revised process

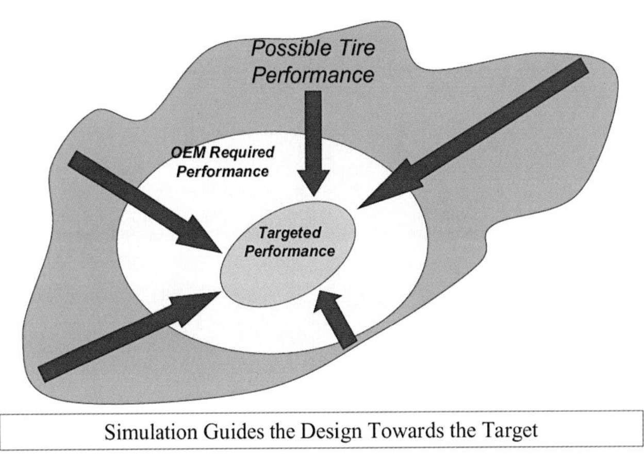

Fig. 9. Use of Simulation to Guide Development

This virtual tire-vehicle evaluation process was applied at the start of Goodyear's involvement in EREV tire development. The tire supplier initially collected details on the proposed vehicle design including vehicle mass, inertia and suspension kinematic and compliance properties. This enabled a true understanding of how the proposed vehicle would be using its tires to be developed before any tooling or physical prototypes were constructed. Through this process, design guidelines for tire characteristics such as footprint pressure distributions, spring rates, and force and moment parameters were established. These design targets in turn were used in a tire mold cavity shape and construction DOE that served to dramatically reduce and center the design space for the first submission. The outputs from this process were the first submission hardware and the initial constructions provided to the OEM. No additional tire constructions were constructed for internal development work.

Initial EREV Tire Submission Results

The results of the first tire submission to the OEM were very positive. Goodyear was successful in providing a tire that not only met the aggressive rolling resistance target set by GM, but also featured the subjective handling or "fun to drive" performance that GM was looking for. This allowed further time and development resources to be applied to refining the tire. It should be noted that the preferred construction was not the tire with the lowest possible rolling resistance. Lower rolling resistance tires were developed virtually and constructions were produced for GM to evaluate. Several other tire suppliers also supplied promising constructions.

Table 1. EREV Tire Selection - Round 1 Results

Characteristic	Submission 1		
	Const. A	Const. B	Const. C
Rolling Resistance	118%	115%	121%
Cornering Coeff.	64%	76%	69%
Dry Traction	97%	97%	101%
Fun to Drive Factor	2	3	2

Table 1 shows the normalized objective data for Goodyear tire constructions produced for Submission 1. The data is color coded with red meaning not within 95% of the goal, yellow meaning within 95% but not meeting the goal, and green meaning meeting or exceeding the goal. Also, the Subjective Fun to Drive Factor rating summarizes on a 1 to 5 scale, with 5 being the best, several aspects including cornering power, linearity, progressivity and on-center steering feel.

Feedback from the first submission was used along with modeling to further refine the design for the second submission. While GM enjoyed the controllability and progression of handling at the limit, GM also indicated that additional lateral and cornering stiffness would be beneficial for handling.

Table 2. EREV Tire Selection - Round 2 Results

Characteristic	Submission 2	
	Const. B	Const. C
Rolling Resistance	105%	105%
Cornering Coeff.	86%	74%
Dry Traction	97%	97%
Fun to Drive Factor	4	3

Table 2 shows the results of the 2nd Submission. The Cornering Coefficient performance improved significantly during this round.

Having shown great promise in meeting both the Rolling Resistance and the Braking Traction goals and simultaneously providing a Fun to Drive tire, the goals were altered slightly for further development. The new Rolling Resistance goal was reduced by 4%, the Dry Traction was held at the original level and the cornering coefficient was reduced slightly. Table 3 shows interim tire ratings after the specification was revised.

Subsequent EREV Tire Submissions and Development

OEM and supplier engineers performed joint rides on development constructions throughout the completion of the EREV program. This close interaction between supplier and OEM allowed an open forum on how the final tire could best benefit the program goals. The processes followed and discussions between companies led to a final, optimized product that incorporates features which collectively create a low rolling resistance, tractive, and sporty all-season tire (see Tables 3 and 4).

Table 3. Interim Supplier Tire Constructions, Submission 2A

Characteristic	Sub 2A Const. B	Sub 2A Const. C	Sub 2A Const. D
Rolling Resistance	92%	94%	105%
Cornering Coeff.	110%	97%	92%
Dry Traction	102%	103%	100%
Fun to Drive Factor	4	4	4

Table 4. Final Supplier Tire Constructions, Submission 3

Characteristic	Sub. 3 Const. J	Sub. 3 Const. K	Sub. 3 Const. L
Rolling Resistance	100%	98%	102%
Cornering Coeff.	90%	93%	98%
Dry Traction	104%	97%	100%
Fun to Drive Factor	4	4	4

EREV Tire Technology

The Assurance Fuel Max tire fitted to the EREV, Submission 3, Construction L shown above, incorporates several technologies in order to achieve its performance. The two performance areas that traditionally conflict with rolling resistance are traction and tread wear. With a traditional tread compound using carbon black as a filler material, it is generally only possible to avoid trade-offs in two of these areas. Silica was introduced as a replacement for carbon black that enables this trade-off to be reduced but not eliminated. This tire along with many other Fuel Max tires is notable in its use of functional polymers. These polymers serve to improve the dispersion of the filler material and further reduce this trade-off (see Figs. 10 and 11).

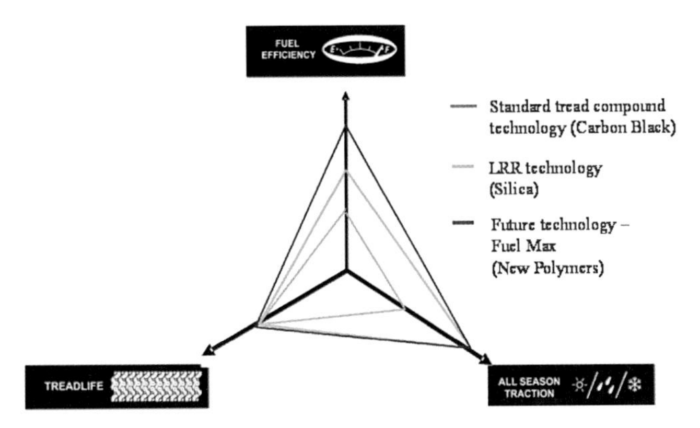

Fig. 10. Tire performance balance

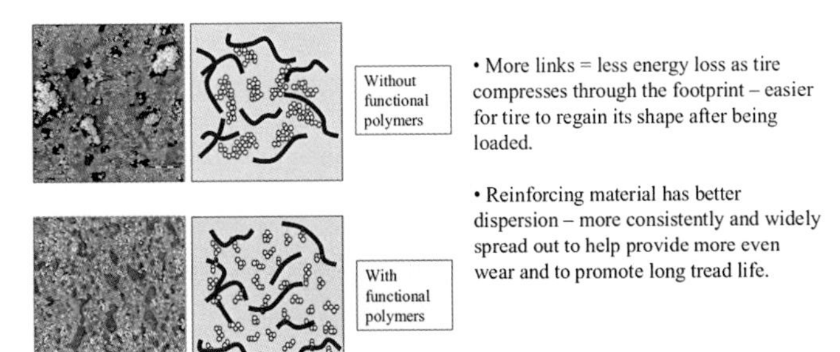

Fig. 11. Fuel Max technology

Rolling resistance modeling revealed that, while the tread compound is the primary factor, mold cavity shape and construction design also play important roles. The tire features a unique cavity shape designed to minimize weight and allow the tire construction to operate as efficiently as possible, both of which serve to further reduce rolling resistance. It also features an envelope-ply construction. Virtual chassis evaluation and first submission input from GM uncovered that while the EREV would benefit from the improved handling potential of a dual-ply construction, it would also require the low rolling resistance of a mono-ply. The solution was to incorporate an envelope-ply, a hybrid construction featuring additional sidewall reinforcements. This combination of mold cavity shape, construction, and tread compound enables the Assurance Fuel Max to help provide a high level of performance not typically seen in a current generation hybrid or electric vehicle.

Delivered EREV Vehicle Performance

As described previously, the EREV vehicle dynamics performance goals were originally established to make the EREV competitive with any other car in its size class and not be limited to competing with just hybrids and electric cars. The EREV tires are an important component in meeting those performance goals. Over 2 years, the OEM and tire supplier worked jointly to develop the tires with the best tire of the final submission of constructions providing the performance summarized in the tables below, which highlight the excellent performance of the vehicle.

Consumer Reports is a well respected magazine that has generated their own double lane change Avoidance Maneuver where the lanes are specified in length and width. The maximum entry speed through the maneuver without hitting cones is the registered speed (see Table 5).

Table 5. Consumer Reports-type test Avoidance Maneuver, OEM measurement

CR Avoidance Maneuver Speed	53.6 mph

In addition to the double lane change Avoidance Maneuver, Consumer Reports also tests the minimum stopping distance from 100kph. While the ABS controls and brake hardware are important, the tire plays a critical role in the stopping ability of the vehicle since this is the sole link between the brakes and the ground (see Table 6).

Table 6. Consumer Reports-type test ABS Stopping Distance, OEM measurement

CR 100kph-0 Stopping Distance – Dry Asphalt	42.6m
CR 100kph-0 Stopping Distance – Wet Asphalt	45.4m

Another specification is the Rollover Resistance Rating generated by NHTSA. It relies heavily on the Center of Gravity of the vehicle and Track Widths and is an indication of overall vehicle stability (see Table 7).

Table 7. NHTSA Rollover Resistance Rating

Rollover Resistance Rating	5 Stars, No Tip Up

Finally, there are many metrics that are subjective in nature and measured relative to a target competitive vehicle. The OE manufacturer uses a 10 point scale where 10 is best. The spider chart below shows the differences on the scale for each metric relative to the target vehicle as measured by experienced OEM evaluators (see Fig. 12).

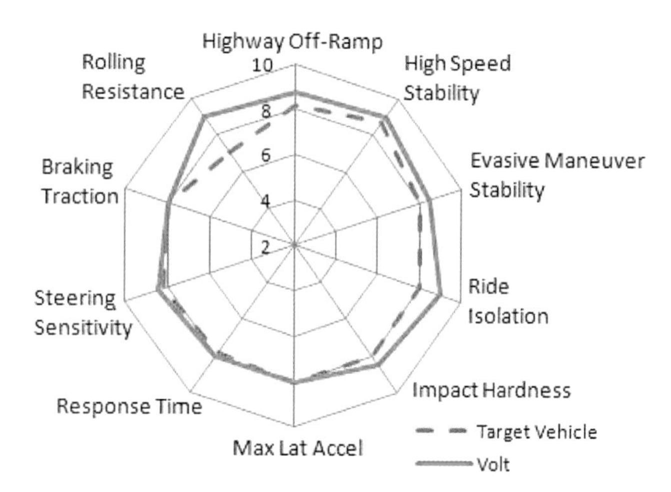

Fig. 12. Performance balance

SUMMARY/CONCLUSIONS

As an innovative electric vehicle with some new approaches to energy usage and vehicle performance balance, the Chevy Volt extended range electric vehicle (EREV) tire development required a special relationship between the OEM and tire supplier community. By establishing early, simulation-based approaches to both vehicle and tire design as well as early connections between OEM and a tire supplier, a tire component contributing significantly to delivered, competitive vehicle performance was achieved. Analytical techniques and processes to improve early exchange of simulation data between OEM and supplier can extend this learning to other vehicle programs and applications.

The EREV program tire development represents yet one more example of how the tire is an increasingly critical component in providing alternative propulsion and other high performance vehicles with world-class fuel economy, vehicle dynamics, noise and vibration, and overall balanced vehicle performance attributes.

This increased focus on tire requirements by OEMs and pressuring to move past historical tradeoffs continues to drive tire supplier application of increased design sophistication and advanced technologies to deliver these new levels of performance.

CONTACT INFORMATION

Dean Degazio
Global Technical Integration Engineer
Hybrid Controls Validation
General Motors LLC
dean.degazio@gm.com
Phone: (248)520-4752

David Hubbell
Global Chief Engineer
Life Cycle Mechanics
The Goodyear Tire & Rubber Company
dave.hubbell@goodyear.com
Phone: (330)319-5099

ACKNOWLEDGMENTS

To all the engineers through the program at GM and Goodyear who made the project a success. Special thanks to the following engineers for providing supporting material and guidance for this paper:

Valarie Boatman, Development Engineer, General Motors

Thomas Long, Tire & Wheel Engineer, General Motors (ret.)

Chuck Lantz, Tire & Wheel Engineer, General Motors

David Cowger, Engineering Group Manager, Tire & Wheel Systems, General Motors

The Engineering Meetings Board has approved this paper for publication. It has successfully completed SAE's peer review process under the supervision of the session organizer. This process requires a minimum of three (3) reviews by industry experts.

ISSN 0148-7191

SAE Customer Service:
Tel: 877-606-7323 (inside USA and Canada)
Tel: 724-776-4970 (outside USA)
Fax: 724-776-0790
Email: CustomerService@sae.org
SAE Web Address: http://www.sae.org
Printed in USA

CHAPTER EIGHT:

A new role for the ICE

Volt's modified Family Zero inline four is along for the ride—until it's needed.

The 2007 Volt concept used a three-cylinder engine, but engineers opted for a modified four-cylinder 1.4 L from the same Family Zero portfolio for the production car. It was the smallest engine in GM's global portfolio capable of delivering the necessary output for Volt. It is shown here with the new 4ET50 electrified transaxle attached.

A new role for the ICE

Judged by its individual components, Volt's 1.4-L four-cylinder combustion engine resembles its nearly identical cousin in the Chevrolet Cruze. It has got an aluminum DOHC head with four valves per cylinder. Its hollow-frame iron cylinder block is a thin-wall casting, making it nearly as light as a comparable aluminum block while retaining iron's inherent strength and noise-damping qualities.

Derived from GM's global Family Zero of small-displacement gasoline inline triples and fours, the Volt engine was designed for efficient operation. Its hollow camshafts are manipulated with two-step variable electrohydraulic phasers. Its lubricant is distributed through a variable-displacement flow-control oil pump to reduce parasitics under light load conditions. The thermostat is map-controlled.

There is no starter motor; the car's generator motor handles that task. And the engine's undersquare bore/stroke promotes good low-and mid-range combustion efficiency, particularly in turbo versions.

But a closer look, and a drive in the Volt, show this is not a typical Family Zero unit (nor is it turbocharged). It is governed to 4800 rpm, the speed at which the engine typically runs when the Volt's powertrain management controller calls for the generator to engage.

There are unique control algorithms (Engine Maintenance and Fuel Modes) that start the engine every 45 days and run it for up to 10 min, in case the owner has been driving solely on battery power, to lube the internal surfaces and run the diagnostics.

It's an off-the-shelf engine almost invisibly modified to serve the Volt's series-hybrid operating mode.

"With our propulsion architecture, the role of this engine is completely different from any other car in the marketplace today, or from what we currently know is coming," said Pam Fletcher, GM's Global Chief Engineer, Hybrid and Electric Engineering.

She noted that the engine is calibrated to operate "completely differently" than engines in the Cruze or other conventional powertrains.

"We did all of our development around maximum efficiency," noted Fletcher, who also serves as the chief engineer of GM's plug-in-hybrid program. "It was a different game than the one we usually play in balancing power and efficiency, because this engine's really set up to run like a generator."

The engine's role in Volt allowed the powertrain development team many degrees of freedom for calibration, as the car's electric generator and traction motors deliver positive and negative torque at a wide speed range. The engine thus does not respond directly to demand from the driver's right foot as in a conventional vehicle, because you are effectively driving the traction motor, not the ICE.

GM has yet to receive an official SAE output rating for it, but the engine produces approximately 63 kW (84 hp) at 4800 rpm. Combined with the full electric drivetrain, total output is 150 hp (112 kW).

"The engine's output numbers are all we need to maintain a minimum battery SOC [state of charge] when the car is cruising at highway speed," Fletcher asserted.

She explained that ideally a purpose-built power unit would optimize the Voltec propulsion system, and likely be more compact and lighter in the car. But opting for the available Family Zero four saved development time, vs. developing a clean-sheet engine to meet the aggressive production timetable.

This was the smallest engine in GM's global portfolio capable of delivering the necessary horsepower for Volt, she said. The engine's 10.5:1 compression ratio was designed to cope with the premium gasoline GM specifies for Volt, due to how the Voltec system is calibrated to handle multiple driving modes in various conditions. Fletcher and other GM engineers are not particularly happy about the premium-fuel spec. They note that an E85-capable version is scheduled for MY2012.

WOT nearly all the time

Integrating the ICE with the car's electric drivetrain consumed much development time, Fletcher said, with multiple iterations of control code developed and tested.

"We put an enormous effort into making the ICE pleasant and unannoying to the customer when it comes on," she said. "We could have chosen to operate it at one fixed speed and load, or perhaps a couple fixed speeds and loads, where you have the balance of maintaining battery charge and not always dissipate charge of the battery, and pick a point that's very efficient—a very specific BSFC [brake-specific fuel consumption] 'island.' But such a powertrain would be very awkward to drive.

"We could still make the propulsion system smooth and responsive to driver input, but from all the driver's sensory perceptions it would be terribly awkward," she said. "We recognized that as being very important. So we made a lot of choices and spent a lot of development to make it a comfortable, smooth, quiet, pleasant-driving vehicle as well as making efficiency a big priority."

Algorithm and control development was a main focus of Fletcher's team, as was working with its colleagues developing the clever new 4ET50 transaxle, based on GM's Two Mode hybrid technology.

"The engine operates at or near WOT [wide-open throttle] pretty much all the time that it's on," she said. "There are a few unique conditions where it doesn't. One of those cases is we do have an 'idle' condition, where we run the engine at less than WOT. To do this, we had to find an rpm point that's pleasing from an NVH standpoint."

The notion of "idle" regarding Volt's operating protocol seems unusual. But Fletcher explained there are times ICE operation is needed when the vehicle is not moving. It may be in the middle of a catalyst light-off, for example. Or it may be in a condition of low battery SOC, where it is necessary to put a little bit of charge back in the battery.

Then there is the aforementioned Engine Maintenance Mode, and Fuel Maintenance Mode. The former is for those drivers who mainly drive electrically and the ICE does not have to come on. GM wants the ICE to be ready for when it is needed, so every 45 days or so if the ICE hasn't come on, the controller starts the ICE in Engine Maintenance Mode and runs it for up to 10 min.

"We actually have a model such that we'll run it so that we get temperature in the oil, to get any accumulated moisture out of it, to basically keep good lubrication in the bearings and keep the gaskets and seals moist and sealing properly. Keep the oil conditioned," Fletcher explained.

The Fuel Maintenance Mode story is a bit different. Volt uses a completely sealed fuel system, its 9.3-gal (35-L) steel fuel tank slightly pressurized, so oxidation really does not come into play. But fuel is blended seasonally, so once every year if a Volt customer has not burned a tank of fuel, the controller will start the engine. This is to combust the fuel in the tank down to the point where the customer will have to add enough fresh fuel to blend it up to an appropriate volatility for the next time it is called to duty.

Veteran GM Powertrain engineer Pam Fletcher considers the Volt program to be a career highlight.

"In reality, people can be 'all-electric, all the time' with this car," Fletcher said. "But you buy a Volt to have peace of mind, to not have range anxiety if you want to travel more than 40 mi.

"We've built in these protection measures just for that. They'll burn a little bit of fuel, but the trade-off is considerable peace of mind."

Optimizing 12 Volt Start - Stop for Conventional Powertrains

2011-01-0699
Published
04/12/2011

Darrell Robinette and Michael Powell
General Motors Company

doi:10.4271/2011-01-0699

ABSTRACT

A cost effective means of achieving fuel economy gains in conventional powertrain is to utilize a 12 volt start/stop (S/S) system to turn the engine off and on during periods of vehicle idle. This paper presents powertrain integration issues specific to a 12 volt S/S system and the powertrain hardware content and calibration strategies required to execute a 12 volt S/S system for start ability, reduced noise and vibration (N&V) and vehicle launch. A correlated lumped parameter modeling methodology is used to determine engine startup profiles, starter hardware and intake cam park position requirements based upon vehicle level response to the startup event. Optimization of the engine startup is reported for a multitude of powertrain configurations, including transverse and longitudinal arrangements with manual, automatic and dual clutch transmissions. Details are provided on the level of hardware and calibration necessary to satisfy a set of prescribed performance metrics, particularly seat track vibration and smooth vehicle launch.

INTRODUCTION

Automotive hybrid systems with only start/stop functionality at idle typically provide a 4 to 10% improvement in fuel economy over a combined EPA cycle, see [1 and 2], and approximately comparable percentage reduction in CO_2 emissions, [3]. Unlike S/S systems with motor-generators running off the engine accessory drive referred to as belt alternator-starters, BAS or BSA, a 12 volt S/S system utilizes a conventional starter on flexplate/flywheel ring gear with no regeneration benefit. The 12 volt S/S system simply shuts off the engine during periods of zero vehicle speed and restarts the engine once the driver performs a lift foot of the brake (automatics - AT's, automated manual transmissions - AMT's and dual clutch transmissions - DCT's) or presses the clutch pedal on manual transmissions - MT's. The level of performance required in certain markets will require the 12 volt auto start system to be as transparent as possible, meaning imperceptible N&V at driver interfaces (steering wheel, seat track and interior noise) with no compromise to vehicle launch performance for automated transmission applications.

N&V CONSIDERATIONS

The basic source-path-receiver model for a conventional automotive powertrain startup event is shown in Figure 1. With this basic model, the source is engine torque pulsations from compression and/or combustion. The path is vehicle sensitivity, defined as the ratio of driver seat track acceleration to torque at the powertrain mounts. The receiver is the driver, quantified by the level of peak-to-peak acceleration at the driver's seat. Source level vibrations are produced by the slider crank mechanism of the reciprocating internal combustion engine and are a result of block force reactions to instantaneous cylinder pressures. The engine speed spin-up profile will have a significant influence on the severity of the receiver response at the seat track depending on how the frequency content of the torque pulsations align with vehicle sensitivity.

For 12 volt S/S, either source level vibrations originating from the engine can be reduced or path sensitivity can be reduced to produce auto starts that are nearly transparent to the driver and vehicle occupants. At a source level, the engine speed spin profile can be managed through cranking speed and combustion to avoid dwelling at engine speeds that align with high vehicle sensitivity. Peak cylinder pressures, whether motoring or firing, can be reduced since this will lead to a reduction in net block torque acting on the powertrain mounts, see [5, 6, 7]. However, a cylinder pressure reduction strategy is complicated by other metrics, such as cold start-ability, start-ability at altitude and emissions. Managing engine park position during the auto stop can reduce force input from the powertrain to the vehicle as shown previously by [6]. Without a controllable motoring device, such as in strong or mild hybrid powertrains, this strategy can be difficult to accomplish by relying on the engine to self park at a particular crank angle degree (CAD) that results in a partial compression and minimal forces upon auto restart. Although, statistically, the engine will tend to park itself at a minimum energy point, such as the start compression of any one of the cylinders. The path can be modified through powertrain mount locations, mount property selection (stiffness, damping, passive or active, etc.). Modification to these parameters alter the frequency response as well as help to reduce the seat track acceleration to powertrain block torque (A/T) amplitude, see [7]. The focus of this paper will be on modifications to the source level vibrations, particularly through powertrain hardware and the approach to engine controls during engine startup.

The receiver, driver's pk-pk seat track vibration is an appropriate means of quantifying the level tactile vibration that is tolerable before becoming objectionable for auto starts. Govindswamy et al [6] and Kuang et al [5] have reported and used seat vibration as the metric for assessing auto starts and stops for hybrid vehicle applications. Testing and evaluation by [6] have shown that the disturbance at the seat track is most noted by the driver during engine starts up due mainly to human sensitivity to low frequency vibrations around 10 Hz.

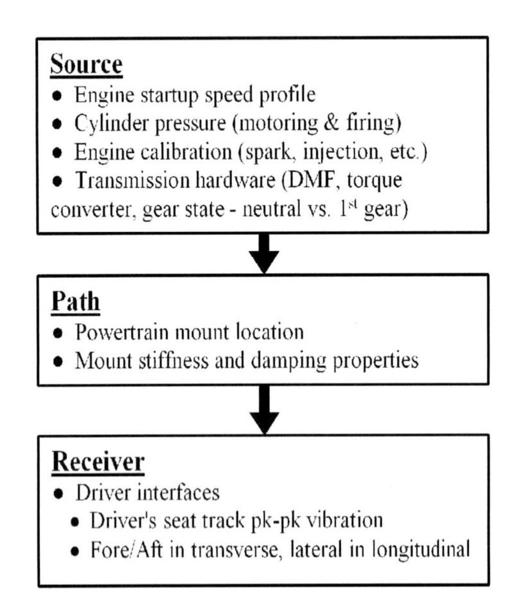

Figure 1. Source-path-receiver model for 12V auto starts.

Figure 2 shows a typical mean engine speed auto start for a direct injected (DI) in line 4 cylinder (L4) engine using a 12 volt S/S system. The startup process is initiated by a 12 volt cranking motor engaged to the ring gear of the flexplate or flywheel attached to the engine crank. After the first compression pulse and the engine controller is synchronized with the position of the crankshaft, fuel can be injected and combusted to accelerate the engine to idle speed. Unlike, port fuel injected (PFI) engines, DI engines eliminate the required manifold mixing dynamics of developing a combustible mixture in the manifold before spark can be added, enabling combustion to start on the second compression as opposed the third or fourth compression, see [8]. This not only reduces time to reach idle and vehicle launch times, but it reduces time spent in an engine speed range that N&V is problematic. For an L4 engine, the speed range that produces N&V issues is generally between 150 and 450 rpm (5 to 15 Hz). This speed range can be separated into two distinct regions, powertrain torsional modes, between 5 and 15 Hz and vehicle modes, between 9 and 15 Hz. The powertrain torsional modes include dual mass flywheel (DMF) resonance, driveline rigid body mode on the half shafts, known as surge, chuggle, or judder and other low frequency drivetrain modes that may occur within the powertrain subsystems. The main vehicle mode is the rigid body mode of the powertrain on the mount system about the axis of the crankshaft. In the case of a transverse powertrain configuration, the fore-aft direction of with respect to the vehicle is of most concern in terms of the powertrain rigid body mode of vibration. For longitudinal powertrains, it is the lateral, or cross car direction that is of most concern for causing objectionable vibration in vehicle and at the driver interfaces during auto starts.

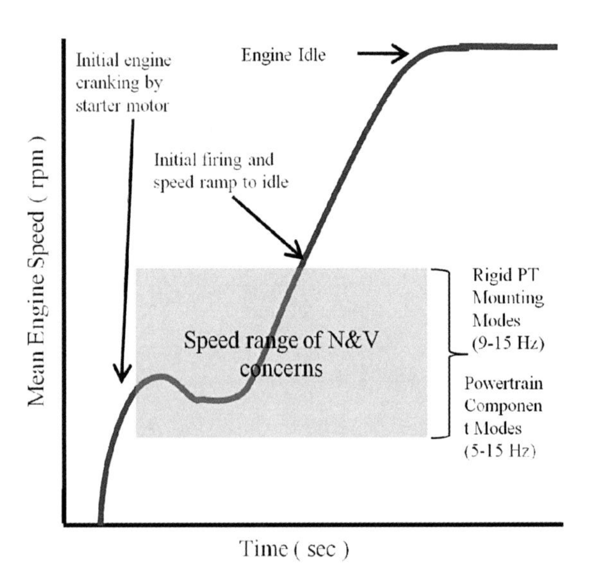

Figure 2. Typical L4 DI engine speed spin up profile for auto start for 12 volt S/S system.

The speed range that N&V is a concern cannot be avoided during a 12 volt auto start, but the time spent operating within this speed range can be reduced. Figure 3 shows engine speed profiles for auto starts events, one a 42 volt-belt-alternator starter (BAS) and the other two 12 volt S/S configurations.

Figure 4 contains the corresponding time history of seat track fore/aft vibration. For the 42 volt BAS system, an engine speed of 450 rpm is achieved in approximately 100 ms, spending little time in the N&V zone, resulting in a seat vibration of 0.4 m/s^2. The ability to spin up the engine with this profile is accomplished with a 5 to 7 kW electric machine with advantageous ratio through the engine accessory drive, providing peak cranking torques of 100 to 120 Nm; see [2 and 3]. Twelve volt starter motors in the 1 to 2.5 kW range are typical in passenger vehicle applications. Peak cranking speeds tend to be around 200 to 250 rpm, as shown in Figure 3 for the two 12 volt S/S applications. For the L4 DI engine, about 250 ms is spent in the N&V zone, cranking to just over 200 rpm and firing occurs on the second compression pulse. With the engine controls minimizing peak combustion pressure through spark retard engine speed drops after the first combustion, before increasing to 80% of engine idle speed after the second combustion. The measured seat track vibration was 0.6 m/s^2. The L4 diesel with DMF measured highest of the three applications at 2.8 m/s^2. This was due to the time spent in the N&V zone, higher geometric compression ratio and high combustion pressures leading to higher block reaction forces to the powertrain mounts.

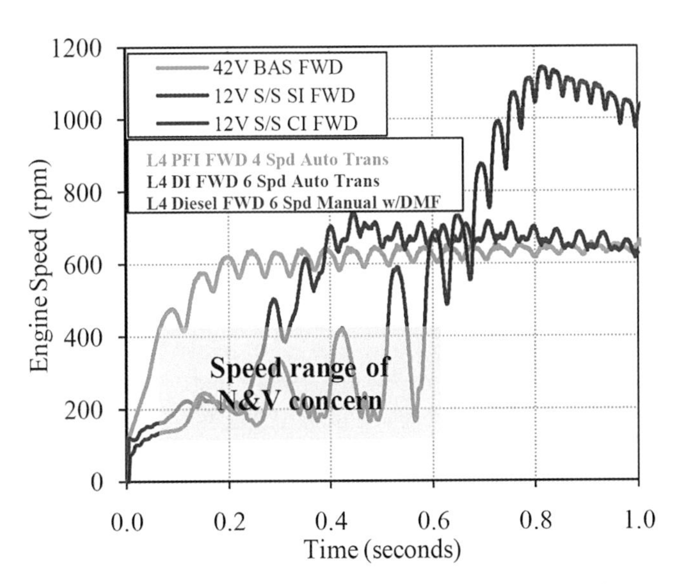

Figure 3. Typical engine speed auto start profiles for various mild hybrid auto starts.

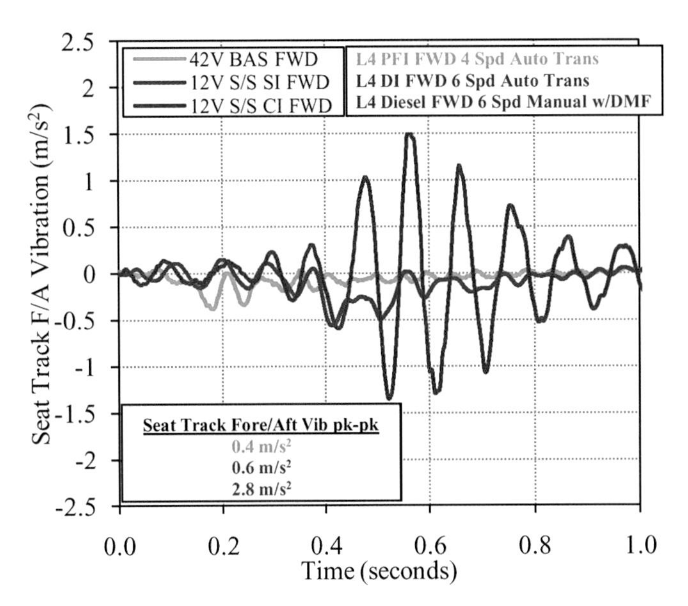

Figure 4. Seat track fore/aft vibration pk-pk for various mild hybrid auto starts.

Further insight into the mechanics behind high seat track vibration during 12 volt auto starts by examining the time history traces of engine speed, seat track vibration and cylinder pressures for the L4 diesel, 6 speed FWD manual with a DMF, shown in Figures 5 and 6. Figure 5 contains engine speed and seat track fore-aft vibration for the 12 volt auto start and Figure 6 engine speed and individual cylinder pressures. A peak cranking speed of 190 rpm is achieved, with cranking pressures in 3 MPa and seat track pk-pk vibration already exceeding the 0.3 to 0.5 m/s^2 threshold of acceptable disturbance. To accelerate the engine from 190 rpm cranking to 800 rpm idle target, fuel is gradually phased in over six combustion events of increasing pressure 5 to nearly 8 MPa, overshooting idle by 300 rpm. The maximum pk-pk seat track vibration of 2.3 m/s^2 occurs after the first combustion event where engine speed peaks at 600 rpm and drops back to the cranking speed of 190 rpm. The drop back to cranking speed is a direct result of the motion of the DMF secondary mass having high angular displacement due to the engine operating at or near DMF resonant speed. The second combustion event elevates engine speed beyond the N&V zone. The amplitude of seat track vibration decreases only when peak combustion pressures reduce to around 4 MPa and steady state idle speed is approached. The combination of high motoring and firing cylinder pressures along with engine speed oscillating at a region of high vehicle sensitivity between 190 and 400 rpm during cranking produce the excessive disturbance at the seat track. Later discussion will show the direct influence of DMF secondary mass motion on engine speed.

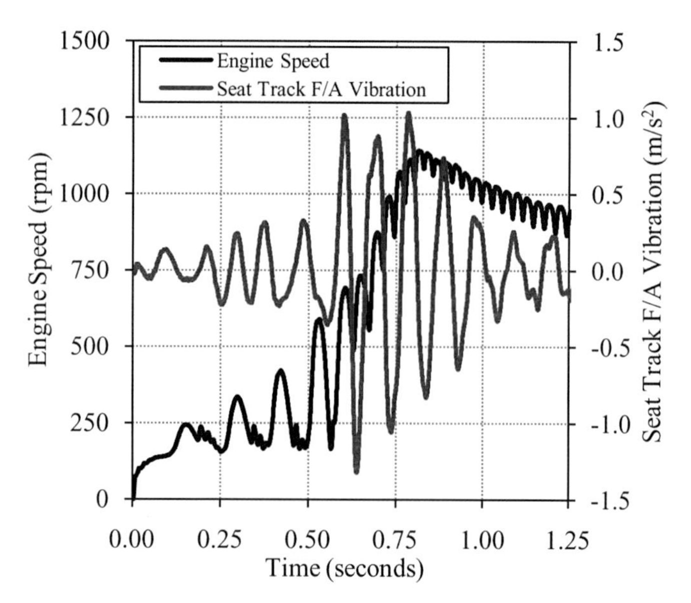

Figure 5. Engine speed and seat track fore/aft vibration.

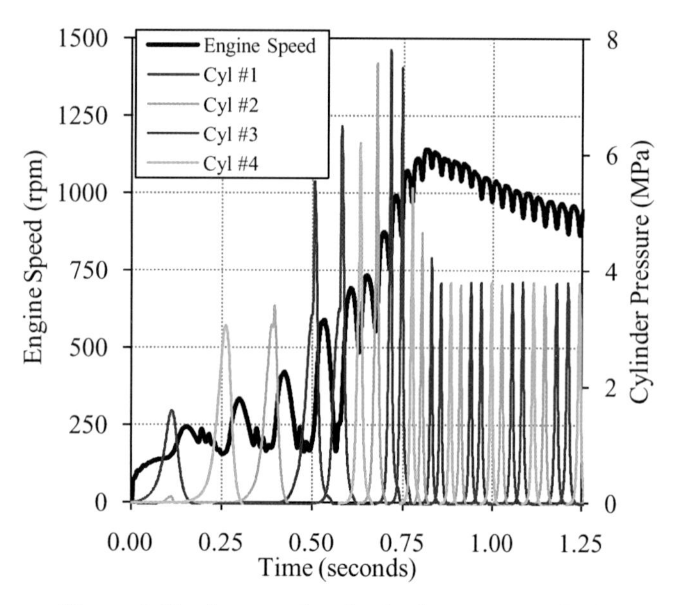

Figure 6. Engine speed and cylinder pressures for L4 diesel, 6 speed FWD manual transmission 12 volt auto start.

START/STOP MODELING TECHNIQUE

To explore the design space of powertrain hardware and calibration changes that improve 12 volt S/S N&V and vehicle launch performance a lumped torsional modeling technique was developed that utilizes measured cylinder pressures from existing engine hardware or analytically generated cylinder pressures based upon future engine hardware designs. Measured cylinder pressures enable validation of the modeling technique and more detailed analysis of a particular 12 volt auto start profile. Analytical cylinder pressures allow the flexibility to investigate effects such as effective compression ratio, peak cranking speed, cranking time and overall calibration of the 12 volt S/S event. The engine model, when driven by analytical cylinder pressures operates off a throttle-opening request and a motoring-firing Boolean input. Comprehended within the engine model are gas and inertia torques, engine friction and accessory drive loads. The engine is cranked via a conventional starter motor incorporating speed-torque characteristics, internal gearing, the one-way clutch, and tooth counts for the pinion and flexplate/flywheel ring gear. The dynamic response of the starter motor for initial cranking is tuned to produce peak cranking torque in 20 to 25 ms. Engine valve timing is also included within engine model calculations, but assumes a fixed position for intake and exhaust cams. This assumption should be accurate for auto start simulations, as phasing for variable valve timing (VVT) camshafts does not typically occur at or above engine idle speeds.

Downstream of the engine, the torsional model includes, details of the launch device (single mass flywheel inertia, DMF inertia damper properties or torque converter speed ratio dependent properties), For automated transmissions, AT's, AMT's and DCT's, clutch pressure profiles are included to capture vehicle launch performance and assess any adverse driveline torsional dynamics that may occur during auto starts. All reaction torques from the powertrain are summed at common node and are imposed upon a single degree of freedom (SDOF) representation of the powertrain inertia in either the fore-aft or lateral direction on the powertrain mounts. The spring rate of the SDOF powertrain mount is tuned to the natural frequency measured in vehicle or analytically computed in the direction of interest with respect to the powertrain architecture (transverse or longitudinal).

Model correlation for keyed and auto starts for a 12 volt cranking system is contained in Figures 7 and 8, respectively. Measured cylinder pressures drove the correlation models, with engine speed serving as the correlation variable to determine the sufficiency of the model to accurately capture the physics of engine startup. In both cases, reasonably good correlation is achieved with only minor discrepancies in engine speed amplitude and only minor shifts with respect to crank angle.

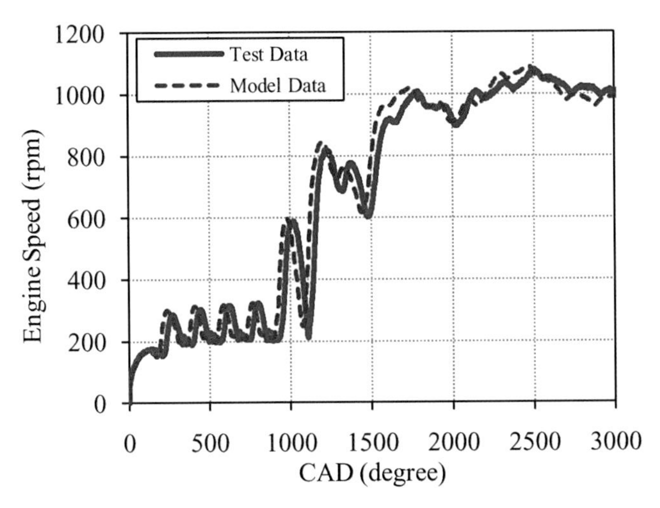

Figure 7. Model correlation for L4 PFI engine, 6 speed FWD MT, 12 volt keyed start.

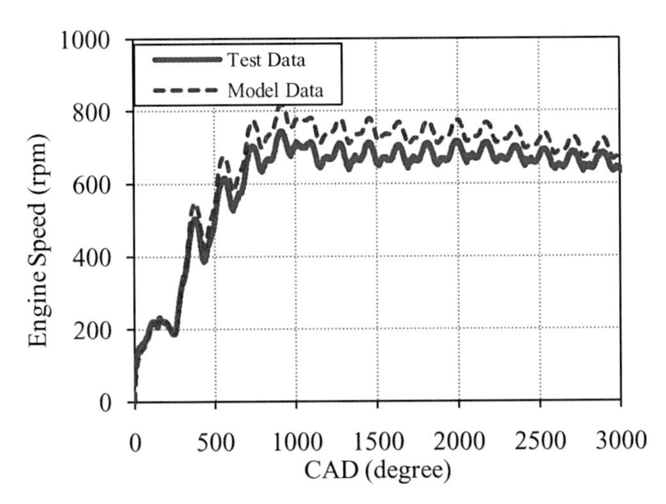

Figure 8. Model correlation for L4 DI engine, 6 speed FWD automatic transmission, 12 volt auto start.

One of the limitations of the modeling technique worth noting, is that the effect on cylinder pressure due to large spark retards is not comprehended accurately unless explicitly accounted for in measured data or analytically generated as such. Thus, modeling results reported using analytical cylinder pressures will not fully comprehend the effects of an aggressive spark retard auto start strategy to reduce firing cylinder pressures and block torque reactions.

POWERTRAIN INTEGRATION

PARAMETER STUDY

The parameters available in the torsional model for tuning 12 volt S/S N&V performance include starter motor performance and gearing, cranking time, peak motoring cylinder pressure, peak firing cylinder pressure, DMF damper properties, torque converter speed ratio dependent properties, transmission clutch torque capacity during startup and vehicle launch. The aim of a parameter sensitivity study is to reduce seat track pk-pk vibration during a 12 volt auto start event and not compromise vehicle launch performance. At a high level, this means utilizing starter motors that reduce the time spent at engine speeds coincident with the frequency of vehicle modes of vibration or powertrain and driveline torsional modes and reducing force input to the powertrain mounts from combustion. To decrease the time spent in the N&V zone for cranking, higher power starter motors off increased torque capability are required. Favorable gearing, both internally to the starter planetary gear set and external ratio of the starter pinion and flexplate/flywheel ring gear can increase peak cranking speeds. At the engine source level, geometric compression ratio, intake cam park position and calibration of spark can all be engineered to produce lower cylinder pressure and thus reduce force inputs to powertrain mounts. However, a number of complex trade-offs exist with modifying cylinder pressures and must be balanced, such as start-ability and emissions. The following sections provide detailed discussion of the effect of each of the available tuning parameter on 12 volt S/S N&V behavior as well as vehicle launch performance for automatic and dual clutch transmissions (DCT's).

For this study path, parameters associated with the vehicle will not be investigated. Instead, vehicle sensitivity data for particular vehicle architecture are incorporated into the torsional models. The summed reaction torques acting on the powertrain structure imposed upon the mounts is input into the vehicle A/T sensitivity to get a prediction of pk-pk seat track vibration in the fore-aft direction for FWD and lateral direction for RWD powertrain configurations. This is the primary metric to be used for assessing any N&V performance gains by varying any powertrain hardware or calibration in the torsional model.

ENGINE

Cranking Speed

Maximum cranking speed is defined as the peak engine speed achieved after a compression pulse. For a 12 volt auto start with a DI gasoline engine this occurs after the first compression pulse, as noted in Figure 3 at roughly 200 ms. The effect of maximum cranking speed on seat track pk-pk vibration is shown in Figure 9 for an L4 DI engine with a 6 speed RWD manual transmission containing a DMF. Figure 9 indicates that the loss function with respect to seat track vibration is exponential function of maximum cranking speed for a given vehicle sensitivity and peak motoring pressure of 1400 kPa and a first combustion event with a peak pressure of 2200 kPa. To achieve the various peak cranking speeds the starting motor speed torque characteristic were modified along the internal and external ratios of the starter. Peak power of the starters varied between 1 and 4 kW, the internal ratio of the starter was varied between 3.64:1 and 5.67:1, while the starter pinion tooth count varied between 7 and 15 teeth. The tooth count of the ring gear on the DMF primary mass remained fixed at 135 teeth.

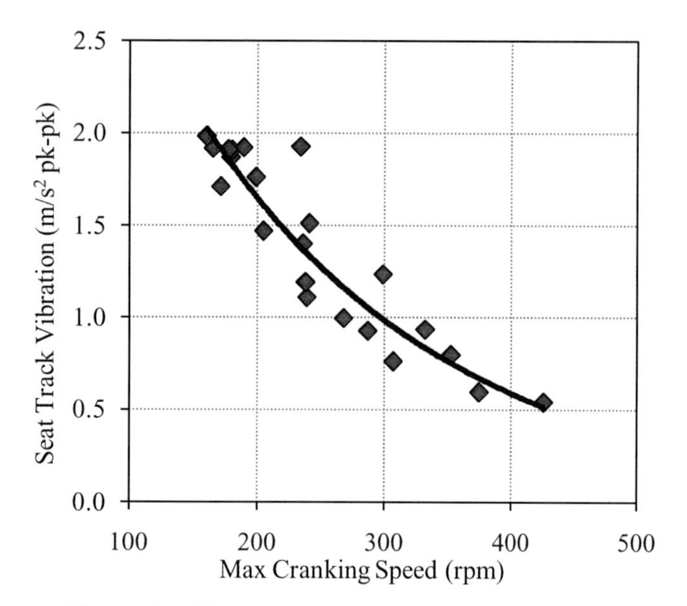

Figure 9. Effect of max cranking speed on seat track vibration, L4 DI, 6 speed RWD manual with DMF.

Maximum cranking speeds in the 300 to 400 rpm range should be a sufficient target for 12 volt S/S applications since this is approximately the point of diminishing returns. The engineering challenge, however, for such a high cranking speed is developing a starting motor capable of such speeds without excessive current draw or voltage drop and packages within the envelope current starter systems. One item to make note of is that the loss function shown in Figure 9 is for a particular powertrain and vehicle combination. Although the general shape of the curve should be similar for different applications, the intersection of the loss function and acceptable seat track vibration may occur at higher or lower maximum cranking speeds.

Peak Motoring Cylinder Pressure

The effect of peak motoring pressure on 12 volt auto start seat track pk-pk vibration is shown in Figure 10 for an L4 DI gasoline engine and 6 speed RWD automatic transmission. Four separate maximum cranking speeds were simulated for peak motoring pressures of 1100 to 1800 kPa. Changes to motoring cylinder pressure are principally achieved by changing intake camshaft timing or phasor park position in the case of VVT engines; retarded park for lower pressures and advanced park for increased pressures. The results in Figure 10 suggest that the seat track vibration loss function is a 2nd order function with peak motoring pressure which flattens out for the range of pressure simulated as maximum cranking speed increases. A large reduction in seat track pk-pk vibration can be noted when maximum cranking speed increased from 240 rpm to 340 rpm, which is consistent with the findings in Figure 9 and reaffirms the notion that cranking to engine speeds higher than the peak of the vehicle sensitivity curve corresponding to the powertrain rigid body mode is advantageous. At cranking speeds in the 300 to 400 rpm range for an L4 engine, the system is desensitized to peak motoring pressures the need to require low geometric compression ratio or retarded intake cam park position is mitigated and base engine design can be focused on meeting other key performance metrics such as torque, emissions and efficiency.

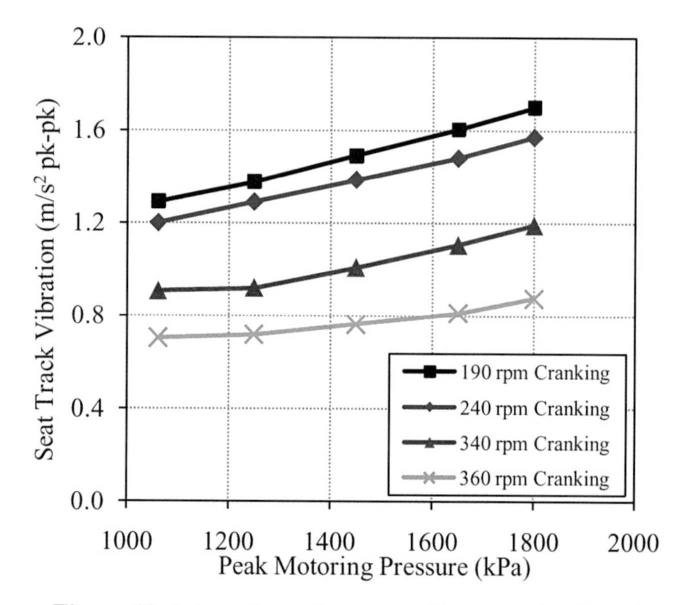

Figure 10. Interaction of max cranking speed and peak motoring cylinder pressure on seat track vibration for peak first firing pressure of 2600 kPa, L4 DI, 6 speed RWD AT.

Peak Firing Cylinder Pressure

As shown previously in Figures 5 and 6 for an L4 diesel, 6 speed FWD manual 12 volt S/S application, the peak cylinder pressure of the first few combustion events during the auto start can significantly affect the pk-pk seat track vibration. To investigate this further with the torsional model, an L4 DI engine with a 6 speed automatic transmission is assumed to peak motoring pressure of 1400 kPa. Peak firing pressure of the first combustion event is varied between 2200 kPa and 5100 kPa for four different maximum cranking speeds. For a given maximum cranking speed the effect on seat track pk-pk vibration is approximately a linear function with respect to the amplitude of first firing event peak cylinder pressure, as seen in Figure 11. Directionally, lower peak firing pressures for the first few combustion events during an auto start can reduce tactile vibration in vehicle. This trend is consistent for all four maximum cranking speeds simulated. As with peak motoring pressure the biggest reduction in seat track vibration occurs once the maximum cranking speed gets above 300 rpm. Maintaining peak cylinder pressures below 3000 kPa during the first few combustion events while balancing emissions, maintaining start-ability and preserving combustion quality are the essential engineering challenges with this strategy.

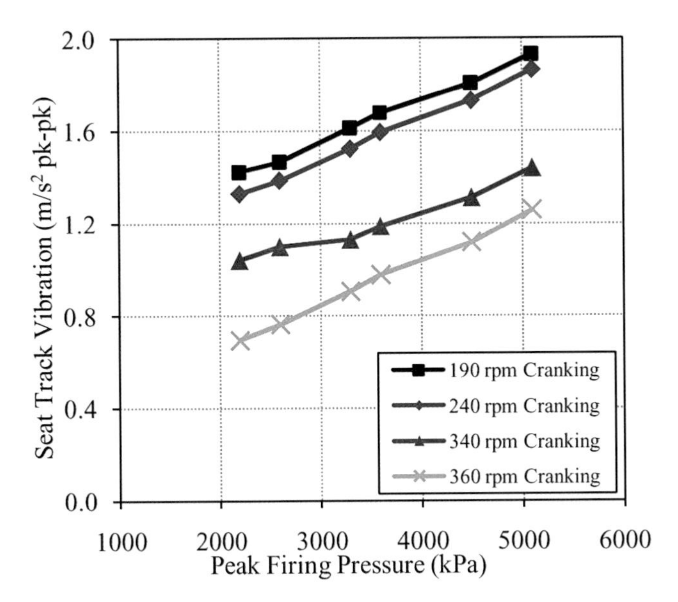

Figure 11. Interaction of maximum cranking speed and peak first firing pressure on seat track vibration for peak motoring pressure of 1400 kPa, L4 DI, 6 speed RWD AT.

The combination of elevated cranking speed, retarded intake camshaft timing/phasor park position and reduced firing cylinder pressures for the first few combustion events produces the best 12V auto start N&V response. A 12V S/S system that utilizes a cranking system with elevated cranking speeds will require less combustion energy to accelerate the engine to the targeted idle speed. Additionally, such an auto start profile will lead to reduced time to idle speed and quicker vehicle launch times as was shown in Figure 3 with the 42V BAS system compared to the 12V system with a conventional cranking speed starter.

MT, AMT AND DCT TRANSMISSIONS

For MT, AMT and DCT transmissions, 12 volt S/S applications there are a few integration issues that can influence the tactile vehicle response and vehicle launch performance during auto starts. For manual transmissions there is no automated clutch apply during or following the engine auto start, the transmission and driveline are decoupled, so the main concerns are powertrain block force reactions, originating from the engine (see previous section) and damper resonance of the launch device. For single mass flywheel 12 volt S/S applications there is no damper resonance since the clutch is released during the auto start, the damper springs are decoupled from the engine. For DMF manual and DCT applications, attention must be given to the primary and secondary inertia's, spring rates, hysteresis and special friction packages of the DMF damper assembly. On DCT applications, launch clutch capacity is controlled during or slightly after engine auto start to launch the vehicle so care must be taken not initiate clutch stick-slip and induce high driveline torsional vibration, resulting in vehicle judder.

DMF Resonance

The driveline and gear rattle isolation afforded by DMF on manual or DCT transmissions is a result of relatively large primary and secondary mass inertia's coupled by a sufficiently low spring rate, long travel arc springs. The first natural frequency of the spring-mass system of the DMF occurs at relatively low engine speeds, on L4 engines for example in the 150 to 300 rpm range. On engine startup, consideration must be given to the DMF's resonant speed and it should be avoided to ensure DMF durability and start-ability of the engine, see [9] for more details. Figure 12 shows a measured keyed start for a DMF equipped L4 DI engine with a DMF mode at 190 rpm. The maximum cranking speed achieved is 175 rpm. DMF travel is relatively low during cranking, but with a 4 MPa peak first firing event (Figure 13), the DMF travel is roughly 54 deg pk, or 104 deg pk-pk. The high angular displacement and inertia torque associated with the secondary mass, cause engine speed to come back to the cranking speed, which is followed by another strong combustion event at 3.2 MPa. After which engine speed peaks at 800 rpm, but is pulled back down to 400 rpm by the inertia torque of the secondary mass, at which point the engine experiences a weak combustion event at 1.2 MPa, because engine speed was trending to overshoot idle. This behavior continues for three more cycles until mean engine speed is sufficiently high to be far removed from the DMF resonant speed and the control system can recover.

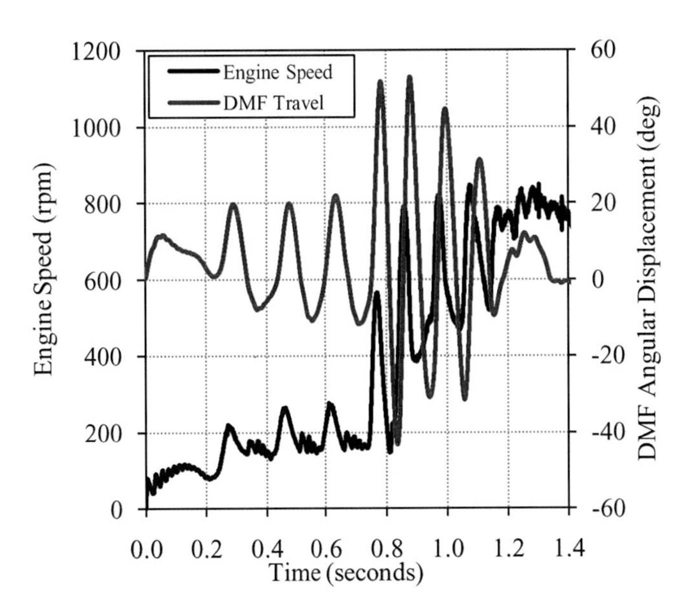

Figure 12. Influence of DMF resonance and high angular displacement on engine speed during keyed start.

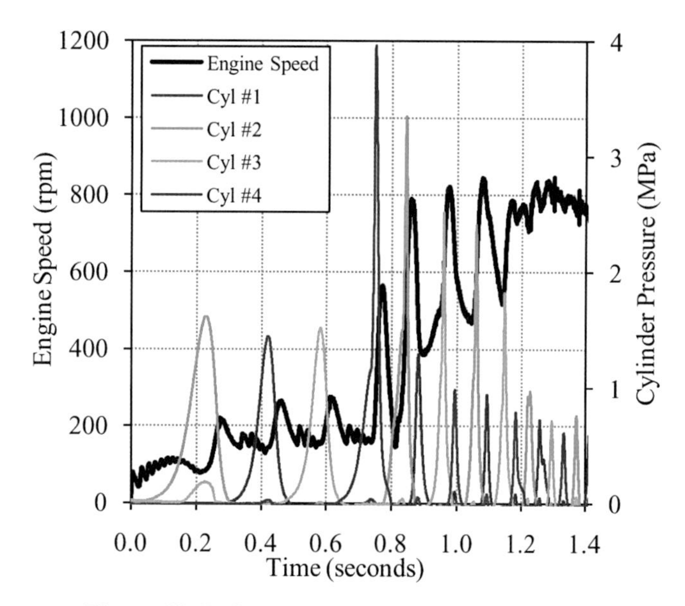

Figure 13. Influence of DMF resonance and high angular displacement on engine combustion and controls during keyed start.

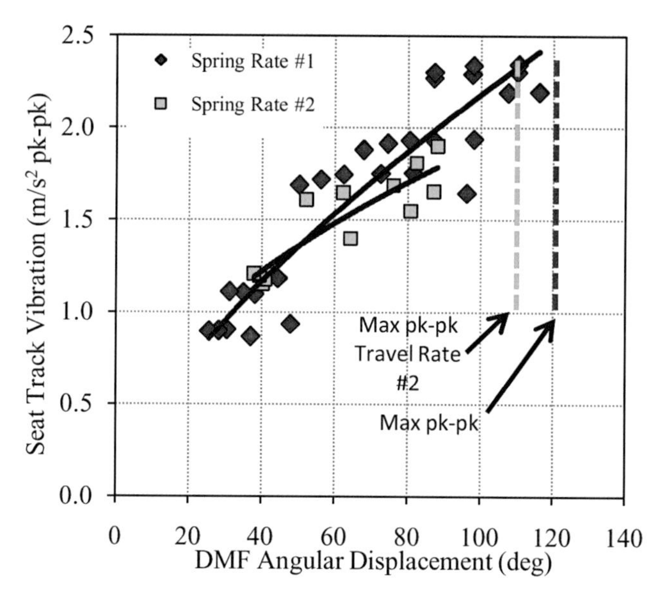

Figure 14. Effect of DMF angular displacement on seat track vibration during auto start for L4 DI, 6 speed RWD MT.

The measured data shown in Figures 12 and 13 illustrates the importance of limiting DMF during auto and keyed starts. Maintaining low pk-pk levels of DMF angular displacement reduces the risk of secondary mass inertia torque influencing engine speed to stay within the N&V zone of a four cylinder engine and cause unacceptable levels of seat track pk-pk vibration. Figure 14 contains simulation data for an L4 DI engine with a 6 speed RWD manual transmission with DMF under various auto start conditions of maximum cranking speed, peak motoring pressure and peak first firing pressure. The spring rates of the DMF damper assembly were also varied along with adding a friction disk with friction varied between 10 and 40 Nm. The base and dynamic hysteresis of the DMF were unchanged. Spring rate #2 DMF has a shorter travel, stiffer first stage rate and softer longer travel second stage rate than spring rate #1 DMF.

The intent of this design of experiment was to develop a loss function for seat track pk-pk vibration and DMF angular displacement. As Figure 13 shows, the general trend for either spring rate is a simple power function, with low DMF travels associated with elevated cranking speeds and low peak motoring and firing cylinder pressures, which is consistent with previously discussed parameters. The tendency with large DMF travels near the pk-pk travel limits results in engine speed oscillating back in the N&V zone as combustion takes place, similar to the data shown in Figures 4 and 11.

AMT and DCT Auto Start Clutch Capacity

The main difference between a MT with a DMF and AMT's or DCT's with a DMF for 12 volt S/S is the automated clutch apply to launch the vehicle. It is desirable to perform the 12 volt auto start with the prescribed methodology discussed, while also properly managing the rate of clutch apply and hence clutch torque capacity to execute vehicle launch that is as transparent as possible to the driver and vehicle occupants. Four clutch apply rates following a brake lift foot 12 volt auto start are shown in Figure 15 for a brake lift foot. The rates of clutch apply and overall capacities vary, producing the engine speeds, transmission input shaft speeds (input shaft for applied clutch) and vehicle acceleration traces shown in Figure 16. For a very slow clutch apply to a low capacity (green traces in Figure 15 and 16), engine speed stabilizes to idle creep smooth, as well as transmission input, resulting in smooth vehicle acceleration. This translates to a very transparent launch to the driver (seat track vibration), but will give the perception of poor vehicle performance since acceleration will most like miss targeted time to reach an observable acceleration. Aggressive clutch applies, either too much capacity too soon (blue traces in Figures 15 and 16), or too rapid apply too late (red traces in Figure 15 and 16), can cause excessive input shaft oscillations and vehicle fore-aft acceleration as engine speed is pulled down to near the DMF's resonant speed as clutch capacity is brought on. Launches of this nature will lead to high seat track pk-pk vibration as engine speed operates for a longer time window within the N&V zone. A blend of capacity apply rate and capacity are needed to provide a balanced 12 volt auto start and launch with a DCT as illustrated by the black traces in Figures 15 and 16. Further investigations are planned to understand the benefits of higher cranking speeds and clutch capacity apply rate on 12 volt auto start N&V behavior.

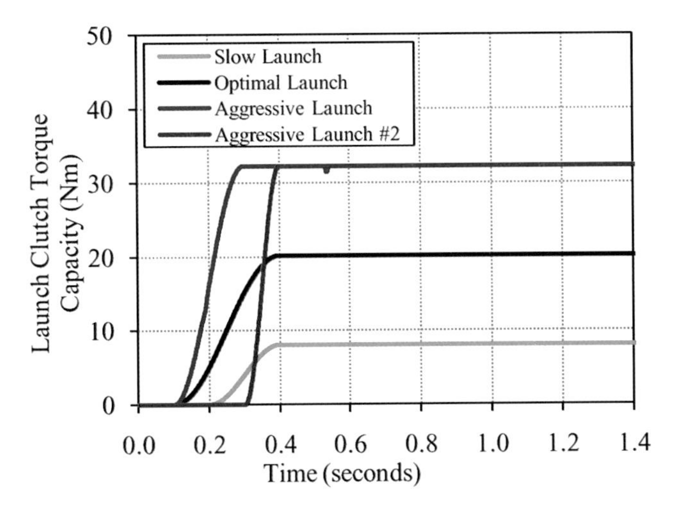

Figure 15. DCT launch clutch torque capacity profiles for brake lift foot 12 volt auto start.

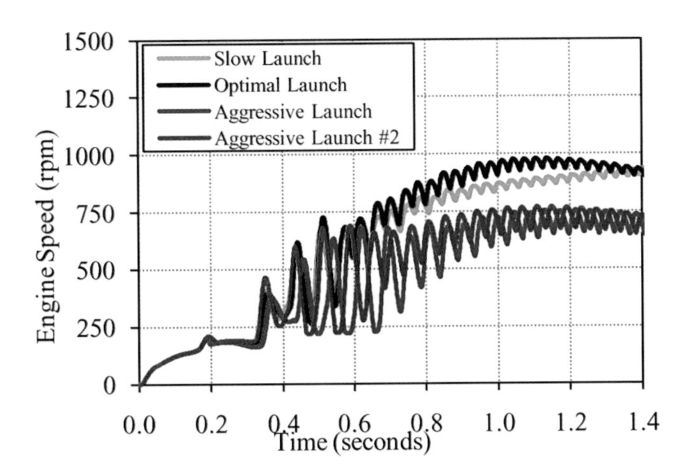

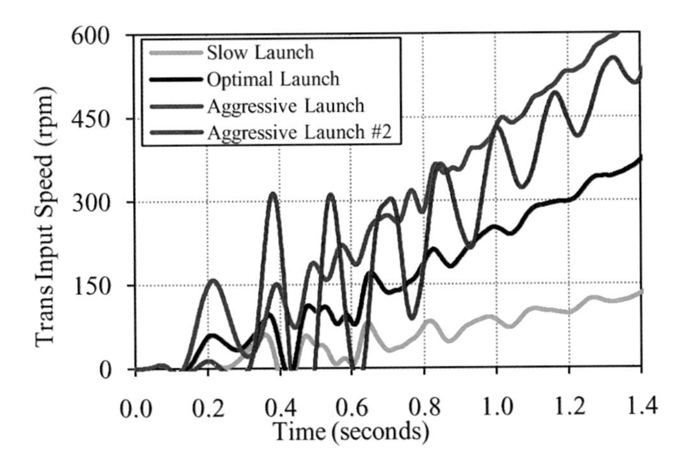

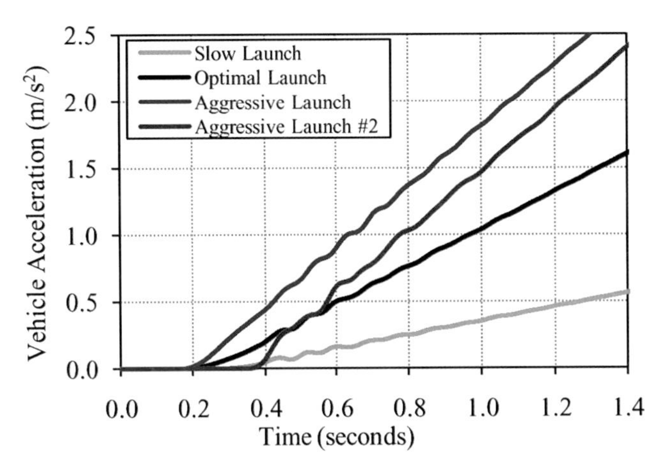

Figure 16. Engine speed, transmission input shaft speed and vehicle acceleration for various DCT launch clutch torque capacities following 12 volt auto start.

AUTOMATIC TRANSMISSIONS

Unlike manual and DCT transmissions, there is no damper resonance to crank through for automatic transmissions, instead the integration issues to consider relate primarily to vehicle launch performance and rate of torque application to the transmission and driveline during 12 volt auto starts. Since the powertrain is most likely to be calibrated to couple the driveline via the transmission launch clutch for first gear, any rotary motion of the engine during cranking and initial combustion is passed to the drivetrain via the torque converter. A balance of torque converter properties, final drive ratio and the engine speed profile during 12 volt auto starts must be achieved so that vehicle startup and launch are as transparent as a non 12 volt S/S vehicle.

K-factor and Final Drive Ratio

For automatic transmissions, the selection of torque converter strongly influences powertrain performance namely through the speed ratio, SR, dependent properties of K-factor, K, and torque ratio, TR. Since the torque converter regulates engine speed and sets engine torque for a given throttle condition at any vehicle speed when operating strictly as a hydrodynamic unit it strongly influences vehicle launch characteristics. Similarly, selection of final drive ratio, FDR, also influences vehicle performance, although not as strongly as the torque converter. To understand the tradeoffs between these two parameters on perceived vehicle launch performance for a 12 volt auto start a simple matrix of torque converter K's (equivalent stall TR) and FDR's were modeled for an L4 DI engine and 6 speed FWD automatic transmission. Table 1 summarizes the results for time to reach 2 kph and maximum vehicle chassis acceleration for a lift foot off brake auto start maneuver. Improved vehicle launch feel can be achieved by selection of lower K torque converters for a given diameter class and a numerically higher FDR. However, selection of these two parameters should not be based solely on launch, but other performance metrics such as fuel economy, 0 to 60 and gradability.

Table 1. Effect of FDR and K on time to 2 kph and peak vehicle acceleration for lift foot auto start for L4 DI, 6 speed FWD.

	FDR	
Torque Converter	2.89	3.23
200 K	2.41 sec	1.77 sec
	0.814 m/s^2	0.952 m/s^2
180 K	1.26 sec	1.07 sec
	1.05 m/s^2	1.21 m/s^2

Idle Speed Overshoot

With the transmission first gear launch clutch at full torque capacity during a 12 volt auto start, engine torque will be multiplied through the torque converter and ratio of the transmission to the driveline resulting in vehicle motion. During the auto start, the torque converter is operating at stall, zero turbine speed, providing maximum torque multiplication. Since the torque converter regulates engine speed and torque relationship shown in Equation 1, referred to as K-factor, K, higher engine speeds lead to higher engine torque, leading to tactive effort at the drive wheels.

$$K = \frac{N_p}{\sqrt{T_p}} \tag{1}$$

Allowing significant overshoot in engine speed can result in rapid and unexpected vehicle acceleration and excessive vehicle speeds during auto starts. This can be an undesirable condition in congested traffic situations of a stop-and-go nature. Figure 17 shows engine and turbine speed along with vehicle speed and acceleration for an L4 DI engine with a 6 speed FWD automatic transmission under for two auto start profiles. As can be noted, an auto start profile with a significant amount of idle speed overshoot, 600 rpm in this case, results in achieving higher vehicle speed and acceleration than a profile with 250 rpm overshoot.

One feature of the turbine speed trace to take note of following the first combustion event in Figure 17 is oscillations of approximately 50 to 75 rpm. These oscillations are a result of torque loading the driveline and lash take up immediately following brake release and initiation of the auto start event. Similar trends in auto start driveline disturbances were reported by [1, 5 and 6]. Turbine speed does not begin to oscillate until after the first combustion when the torque converter's torque capacity increases and the drivetrain responds to the step change in applied torque. Looking at the vehicle chassis acceleration traces in Figure 17, it can be noted that the fore-aft acceleration of the vehicle is not smooth as a result, which will translate into perceived vibration at the seat track and perception of a poor startup and launch quality. For the small idle speed overshoot, the oscillations in turbine speed reduce in amplitude due to lower engine torque and driveline loading at initiation of combustion.

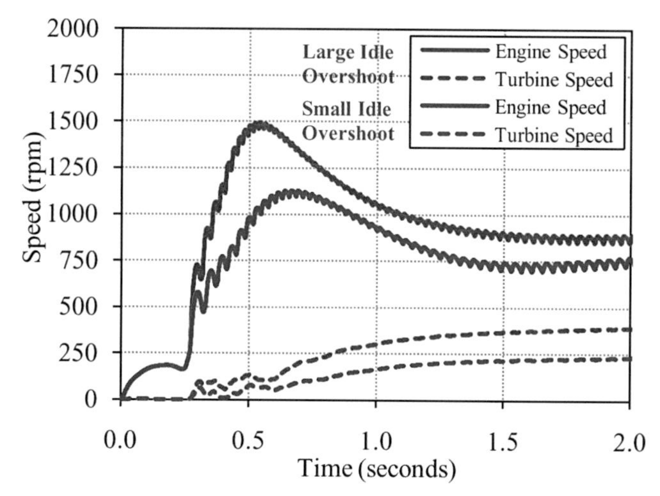

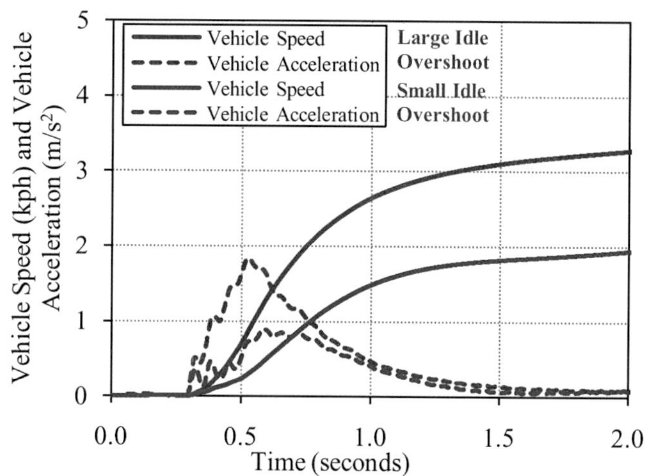

Figure 17. Auto start and vehicle launch performance for large (red) and small (blue) idle speed overshoots, L4 DI, 6 speed FWD AT.

Figure 18 contains model data for a V8 DI engine and 6 speed RWD automatic transmission under an auto start condition for two engine speed profiles, one with a large idle speed overshoot typical of keyed starts and the other a more managed idle speed overshoot. For large displacement, high torque output engines coupled to automatic transmissions, torque converters tend to be large in diameter and have numerically low K's. This results in considerable torque capacity even at low engine speeds such as idle. For the powertrain modeled in Figure 18, an overshoot of idle speed like that of a keyed start can lead to extreme vehicle acceleration and a very rapid change in vehicle speed. With a 400 rpm overshoot the vehicle goes from 0 to 6.5 kph in little over 500 ms and reaches a peak chassis acceleration of near 5 m/s2 (0.5 g's). Managing the engine idle overshoot to around 200 rpm and targeting idle around 600 rpm reduces the initial rise in vehicle speed to 3.5 kph in 500 ms, and a peak chassis acceleration of just over 2 m/s^2. The turbine speed disturbance is also reduced and the irregularity in chassis acceleration is reduced by limiting idle speed overshoot, leading to improved tactile feel of the auto start and launch.

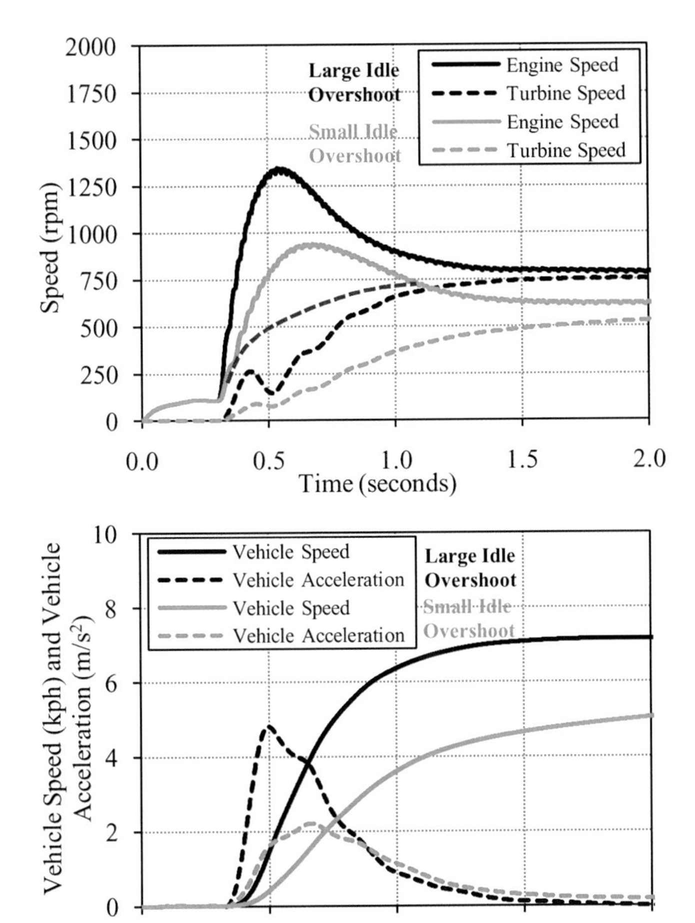

Figure 18. Auto start and vehicle launch performance for large (black) and small (green) idle speed overshoots, V8 DI, 6 speed RWD AT.

Ideally, the engine speed profile during startup would follow the red dashed line in Figure 18, further reducing, turbine speed oscillations and chassis acceleration below 2 m/s^2. Refinements to the model to handle large spark retards to reduce cylinder pressure of the first few combustion events would improve the model's predictive capability.

Transmission Launch Clutch Capacity

Neutral idle or limiting the torque capacity of the transmission 1st gear launch clutch can be an enabler to smoother 12 volt auto starts by not torque loading the driveline during cranking and initial combustion as shown previously to cause disturbances in turbine speed. Figure 19 contains results comparing the launch performance of an L4 DI engine and 6 speed FWD automatic transmission with full torque capacity of the transmission launch clutch (black), limited capacity (green) and a neutral auto start capacity (red). By limiting the transmission launch clutch capacity turbine speed flares to 500 rpm as engine speed increases to idle. This requires precision capacity control of the clutch apply pressure, between 0 and 60 Nm. During a neutral auto start, turbine speed flares to 1000 rpm when the launch clutch is brought to full torque capacity at 1 second. Both limited capacity and neutral auto start launches reduce the driveline torsional oscillations, but add 500 and 1000 ms delays, respectively. Either delay will result in unacceptable launch time with the perception of a sluggish start, especially if there is an immediate driver tip-in request. Further, since most transmission launch clutches are designed at a large radius for maximum capacity, the gain of the clutch will prevent enough control resolution unless special hardware or software provisions are made for the transmissions.

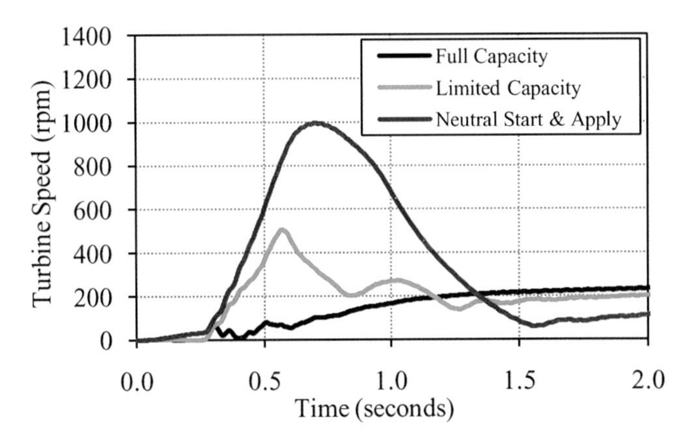

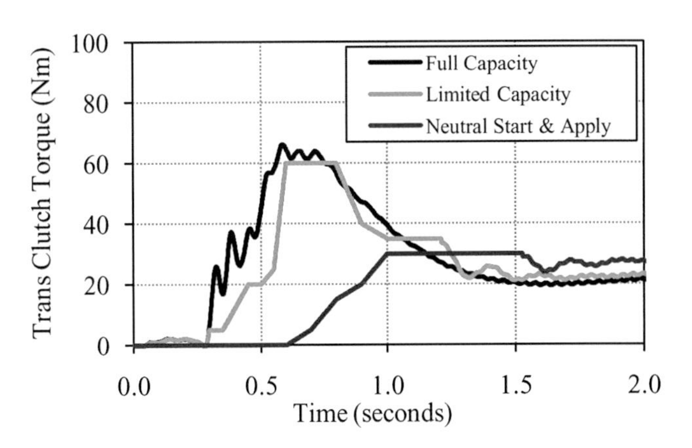

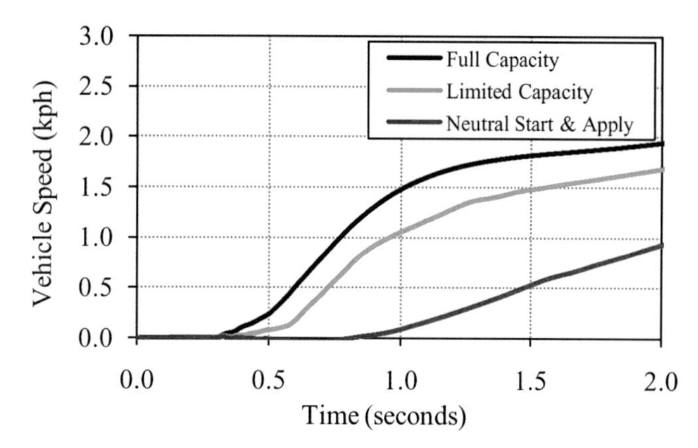

Figure 19. Effect of transmission launch clutch capacity and neutral auto starts on vehicle launch performance, L4 DI, 6 speed FWD AT.

OPTIMIZED AUTO START

Table 2 summarizes the hardware content and calibration strategies found to minimize seat track pk-pk vibration without degrading vehicle launch performance.

Table 2. Summary of hardware and calibration for 12 volt auto start N&V and launch performance

Parameter	Value	Comment
Cranking Speed	300 to 400 rpm	Requires optimize speed-torque characteristics of starter motor and gearing
Motoring Pressure	Maintain start-ability	Desensitized with high cranking speeds
Firing Pressure	< 2000 kPa	First few combustion events
Launch Clutch Capacity	Maintain engine speed drop above DMF mode	AMT's and DCT's
Idle Speed Overshoot	< 20%	AT's

CONCLUSIONS

A lumped parameter torsional modeling technique was used to determine powertrain hardware and calibration strategies necessary for minimizing seat track pk-pk vibration and preserving vehicle launch performance for 12 volt S/S auto starts. Data acquired on developmental 12 volt S/S vehicles was used to show model correlation and illustrate key issues related to improving N&V including trends in engine speed, cylinder pressures and DMF angular displacement. In general, it was found that minimizing time spent at engine speeds aligning with the rigid body modes of the powertrain on its mount is a key enabler to reducing vibration amplitude felt at the driver's seat track. A cranking the engine to elevated speeds of 300 to 400 rpm, reduces time spent at vehicle sensitive vibration modes while reducing system sensitivity to peak motoring and firing cylinders pressures. Maintaining combustion pressures below 3000 kPa for the first few combustion events lessens seat track vibration, while also reducing overshoot of targeted idle speed. This strategy derives further benefit in terms of lessening the initial torque step and resulting driveline torque oscillations for automatic transmission that would lead to objectionable vehicle fore-aft acceleration. DMF angular displacement is minimized with the auto start strategy presented for manual and DCT transmissions. For DCT's, clutch capacity and rate of apply need to be balanced to achieve favorable launch performance without disturbing driveline torsional modes or stalling the engine. Future work and studies planned for 12 volt S/S integration include N&V performance for auto stops and driver change of mind starter re-engagement and cranking.

REFERENCES

1. Bishop, J., Nedungadi, A., Ostrowski, G., Surampudi, B. et al., "An Engine Start/Stop System for Improved Fuel Economy," SAE Technical Paper 2007-01-1777, 2007, doi: 10.4271/2007-01-1777.

2. Canova, M., Sevel, K., Guezennec, Y., and Yurkovich, S., "Control of the Start/Stop of a Diesel Engine in a Parallel HEV with a Belted Starter/Alternator," SAE Technical Paper 2007-24-0076, 2007, doi:10.4271/2007-24-0076.

3. Wagner, J., Mencher, B., and Keller, S., "Bosch System Solutions for Reduction of CO_2 and Emissions," SAE Technical Paper 2008-28-0005, 2008, doi: 10.4271/2008-28-0005.

4. Tamai, G., Hoang, T., Taylor, J., Skaggs, C. et al., "Saturn Engine Stop-Start System with an Automatic Transmission," SAE Technical Paper 2001-01-0326, 2001, doi: 10.4271/2001-01-0326.

5. Kuang, M.L., "An Investigation of Engine Start-Stop NVH in a Power Split Powertrain Hybrid Electric Vehicle," SAE Technical Paper 2006-01-1500, 2006, doi: 10.4271/2006-01-1500.

6. Govindswamy, K., Wellmann, T., and Eisele, G., "Aspects of NVH Integration in Hybrid Vehicles," *SAE Int. J. Passeng. Cars - Mech. Syst.* **2**(1):1396-1405, 2009, doi: 10.4271/2009-01-2085.

7. Komada, M. and Yoshioka, T., "Noise and Vibration Reduction Technology in New Generation Hybrid Vehicle Development," SAE Technical Paper 2005-01-2294, 2005, doi:10.4271/2005-01-2294.

8. Kataoka, K. and Tsuji, K., "Crankshaft Positioning Utilizing Compression Force and Fast Starting with

Combustion Assist for Indirect Injection Engine," SAE Technical Paper 2005-01-1166, 2005, doi: 10.4271/2005-01-1166.

9. Mohire, S. and Burde, R., "Evaluation of Interdependent Behavior of Dual Mass Flywheel (DMF) and Engine Starting System," SAE Technical Paper 2010-01-0188, 2010, doi: 10.4271/2010-01-0188.

DEFINITIONS/ABBREVIATIONS

A/T
Acceleration to torque sensitivity

AMT
Automated Manual Transmission

AT
Automatic Transmission

BAS
Belt Alternator Starter

CAD
Crank Angle Degrees

DCT
Dual Clutch Transmission

DI
Direct Injection

DMF
Dual Mass Flywheel

F/A
Fore/Aft

FDR
Final Drive Ratio

FWD
Front Wheel Drive

K
K-factor

kph
Kilometers per Hour

MT
Manual Transmission

PFI
Port Fuel Injection

pk
Peak

pk-pk
Peak to Peak

N&V
Noise and Vibration

RWD
Rear Wheel Drive

SDOF
Single Degree of Freedom

SR
Speed Ratio

S/S
Start/stop

TR
Torque Ratio

VVT
Variable Valve Timing

The Engineering Meetings Board has approved this paper for publication. It has successfully completed SAE's peer review process under the supervision of the session organizer. This process requires a minimum of three (3) reviews by industry experts.

ISSN 0148-7191

SAE Customer Service:
Tel: 877-606-7323 (inside USA and Canada)
Tel: 724-776-4970 (outside USA)
Fax: 724-776-0790
Email: CustomerService@sae.org
SAE Web Address: http://www.sae.org
Printed in USA

Optimizing ICEs for hybridization

Automakers and powertrain R&D specialists are developing combustion engines specifically for series-hybrid, EREV, and PHEV applications. Top engineers explain the opportunities involved.

ABOVE: Chrysler unveiled extended-range electric vehicle concepts at the 2009 Detroit auto show. The system's internal-combustion engine was a parallel-twin-cylinder unit derived from the company's 2.4-L inline four. The future of Chrysler's hybrid program currently is unclear.

The industry's growing interest in series hybrids and various types of plug-in and extended-range electric vehicles (PHEVs and EREVs) is giving engine designers a rare opportunity—the clean-sheet optimization of the internal-combustion engine (ICE).

While the pioneering 2011 Chevrolet Volt and Fisker Karma will launch with off-the-shelf ICEs to minimize cost and speed development, their next-generation successors are expected to feature dedicated, more efficient, and even unconventional power units.

"The series-type hybrid has created a huge opportunity to explore many solutions in various directions for ICEs," explained Paul Najt, Group Manager of the Propulsion Systems Lab at General Motors Research. "For the first time, we can develop engines which don't have to perform from idle to 6000 rpm and do everything in terms of powering the vehicle."

Instead the ICE's sole role in these configurations is to power a generator, which charges the vehicle's battery when its state-of-charge (SOC) drops below a predetermined level. The engine runs steady-state at various fixed rpm levels, depending on battery charge requirements and control strategy.

When engaged, the engine/generator sustains a minimum battery SOC, extending the vehicle's range for potentially hundreds of miles beyond its initial charge. This allows it to reach a place where it can be fully recharged by plugging into the electric-power grid.

"The engine serves very much like a small APU [auxiliary power unit]," noted Najt, a veteran of more than 30 years of advanced engine R&D at GM. "Because it's decoupled from the drive wheels, we can design the engine around a very narrow set of operating parameters," that best suit its thermal efficiency and brake-specific fuel consumption map.

The series-hybrid system typically allows cylinder displacement to be significantly reduced, as it does not have to handle the peak loads incurred during acceleration. Most series hybrid/PHEV testbeds and prototypes currently range from 800 cm^3 to 1.2 L, with 1.0-L emerging as a popular size based on average and peak power requirements, and the balance of packaging, mass, internal friction, and pumping losses.

Najt noted that future ICEs dedicated to series hybrid/PHEV propulsion will be tailored to the vehicle's intended duty cycle. To spin the generator in a small city car requiring perhaps 15 kW (20 hp), for example, the ICE will be sized and equipped differently than one designed for heavier vehicles or sports cars, where far greater peak power is needed.

Where 90 hp/L (67 kW/L) would be a typical specific-output bogey for a conventional engine application, 70 hp/L (52 kW/L) would probably be more than adequate for the series hybrid/PHEV, noted the experts interviewed for this article. They also expect the engines to be gasoline fueled, diesels presenting a cost challenge.

For the first-generation Volt, GM engineers chose an existing 1.4-L Ecotec designed for B/C-segment front-wheel drive vehicles. Simulations and early mule tests showed 53 kW (70 hp) at 3000-4000 rpm would be required for the EREV application. With this system, GM engineers admit their engine selection was conservative for this vanguard vehicle. Experts note that typical four-stroke gasoline engines operate most efficiently between 2500 and 3500 rpm.

"It comes down to how big a battery do you want to carry and how powerful do you want your electric motors," Najt explained. Both factors have a significant effect on the engine-control strategy and specification of the engine/generator.

Cost reduction through design

"When you're looking at designing a range-extender engine, everything's up for grabs—within the confines of cost," observed Neil Fraser, Senior Principal R&D Engineer at Mahle Powertrain, which recently completed a detailed study of optimizing current production engines for series hybrid/PHEV use.

"You're not designing for typical 6000 rpm and 100-bar peak cylinder pressures, so you can really start to minimize friction," he said. "You can reduce valve spring pressure, for example, and size all internal dimensions for what your boundary conditions and ultimate loadings are."

In some cases, main bearing journals could be downsized, he said, as long as durability at least equivalent to a conventional ICE is maintained. The potential for long periods of vehicle operation without the ICE in use, as well as the requirement to go from "off" to peak load when the battery calls for a charge, requires rethinking lubrication systems and bearing robustness.

Some engineers are investigating assembled crankshafts with rolling-element main and con-rod bearings. Electric oil pumps that can be switched on and off also are under consideration.

In terms of dedicated clean-sheet designs, however, various OEMs and the major powertrain design houses are exploring an array of engine

FEV also developed a Wankel-based power generator, installed for European testing in a Fiat 500. Visible in this photo of the underside of the car (from left to right) is the front-mounted electric traction motor that drives the wheels, the enclosure for the battery pack (center), and the cleverly packaged Wankel-APU slightly to the left of the spare-tire well.

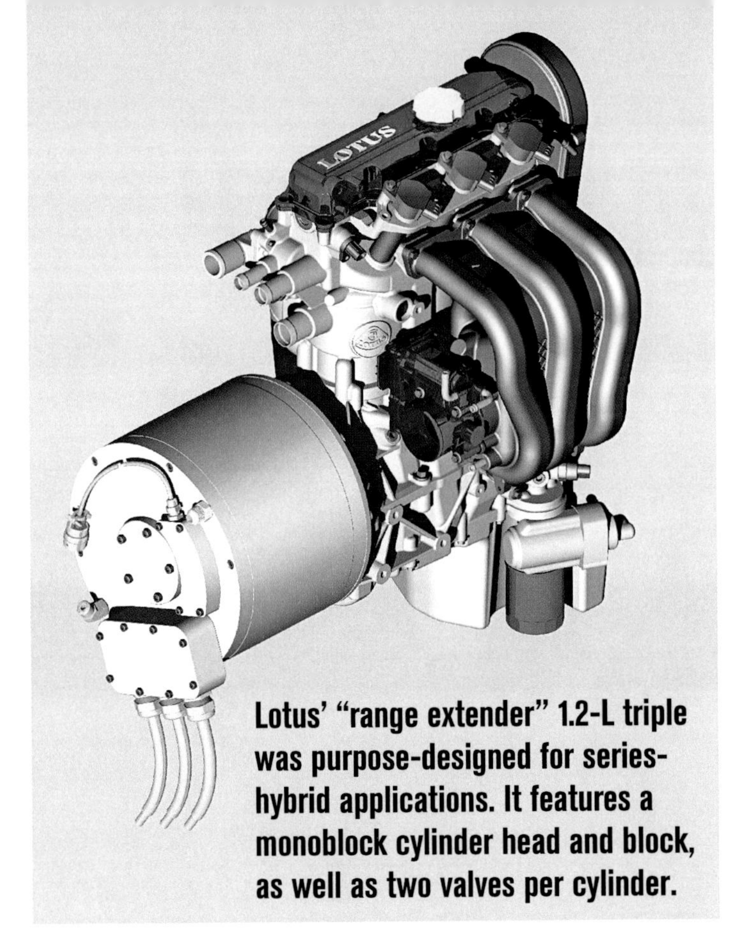

Lotus' "range extender" 1.2-L triple was purpose-designed for series-hybrid applications. It features a monoblock cylinder head and block, as well as two valves per cylinder.

configurations. The list of active programs includes parallel- and V-twins; inline triples, fours, fives, and sixes; Wankel rotaries; opposed-piston alternative fuel types (aimed at military hybrid applications), and even hydrogen fuel cells and miniature gas turbines. All offer an equally broad range of capacities, outputs, complexity, and cost.

Responding to growing demand from industry customers, FEV has developed 10 different driveable EREVs for evaluation within the past year, noted Dr. Jochem Wolschendorf, Chief Technology Officer. The company executed vehicle integration on the prototypes as well as some engine designs.

"The bandwidth of generator output in these units spans 15 to 200 kW (20 to 270 hp) to cover small cars through large military vehicles," he said. Design modularity enables FEV to minimize complexity and cost among power ranges and engine families, he explained.

Two of the key opportunities in designing dedicated ICEs for hybrids are simplifying the engine architecture and decontenting components where possible. GM's Najt noted that SOHC with two valves per cylinder would help reduce friction and could be sufficient, as long as an efficient combustion chamber is retained.

FEV's Wolschendorf noted that variable valve timing systems might not be needed, due to the limited rpm range, although he added that a simplified VVT could be retained for start-up emissions control. Doing away with VVT would also reduce the oil pump requirements.

At the 2009 Frankfurt Motor Show, Lotus Engineering unveiled a purpose-designed range extender engine that is being tested in a Jaguar XJ with EREV powertrain. (See AEI November 2009, page 19). The 1.2-L lightweight triple features a monoblock cylinder block/head, simplified valvetrain, and other elements designed to minimize cost and optimize efficiency.

Managing NVH a priority

Designing for low NVH is one of the major challenges in developing ICEs for series hybrid/PHEV use, according to experts. When the vehicle switches from charge-depleting mode, where it is running smoothly and quietly on battery power alone, to charge-sustaining mode when the ICE engages, the transition can be surprising if not carefully managed.

"It is one of our top priorities in developing engines for these applications," said Wolschendorf. "When the engine kicks in, suddenly it's a noisy vehicle—no matter what type of ICE you are using."

Indeed, AEI discovered this during a November 2009 test drive of the Volt—the first automotive publication to sample the car in charge-sustaining mode. Even in prototype form, the car is extremely refined and quiet as a battery EV. When the ICE engaged during our drive at GM's Milford proving ground, it immediately revved to around 3000 rpm as called for by the electric powertrain's control software.

It was a comparatively loud intrusion into what had been a nearly silent ride. Going from "off" to high load in a split second is a unique trait of series hybrid/EREV engines, and it is a challenge for developers.

"The best scenario is to make sure the customer doesn't know the engine is running," explained Wolschendorf.

Volt engineers are revising the car's control software and they've upgraded engine mounts to smooth the transition when the generator kicks in. But they admit overall NVH could be much reduced with an optimized ICE. (See AEI Online for the complete Volt test story.)

NVH concerns present a particular challenge to twin-cylinder engines for series hybrid and EREV applications, despite the obvious attractions of reduced mass, compact packaging, and lower unit cost.

"Some of our customers have been interested in twin-cylinder engines, both inline twin or Vee, for series-hybrid applications in smaller vehicles. We've found that's very difficult due to mechanical vibrations mostly, and due to noise," said Paul Whitaker, Chief Technologist for Spark-Ignition Engines at AVL.

He and Jerry Klarr, AVL's Director of North American Hybrid development, noted the primary concern with twins, even those with balance shafts, is NVH at start-up.

Sophisticated (and costly) active engine mounts are one solution, they said. Also, electric oil pumps capable of being switched on and off have the ability to maintain a level of oil pressure for start-ups, which would help benefit bearing wear while also mitigating timing-chain noise at start-up, Whitaker said.

Game-changing opportunities

Minimizing engine mass is a key design bogey for series-type hybrids/PHEVs, because the ICE is simply "along for the ride" when the vehicle is operating as a battery EV. The engine adds mass to a system that is already challenged with the weight of the battery pack and e-motors.

Powertrain-technologists AVL and FEV each developed city-car demonstrators to showcase the packaging, mass, specific-output, and NVH benefits of using Wankel rotary engines in the series-hybrid/PHEV configuration.

Above: FEV's Dr. Jochem Wolschendorf (right) with Manager of Electronics and Controls Kevin Rzemien and one of the company's ICE-generator concepts.

Below: Jerry Klarr (left) and Paul Whitaker of AVL North America in one of the company's dyno rooms.

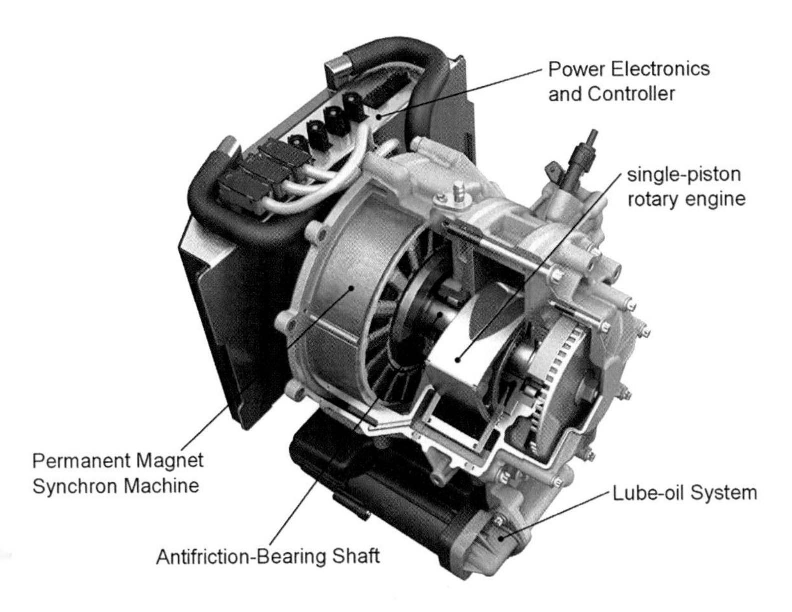

AVL's engine-generator uses a single-rotor Wankel rotary engine. The super-compact, modular unit is now under test in a Mini in Europe.

The arch-competitors each chose single-rotor engines for their compactness, low mass, and scalability in their vehicles—a modified Mini from AVL and a Fiat 500 from FEV. Both cars are busy running real-world test cycles in Europe.

While AVL installed its Wankel generator in the Mini's traditional engine compartment, FEV chose a radical layout—nestled within the Fiat's floor pan underneath the front seats (see photo on p. 9).

"This installation shows the packaging advantages of the series-hybrid configuration," noted Wolschendorf. "Even in a car as small as the Fiat 500, the compact generator unit can be fitted underneath the passenger compartment."

Such flexibility offered by dedicated designs could help create new opportunities in interior packaging and frontal-crash management.

A question for the future is whether the ICE in a series hybrid or PHEV will be a main purchase criteria for the end customer. But at this early stage of the development curve, engineers are focused on making their operation transparent.

In the past year, FEV has integrated 10 prototype series-hybrid/EREVs into customer test programs. This Chrysler application (in a Dodge Caliber) used a Mercedes smart 800-cm³ triple.

CHAPTER NINE:

Flogging a Mule

The first drive of a Volt prototype in charge-sustaining mode revealed a lot about the car's development pace 11 months before production.

The author samples a pre-production Volt with the full Voltec propulsion system in operation—Milford Proving Ground, November 18, 2009. (Photo by John F. Martin for Chevrolet)

Flogging a **Mule**

During *AEI*'s late 2009 prototype drive, Volt's transition from nearly-silent electric drive to charge-sustaining mode was a bit ragged. At the time, GM engineers were still designing the software set of rules that enable a near-seamless "feathering" from EV mode to charge-sustaining mode. By start of production a year later, Volt had received more than 20 VESCOMs (major vehicle control software calibrations), plus many more iterations (v.1, v.2, etc.). (Photo by John F. Martin for Chevrolet)

The moment of truth is approaching, I think to myself, and mash the accelerator pedal to the floor. As I point the Chevrolet Volt development mule toward the steep hill in front of me, the "range to empty" display on the car's cluster signals less than one mile of battery range remaining. Within seconds, the car's power controller will sense the minimum state of charge and call for the combustion-engine-powered generator to kick in.

That's the moment I'm waiting for. It's why I'm here at GM's Milford Proving Ground on a chilly day in early November 2009. *Automotive Engineering International* is the world's first industry publication invited by GM to sample the full performance of Volt's extended-range hybrid-electric powertrain.

Like other media, I've driven Volt mules in battery-only EV mode in short spurts around the flat Warren (Michigan) Tech Center. But today I'll experience what Volt sounds and feels like when its four pistons chime in to sustain battery charge.

Will the 1.4-L four-cylinder be annoying in this otherwise nearly silent electrified car? Will it roar, rather than whisper, at its point of engagement? What about the sensation of pushing the "gas" pedal, which has no direct connection to the engine's throttle and thus does not control rpm?

For my test at Milford, the engineers intentionally serve up a Volt with a nearly depleted battery. GM engineers say a fully charged Volt is capable of 40 mi (64 km) of purely electric driving before the controller calls for the 53-kW (71-hp) generator to start and sustain the battery's minimum charge level—the "extended range" operating mode.

Riding shotgun with me on the twisty, undulating 3.7-mile road course is Volt's Vehicle Line Executive, Tony Posawatz. A veteran GM engineer with a big-picture view of vehicle electrification, he's been with the Volt program since its inception. I tell Tony that even though we're not driving the car on remote public roads, without a place to stop and plug in, I still sense every future EV owner's ultimate fear: running out of battery power.

"That's the beauty of the E-REV solution," he replies with a chuckle. "We've eliminated those worries."

As we take a few initial laps, the EV range display counts down toward zero miles. Going up the steep grade with the car under load, I expect the generator to kick in. Instead, we crest the hill and remain on smooth, quiet battery power. Surprisingly, the engine doesn't engage until we're off load and half way down the other side of the hill. Not what I expected, I tell Tony.

I'm impressed that the engine's initial engagement is subdued and seamless. But a few hundred yards later, as we snake through the track's infield section, the engine rpm rises sharply. The accompanying mechanical roar reminds me of a missed shift in a manual-gearbox vehicle. For a moment the sound is disconcerting; without a tachometer, I guess that it peaked around 3000 rpm.

I ask Tony what's going on.

"The system sensed that it's dipped below its state of charge and is trying to recover quickly," he says. "The charge-sustaining mode is clearly not where we want it to be yet."

The engine sound disappears immediately, although the internal combustion engine (ICE) is still spinning the generator. A few times later in my test, the generator behaves in similar fashion—far too abrupt and unrefined for production. But start of production is 11 months away, so there's time for the control and calibration experts to find solutions.

According to Posawatz, at the time of my test drive the Volt engineers are in the process of revising the car's control software. Their goal is to enable "feathering" the transition from the nearly silent all-electric mode to the charge-sustaining mode, when the generator will be operating.

"We're designing a software set of rules, which will just require more seat time for the engineers to finish," he explains. "We have nine months to work this out."

The sound of the generator running at steady highway speeds is something Volt owners, and others who appreciate the flexibility and efficiency of this type of hybrid system, may have to accept. They'll also have to get used to the lack of traditional 'vroom-vroom' feedback. That's because when your right foot presses the Volt's "gas" pedal, the ICE's rpm does not change. The pedal controls only the flow of battery power to the electric drive motor—it has no connection to the generator, which is programmed to run at constant, pre-set speeds.

Unlike many EVs, including the $102,000 Tesla Roadster, the Volt's electric drive has no audible whine. The car feels solid and planted on the road. Clicking the button marked 'Sport' on the dashboard releases a bit more oomph than when in Normal mode. In terms of efficiency, there isn't much difference between the two except at peak power.

The Low mode is unique among EVs. While coasting, this useful feature applies electric motor braking, then smoothly blends in the regular brakes.

Beyond the regenerative function, Low mode offers one-pedal driving in slow speed, stop-and-go, and downhill environments. The Volt's regenerative braking is progressive and predictable. It can be applied using the brake pedal or by pulling the shift lever down into Low mode. This is in stark contrast to the abrupt regenerative braking delivered by BMW's Mini-E, for example.

There is minimal body lean in Milford's tight corners. The low-rolling-resistance Goodyear tires created specifically for the Volt provide excellent grip.

Throughout my test, the mule behaves admirably. I'm extremely impressed by Volt's refinement and overall performance at this stage of development.

CHAPTER TEN:

Charging and connectivity

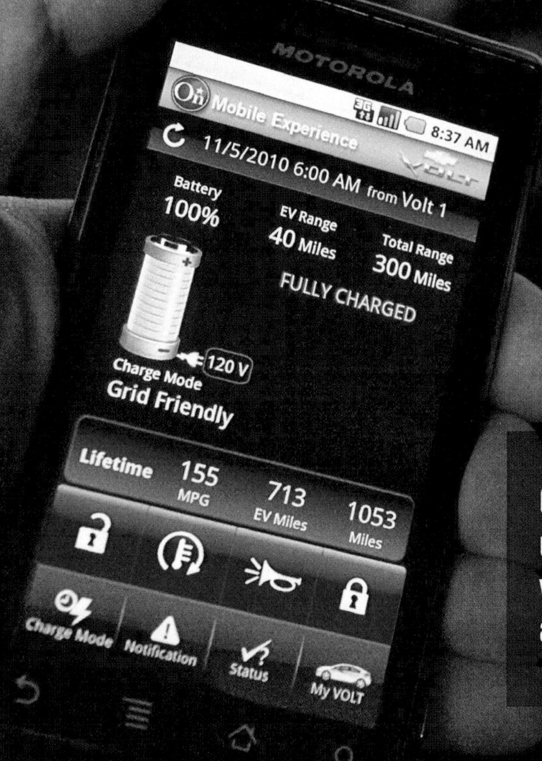

GM's OnStar telematics offer almost unlimited opportunities to create value with Volt and its progeny, according to experts.

GM engineers designed in maximum flexibility for keeping the Volt juiced up and connected —to the grid and to the Internet.

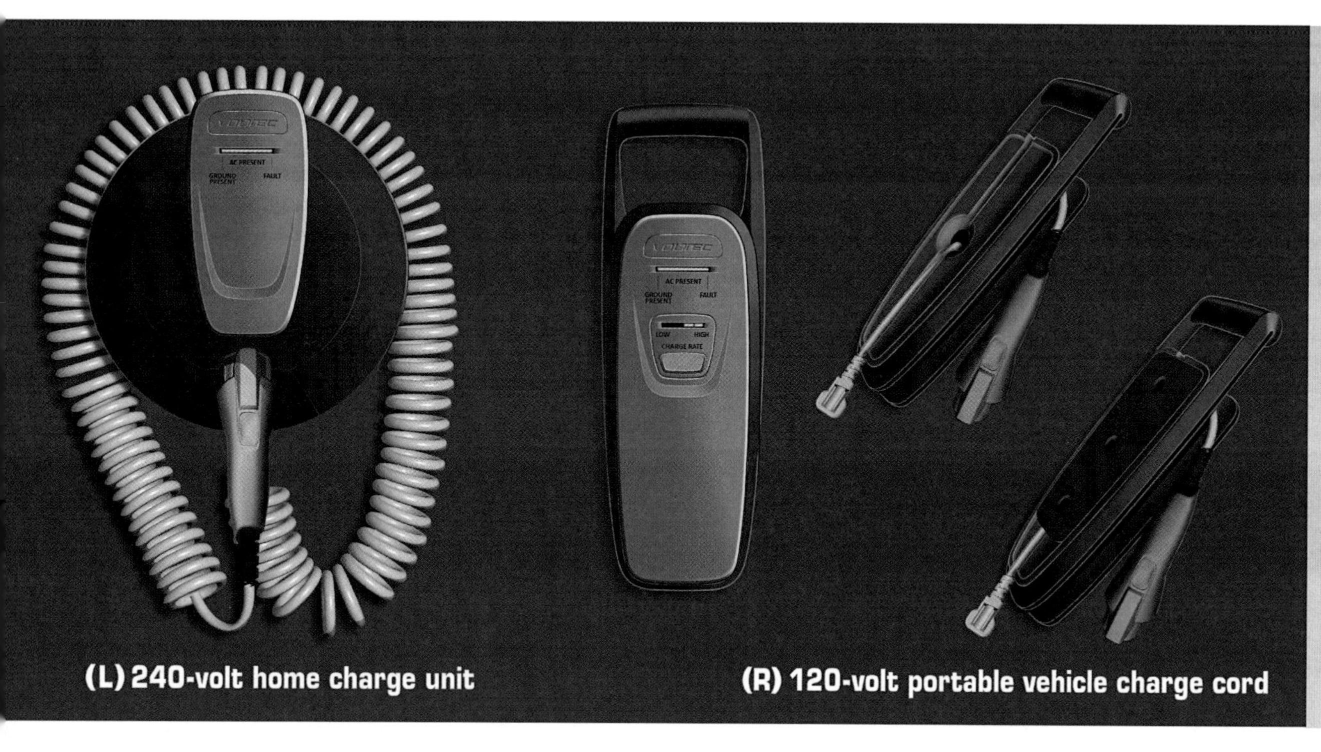

The optional (left) and standard charging sets in Volt's EVSE. The standard 120-V set is stowed in the rear luggage area.

To know Volt, you must learn a new language. Vehicle electrification brings with it almost Defense Department levels of arcane terms and acronyms, particularly those related to vehicle charging and telematics. PEV (plug-in electric vehicle) and PHEV (plug-in hybrid) engineers have a head start with the new lexicon. For others, it is time to get smart.

EVSE is electric vehicle supply equipment—the off-board hardware needed to supply charge energy to the vehicle. Volt's EVSE includes the vehicle's charging cord, residential or public charging stands, attachment plugs, power outlets, and the vehicle connector.

Charging with ac Level 1 uses 110/120 V ac from standard 15 A or 20 A household outlets. Charging with ac Level 2 takes 208/240 V ac up to 80 A. Level 1 is the standard charging used in Volt and the 2011 Nissan Leaf, with a Level 2 option available for both cars. Level 1 and 2 equipment is expected to serve as the workhorse of EV charging for the near-term future, according to experts.

Volt is equipped with an onboard 3.3-kW powered charger supplied by Delta Electronics. According to General Motors Engineering Specialist Gery Kissel, the unit is sized to the recharge requirements GM engineers set for Volt—to replenish the 16-kW·h battery to half its total energy capacity (8 kW·h) within 10 h using the Level 1 power supply. For Level 2, the requirement is a 4-h recharge, depending on efficiencies.

Many safety and durability requirements had to be met in developing Volt's Level 1 and 2 charging sets, explained Vehicle Line Executive Doug Parks. The design had to endure a 10,000-cycle life with exposure to dust, corrosion, and water. Besides complying with SAE International's pioneering J1772 charge-connector standard published in early 2010, it also had to meet various IEC and Underwriters Laboratories standards. The durability tests for Volt's standard 120-V, 20-ft-long (6.1-m-long) charge cord include being driven over repeatedly by the vehicle.

Volt's charging sets were developed by Lear Corp., whose engineers began working with the Volt team in 2008. Best known as an interior systems Tier 1, Lear has steadily expanded its focus on electrified vehicle technology since it acquired UT Automotive in 1999. Last year, more than 60% of Lear's patents were related to EV and hybrid technologies. The shift in R&D investment indicates where the supplier is placing its bets for the future.

Connecting Volt and OnStar

AEI's experience with Volt showed it to be as simple to recharge as a mobile phone. The car is also about as easy and intuitive to operate overall. "It's not a spaceship, and that's by design," explained Dave Lyon, who oversaw Volt's interior design and development as GM's Executive Design Director for North American Interiors.

Indeed, the Volt team leveraged its OnStar telematics technologies and organization to make the ownership experience convenient and seamless. On- or off-board charge programming, for example, can be done via smartphone when the vehicle is plugged in by going through the dedicated MyVolt.com website or through Chevrolet's Volt mobile application powered by OnStar Mylink.

Volt mobile apps offer owners the ability to use grid energy to pre-condition the car's battery before driving, depending on ambient temperatures. It can also remotely control cabin pre-cooling and heating, the latter via a new electric cabin warmer that operates when the combustion engine is not engaged.

"Early in development we realized that OnStar offers endless possibilities for connecting Volt and its owners to the energy infrastructure," noted Micky Bly, GM Executive Director, Electrical Systems, Hybrid & Electric Vehicles and Batteries. He believes Volt and other early EVs will require owners to learn new behaviors regarding both the human-machine interface (HMI) and the human-vehicle interface (HVI)—how we react to the vehicle.

"We're accustomed to charge our cell phones at night, and plug our laptops into docking stations, so having charging devices around us is a natural thing today," he said. "But we're not yet used to doing that with vehicles," he said.

"Early on, you'll want to check to see that the car's plugged in at night—maybe

Telematics and the Volt

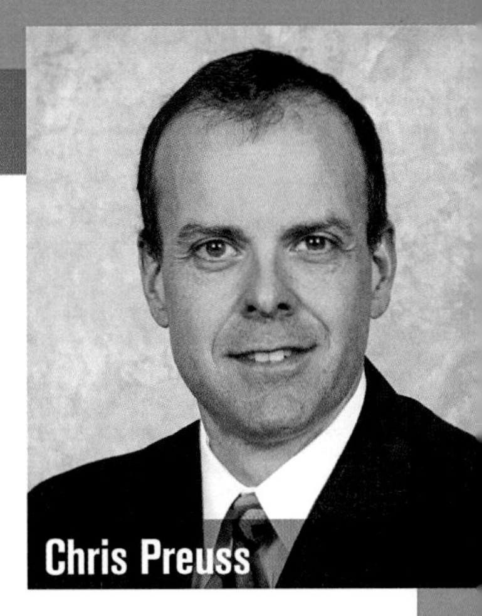
Chris Preuss

Electrified vehicles offer unlimited technology and business opportunities to link the vehicle, customers, and the energy and communications infrastructures. OnStar CEO Chris Preuss recently spoke with *AEI* about his company's developments in this fast-growing sector.

Q: What's the level of importance of the Chevrolet Volt as it relates to the development of OnStar?

Preuss: It's huge. Because as much as Volt is a demonstration platform for the best of what propulsion technologies can do, it's also an example of the best of what telematics can do. That's how OnStar sees it. We believe 'smart grid,' and being present in the smart-grid area, is an enormous business opportunity for us.

Secondly, it's going to be a demand from consumers as they use these technologies to manage their lives. Nobody has really gotten into this space substantially, so a lot of our IT infrastructure development right now around the Web-based applications, and around how we're aggregating data from the vehicle, is focused on interacting with smart grid. We're on the forefront of this. And we're working with a lot of outside companies to tap that potential.

Q: GM engineers involved with vehicle electrification are spending greater amounts of time working with their counterparts at the energy utilities and with those developing charging hardware. How do you expect OnStar will leverage these activities?

Preuss: As we talk about future business opportunities for OnStar, we might want to acquire technologies or companies that are in that [charging technology] space. It would be of key interest to us because we think Smart Grid is the future. It will be a big differentiator, and we've been involved from the ground floor.

Q: Is OnStar working with Jon Lauckner's GM Ventures group to seek acquisitions, as well as possibly sell IP or rights to GM and OnStar IP?

Preuss: Without being specific, we are very aggressively looking in this space—both in terms of what we possess and what we want to create. We're seeing more and more of our IT capability as 'core' technology—it defines the value in our products. From my perspective, it's our 'secret sauce' and it's a place where GM and OnStar lead. When I joined OnStar [spring 2010], I did not appreciate the level of development that was already resident in this space.

Verizon Chief Technology Officer Dick Lynch is the architect of LTE—the next-generation 4G cellular Verizon is launching this year. Dick is one of the biggest evangelists for OnStar—not just because he thinks it's a great service, but he knows what we've been able to figure out, how to move massive amounts of data, how to translate and analyze and make it consumer-usable. We do better than anybody out there.

The automotive telematics space is vast. I've spent two-thirds of my time working to understand it. Volt and its successors are just the beginning of what's possible in terms of vehicle-to-grid technologies, and telematics overall. There's some amazing stuff coming.

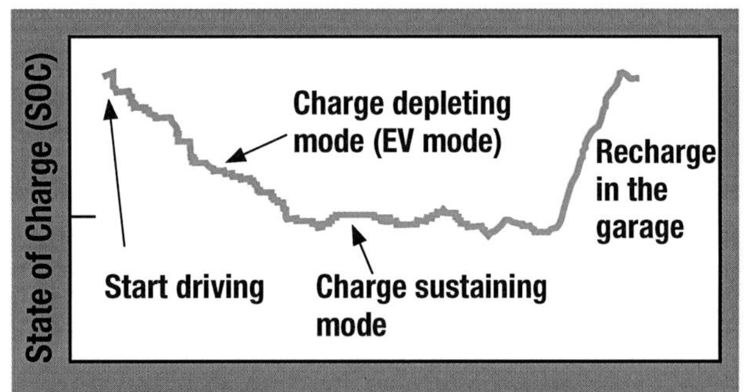

A simplified plot showing Volt's drive modes as related to battery SOC. The point is to rely on grid energy for power.

your kids took the car out, brought it back and didn't plug it in," Bly said. "One of our mobile-app features provides notifications around a variety of events, including battery state of charge. But seemingly more trivial but no less important is, 'Is the car plugged in at night?'"

He said the OnStar mobile apps allow Volt owners to communicate directly with their car, requesting and receiving text notices about charging status.

OnStar's almost unlimited upside, as it relates to keeping Volt owners connected to their vehicles, will come increasingly from remote-server-based tools and information. This technology trend is known as "cloud computing," and Bly believes it will help bring GM's PEVs into the mainstream by making them more user-friendly, capable, and, perhaps most importantly, rapidly upgradable.

Automotive electronics analyst Paul Hansen, publisher of the respected *Hansen Report on Automotive Electronics* agrees. He predicts "a wonderful convergence of 4G connectivity—zero latency, wide-wide bandwidth communications"—being linked to Internet cloud computing. Hansen reckons PEVs like Volt with sophisticated electronics architectures will serve as excellent platforms for this trend.

The physical act of charging Volt goes like this: Pop the hatch to the charge port on the left front fender. The standard 120-V portable charge-cord pack stowed in the rear cargo area enables Normal charging for most situations, and Reduced charging for when electrical current is limited.

Plug the charge cord's coupler into the receptacle until it clicks. The cord's LED charge-status indicators glow green to identify the vehicle can be charged. The LEDs flash red if the cord will not permit vehicle charging due to incorrect voltage or if the electrical outlet lacks a proper safety ground.

The process is basically similar with 240-V charging, a technology space that is already flush with suppliers. Chevrolet collaborated with SPX Service Solutions to offer the latter's 240-V home charging set priced at $490. That is before installation, of course, which must be handled by a certified electrician.

Chief Engineer Andrew Farah reckons a typical garage installation including local permits might cost $1500 to $2000, depending on location and codes. For that amount you could buy a lot of gasoline, EV skeptics argue. But for most homeowners, the first charger installation will be the only one needed.

Setting standards critical to the EV future

The challenge of implementing Volt, **Opel** Ampera, and their progeny becomes evident when one considers the diversity of electric-energy providers worldwide. There are more than 3000 electric utilities in the U.S., and scores in Europe, making the many standards efforts currently under way key to documenting and getting the widespread consensus OEMs need if electrified vehicles are to move beyond their current niche.

"We have to get the technical interface between vehicles today and in the future on the same page as the grid today and in the future—and we've seen some recent progress on this," said Britta Gross, GM's Director of Hybrid and Electric Infrastructure Strategy.

Since the advent of Volt, she has helped build dialogue between GM, policy makers, energy utilities, and various industry stakeholders regarding the emergent EV-charging base. The "Plug-In Cities" plan she spearheaded is a simple one-page outline of the critical steps cities and states should put in place to make vehicle charging possible. "It's a starting point for what these communities should do to make cities plug-in-ready," she said.

Volt's engineering leadership believes home charging will be the most popular means to juice up the car for the foreseeable future. Lack of public charging won't be a problem due to the car's petroleum-fueled range extender, they noted.

Gross agrees. "All the advancements that must be made on the grid side—how to plug in a car; how to communicate between the vehicle and the grid (V2G) or between the utilities and the car; how

As champion of GM's Plug-in Cities plan, Britta Gross sees progress in making electric energy a ubiquitous 'fuel' for vehicles.

we will one day work billing issues—don't have to be figured out on day one," she said. "That's because most of it becomes an issue only when you're talking about large volumes of vehicles. That's when we start to care that they're not all charging at the same time."

Setting expectations correctly is critical, she explained. For example, Gross noted that a few years ago utilities and automakers (including GM) were "not on the same page" in how they were answering PEV-related questions from consumers. "Utilities were talking about secondary use of batteries; they were talking about V2G and V2H [vehicle to home]. We said, 'Wait a second, guys! We've got to get the car right first.'"

Standardizing the EV charge-connectors' dimensions and function through the SAE J1772 standard (along with the wide-ranging J2836 communications standard) was a pioneering step toward getting the industry on the same page. J1772 will help minimize cost for automakers, charge-station suppliers, and energy utilities. It also greatly simplifies consumers' transition to EVs. With the J1772 hardware ready to go for Volt and other makers' PEVs in the pipeline, EV charging locations are expected to proliferate in the U.S. and in other regions.

SAE is developing a new version of J1772 that includes a standard for dc charging (also known as "fast charging"). Gery Kissel, who chaired the J1772 task force, said dc charging is still under development. He explained that the technology uses an off-board charger to connect directly to the vehicle's onboard high-voltage battery bus. Charging via dc allows for extremely high power (>100 kW) transfer and thus promises recharge times in minutes rather than hours.

Interest in faster charging times has sparked investigation of a vehicle charge connector that combines ac and dc capabilities in a single unit. But keeping the size of an integrated ac/dc connector small and lightweight is a technical challenge, Kissel said. Approval for SAE's dc charging standard is anticipated in the late 2011 time frame.

Meantime, Volt will be in the vanguard of what some experts see as a new era of vehicle propulsion and mobility technologies in general. The change from petroleum to electricity is not unlike moving from coal to oil—or shifting from the hay that "fueled" genuine horse power when the automobile era dawned, to gasoline.

"With an E-REV, there's a whole bunch of 'handshaking' between the vehicle and the utility," explained Britta Gross. "It's a new relationship, and we need to make sure it's a smooth experience for consumers."

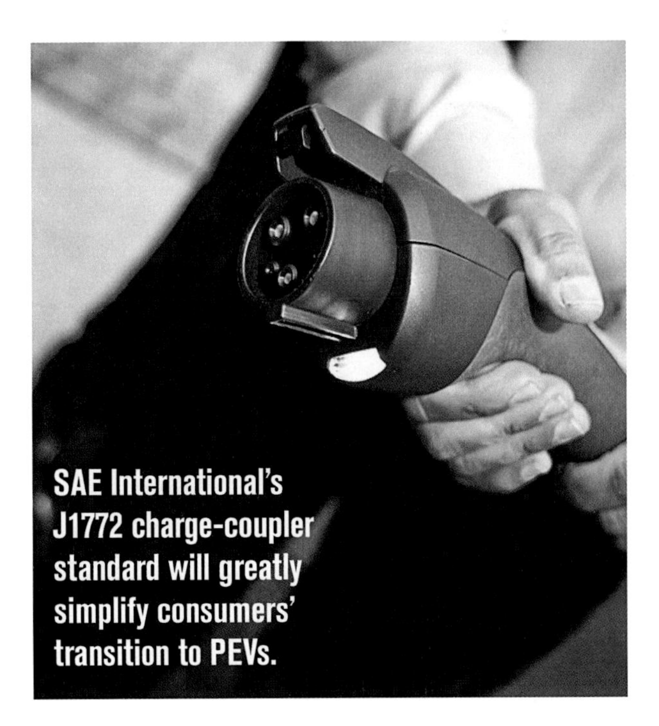
SAE International's J1772 charge-coupler standard will greatly simplify consumers' transition to PEVs.

Chevy Volt debuts

GM's Global A electrical architecture

GM's Micky Bly said the Volt program has helped enable GM's EEs to take new approaches to vehicle electrical architectures.

As General Motors Executive Director, Electrical Systems, Hybrid & Electric Vehicles and Batteries, Micky Bly is one of the busiest engineers at the reborn automaker. Recently, *AEI* Senior Editor Lindsay Brooke caught up with Bly at GM's Warren Tech Center to talk about the 2011 Chevrolet Volt and its impact on GM's electrified-vehicle development moving forward. Following are excerpts from that interview:

Q: Micky, what sort of opportunities does the Volt represent for GM in terms of moving vehicle electrical architecture design forward?

Bly: What's been really good about the Volt program from an engineering perspective is it's allowed us to challenge the architecture team to do things differently, in a manner that's faster and at more of a risk level than traditionally we're used to in this area.

In this sense Volt has served as a rallying point, an enabler, for ways to do things differently—from subsystems design and integration to the manner in which we present the vehicle's energy-use profile to the customer. The design engineers continue to bring us great ideas, and I've had to push back on some features as we get close to the start of production.

Q: Does GM have an internal designation for the Volt's electrical architecture?

Bly: We call it Global A, and we're rolling it out around the world—that's a first for us. Global A features very much re-usable, replaceable, modular-type software and controllers. Developing Global A has shown me the power of GM's global electrical-engineering team. The team in India is doing some of the software; the controls guys up in Milford [GM's Michigan proving ground] are literally speaking the same language around this feature. So it's been good.

Q: Are we at a development stage with electrified vehicles where engineers have to accept design complexity—multiple controllers, complex harnesses, etc.—before they can begin to make things simpler?

Bly: Yes. I think we'll see some up-integration of some controllers, where they make sense from a cost perspective. The main reason is the memory density that's on these controllers, the cost of memory, is so cheap now. Before you'd have a unique box because most of it was in the memory—you didn't want to have a very large box with a lot of memory.

Now we're able to optimize this from a memory point of view with the up-integration. It saved us a lot of money.

Q: Having developed OnStar to its current stage seems to have dovetailed well with GM's development and launch of Volt.

Bly: Yes. The key thing is we have a secure data connection to the vehicle, with a very flexible back office and the notion of the kind of APIs [application programming interface—an open standard approach to how developers write standards to interact with devices] that we make possible to connect to the vehicle. It would be irresponsible to let anybody access the vehicle on their terms.

What we're doing, as part of our experimentation, is allowing those developers to access OnStar with the appropriate messages and then we provide the secure connection to the vehicle. It helps protect the customer and protect the integrity of the operations.

We get a lot of unsolicited ideas that come our way. Over time, we've implemented some of them. We were getting OnStar customer e-mails before e-mail was really part of our normal life. But the appropriate "hands-on-the-wheel" types of technologies are what we want to make sure we're pioneering.

Voltec Charging System EMC Requirements and Test Methodologies

2011-01-0742
Published
04/12/2011

Vipul M. Patel and Donald Seyerle
General Motors Company

doi:10.4271/2011-01-0742

ABSTRACT

With the advent of vehicle manufacturer driven on-board charging systems for plug-in and extended range electric vehicles, such as the Chevrolet Volt, important considerations need to be comprehended in both the requirements specified as well as the test methodologies and setups for electromagnetic compatibility (EMC). Typical automotive EMC standards (such as the SAE J1551 and SAE J1113 series) that cover 12 volt systems have existed for many years. Additionally, there has been some development in recent years for high voltage EMC for automotive applications. However, on-board charging for vehicles presents yet another challenge in adopting requirements that have typically been in the consumer industry realm and merging those with both the traditional 12 V based system requirements as well as high voltage based systems. This paper will investigate those additional EMC standards (used in the consumer electronics industry) that are applicable to on-board charging systems as well as what modifications need to be made to those standards to adopt them to an on-vehicle implementation. Additionally, this paper will explore the specific details of the existing 12V EMC standards and procedures, indentifying which are directly applicable and how the specific test setups may need to be adjusted.

INTRODUCTION

Due to ever demanding fuel economy requirements by governments, environmental performance expectations from consumers, and volatility of fuel prices, vehicles in the hybrid to pure electric vehicles have grown in the market place. In the recent past, hybrids have been offered by many OEM's. These vehicles improve overall fuel economy, but do not tap into the electrical grid for power storage. There is an emerging segment that taps into the electrical grid for some varying amount to energy storage. In this segment, on one end of the spectrum are PHEV's (plug in hybrid electric vehicles). These vehicles typically have the least electrical grid storage capability. On the other end of the spectrum are pure EV's (electric vehicles), which rely only on stored energy. In between these two are EREV's (extended range electric vehicles), such as the Chevrolet Volt, which have a good electric only range performance, yet can go to a back-up on board charger when the stored electrical energy is used up. Key to adaptation of these vehicles that tap the electrical grid for energy storage is ease of use of the charging system. Earlier generations of electric vehicles required a dedicated charging station. Due to the size of these they had to be fixed in location, and could not be carried around in the vehicles. Due to this, charging "on the go" was not easily possible, and drivers needed to have a charging infrastructure to accommodate their travels. With this newer generation of grid electricity charged vehicles, the charging system has moved to a system that is easier to travel with. Standards such as the SAEJ1772 have developed that commonize charging systems. Within the SAEJ1772 there are several architectures proposed. The focus of this paper will be on the EMC aspects AC level 1 and AC level 2 charging system (see figure 1).

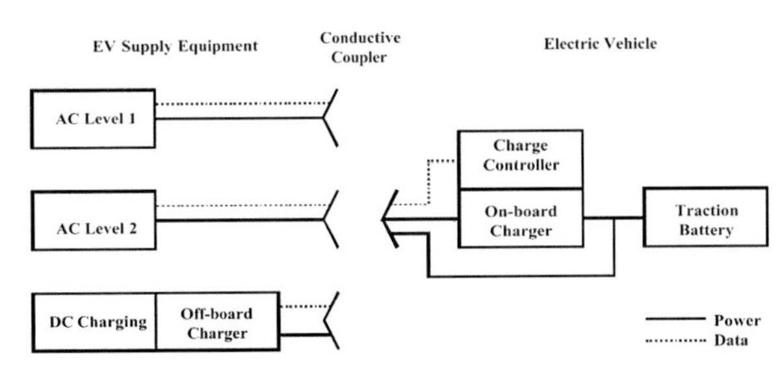

Figure 1. Charging architecture

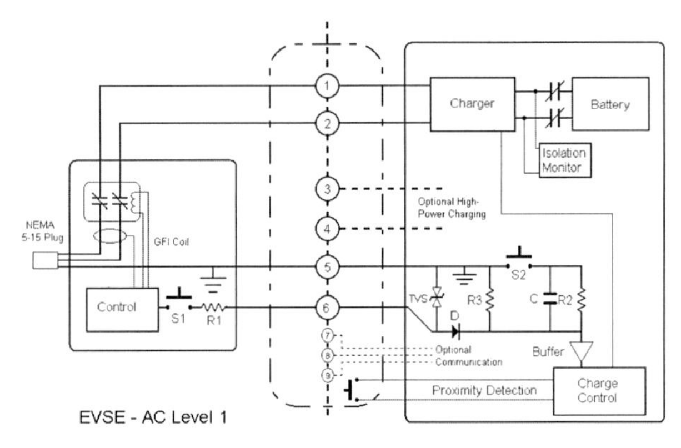

Figure 2. Level 1 charging

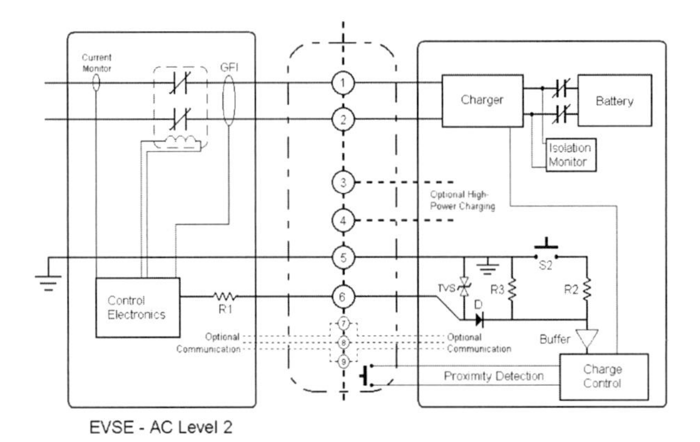

Figure 3. Level 2 charging

The main difference between AC level 1 and AC level 2 (figures 2, 3) is that the level 1 is a lower typical household outlet (120V in the US), the level 2 is a higher voltage and branch current rating. As can be seen from the architecture, there are 3 key electrical components to the system. The first one is the EVSE supply equipment, the role of this component is to check the electrical safety of the system and once it is safe, provide a data signal to the vehicle and then once the vehicle acknowledges that, close the contacts to provide a current path to the on-board charger. This is the second component in the system, the on-board charger. The role of this component is to take in the AC, rectify it, do power factor correction and finally provide a DC level to the battery. The final component is the charge controller, which manages the control of the charge level, time, etc. The topic of this paper will focus on the EVSE and on-board charger, since these are in the ac power path.

As far as the EVSE, there are two types, one for level 1 charging and one for level 2 charging. The level 1 charging is a lower voltage charging and thus does not have the safety restrictions of the level 2 charging. For this reason the level 1 charging EVSE is a small unit that can be taken with the vehicle for charging anywhere there is an ac wall outlet, (see figure 4). The level 2 charging system is a wall mounted electrician installed unit, that is fixed in position, (see figure 5). Both of these units provide ac directly to the same onboard charger.

Figure 4. Travel Cordset

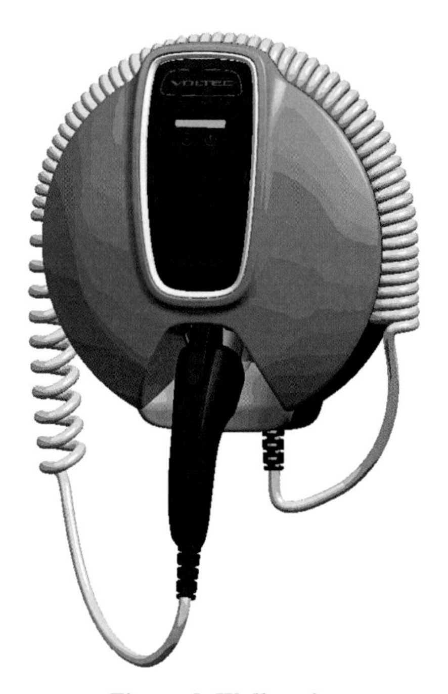

Figure 5. Wall station

Since the EVSE is essentially not on the vehicle, the primary specifications from an EMC standpoint for it would be from the IEC set of standards which are non-automotive. There are some exceptions to that for the few lines that cross into the vehicle (the control pilot and AC lines). The on-board charger would be subject to a mix of both automotive and non-automotive specifications, because it is on the vehicle, yet gets AC from outside the vehicle. There are numerous documents that cover the testing of these devices.

Automotive specifications come from both the SAEJ113 series and OEM specific procedures for which these in turn reference CISPR25 and several ISO documents. For the non automotive requirements, there are documents such as the UL2231 and IEC61851, which in turn reference IEC EMC documents. The next sections will review EMC requirements and methodologies of the EVSEs and on-board charger, and will discuss component test requirements and setups such that each component can be validated on its own. The advantage of that is that many times the suppliers of these components can be different companies and even in the case of the same company developing both, there are different project teams working on each component with their own timing issues.

BODY

EVSE EMC Requirements and Methodologies

As noted earlier the EVSE is not on the vehicle and will have a mix of requirements from the non-automotive specific ones to automotive ones. In general the EVSE plugs into a power line at one connection point and provides this power along with control information to the vehicle at another connection point. Since the wiring layout and setup is an important part of the EMC setup, this paper will propose wiring and grounding setups for the various tests in addition to device setup.

Transients

Since the side that connects to the power grid does not run in the vehicle, the transients on that side would only be ones that are standard that are used on other plugged in devices. In this area there are two specifications that apply that are electromagnetic transient phenomena. They are the surge transient (IEC 61000-4-5) and the fast transient (IEC 61000-4-4). The surge transient simulates transients that occur as a result of lightning induced surges that can occur on power distribution networks. There are various mechanisms that lightning can get to an outlet, one of them is a field induced surge on the overhead lines and another is lightning strikes to the ground which raises that potential. While there are various mechanisms for the surge to get to a receptacle, the IEC 61000-4-5 has a waveform that represents this surge energy and is applied both differentially (line to line) and also common mode (line to ground). The waveform is coupled via a coupling network to the device under test. The fast transient is supposed to simulate the transient that happens when loads on the distribution network are switched on and off. When inductive and capacitive loads are cycled they produce a transient on the power distribution system and these transients show up on the device. While the surge transient is run differential and common mode, the fast transient is just run differential mode.

On the side of the wiring that interfaces with the vehicle we would need to impose the automotive transients. Within the automotive suite of transients tests, there are two fundamental types of tests. One of them is for conducted transients, as the name implies this would be for lines that can have transients that are conducted to lines via a direct battery connection (unregulated/unfiltered lines). The other one is for coupled transients that occur as a result of co-routing of wiring of noisy lines to lines under test. As can be seen from figure 2 and 3, the lines from the EVSE that run within the vehicle do not have battery power line coming in or a pull up to the vehicle battery, so the conducted transients would not be applicable. The coupled transients would be applicable since these lines would be running in the vehicle with other lines. When running the coupled transients test, an important part of running the test would be to duplicate the impedance of these lines on the load side. Running a test with the wires un-terminated would not represent what actually happens on the vehicle, since the AC lines would be terminated by the differential (X) capacitance that is on the charger and the common mode (Y) capacitance that is on the charger. In addition the control pilot line also has some capacitance to ground on the load side.

Electro-Static Discharge

In the case of ESD, there are various tests and modes to be considered. Within ESD, there are tests that are done when a unit is powered up and also in a powered off mode to assess the handling immunity of the device to ESD. Typical automotive module testing for ESD handling is for assembly plant handling of a unit. For this reason, the test levels are not the highest, because assembly plants are a known environment. In the case of the travel cordset however, a customer will be handling it regularly, and thus the ESD levels should be increased to the levels that are done for powered on customer level tests. In addition the wall station should also be subject to the same levels, since it will be installed in a uncontrolled environment.

For the powered off test, both direct contact and air discharges should be done to the 3 lines that connect to the utility, see figure 6. Also this test should also be performed to the surface of both units shown in figures 4 and 5. Finally the output receptacle should also be tested (figure 7). For the powered on case, the ac plug lines are not accessible, but the unit and output receptacle are and should be tested. The only deviation should be to not test contact discharge on the output receptacle for safety reasons, in case this line is live. Normally this line will not be live until it is connected to the vehicle, but it is a safety precaution. The metal contacts on this connector are recessed slightly in the plastic cavity, so an air discharge can be attempted.

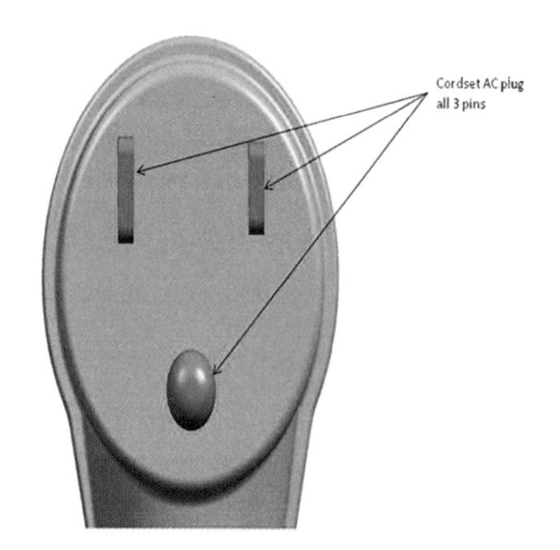

Figure 6. AC plug

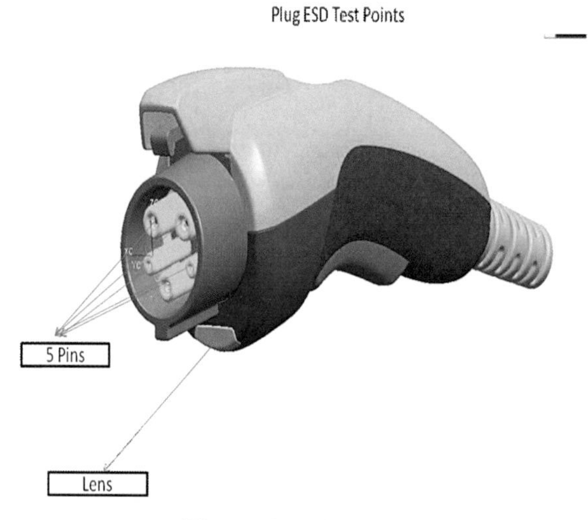

Figure 7. Vehicle plug

Radiated Immunity

Radiated Immunity testing can be accomplished with IEC61000-4-3 and IEC61000-4-6. For low frequencies, IEC61000-4-6 is used. Because of the low frequencies and component chamber size limitations, this method couples rf energy directly on the wiring. When running this test method, it is important to induce energy on the outlet wire, since this has a pilot signal also when in charging mode. In addition, when running charging mode, the load equivalent impedance should also be used. Also, when running this test, a ground plane should be attached to the outlet side safety ground to simulate the vehicle body. This is important for low frequency testing because the ground plane simulates displacement current paths that are present. For higher frequencies where the wavelength is shorter, a direct rf exposure method is done. This is specified in the IEC61000-4-3 procedure.

Radiated Emissions

For radiated emissions, there is the FCC test for unintentional radiators. When this product goes to other regions outside of US, then EN55011 and other local requirements are applicable. Since this device is not on the vehicle it will need to follow the proper certification requirements for applicable markets.

On-Board Charger EMC Requirements and Methodologies

Since the charger is on the vehicle, many of the EMC requirements will come from traditional automotive specifications. Figure 8 has a typically mechanization of an on-board charger. As can be seen in the figure, the unit gets ac power in and provides a high voltage dc (this high voltage dc is shielded typically in a vehicle) and is controlled via a communications bus. Depending on the architecture, there can be a 12v supply also and some discrete lines for control.

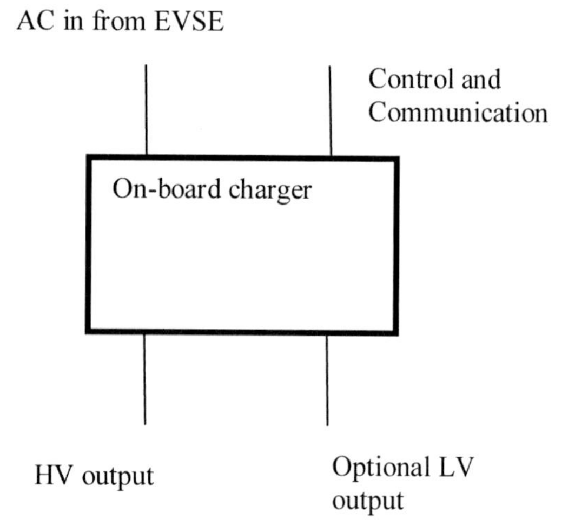

Figure 8. Charger mechanization

Transients

For the transients tests, the primary testing will come from the normal automotive EMC specifications. There will be some supplemental EMC procedures that will apply from the IEC set of tests. As far as the coupled transients tests, the HV DC lines would not be applicable because those lines are shielded and ground terminated at both the charger end and battery end. Typically the power and ground lines are not applicable because they are tested for the conducted automotive transient immunity test. However, in this case these power and ground lines are unique in that they are not tested in the normal automotive conducted transient immunity tests since they are AC power and ground, thus should be tested for the coupled transients test. An important part of the setup is to ensure that there is no second earth ground for the test setup, since the vehicle body is isolated from the ground, see figure 9. As can be seen from figure 9, the ground plane should not be grounded to the facility, as well as any incidental earth ground provided by cooling fans or monitoring equipment. In the conducted transient immunity, any 12v power lines and control lines that are pulled up to 12v need to be tested. Note for these tests there is no new setup procedures, since they are 12v. For the conducted transients on the AC power lines noted in the EVSE section, this test is not absolutely necessary. Since the EVSE needs to protect its electronics from the surge and fast transients, it provides incidental protection for the charger upto the voltage on the clamping device used. As an example, if the EVSE has MOV's rated to 500V, the charger will get protection for transient voltages above that, but the charger would have to withstand transients below that level. In addition, the hardware in the EVSE needs to be understood to ensure the clamping is in parallel to prevent pass through of transients.

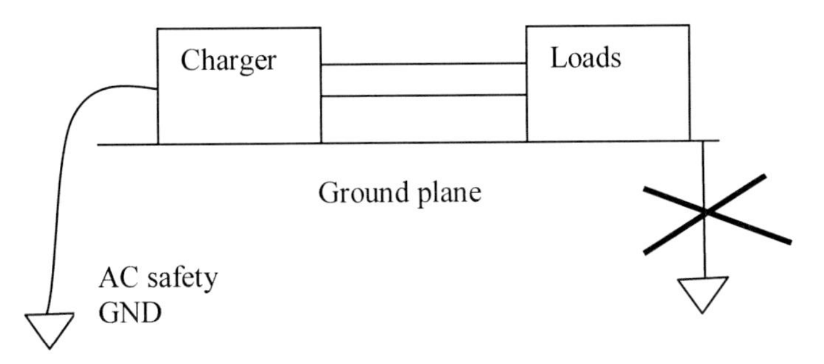

Figure 9. Charger EMC test setup

Electro-Static Discharge

For ESD handling tests, the standard automotive set of handling tests need to be done. One exception to the setup would be to ground the charger to the conductive ground plane. Typically the unit is insulated from the ground plane for automotive because the operator is handling the unit before assembly into the vehicle. The charger is tested in this unique way because it is chassis grounded and once it is installed, an assembly center operator or a service technician can try to reach the charger while inadvertently touching a connector terminal. This represents a worst case scenario, since the charger will have current flow from the pin for ESD discharge. High voltage line shields would be subject to the test, yet the center pin would not be if it is recessed in the connector. AC lines would be subject to this test, since the receptacle on the vehicle has prongs that can be hit with ESD. For the ESD powered on test, there are a limited number of lines that are subject to it since there is no customer control surface on it. The communications lines are subject, since these eventually go to the service connector in the vehicle, and a service center can discharge to it while trying to find the pin.

Radiated Immunity

For the radiated immunity there are two major modes the unit needs to be tested in, one is in charging mode with the AC line attached and the other is in vehicle active non-charging mode. In the charging mode functions that need monitoring are output voltage and current. In addition, internal data parameters such as temperature need to be monitored via serial data. To accomplish this appropriately, suitable load needs to be attached to the high voltage output. This load needs to be able to handle the power coming out of the charger (approximately 3KW for 240 V level 2 charging on the Volt). In addition to this resistive part from HV+ to HV-, there also needs to be equivalent X and Y capacitance that simulate the capacitance in the battery electronics. When running these tests it is important to comprehend the setup and assess the appropriate method to test. As an example, when running low frequency radiated immunity, a injection probe is used to excite the wires in the harness. Examining the setup shown in figure 10, it shows two ground loops. One is from the AC ground to the test ground plane (which is grounded per automotive EMC standards). The other one is between the HV output of the charger to the simulated battery load on the coax shield. When these are included in the injection probe, they are equivalent to a shorted turn on the secondary of this primary to secondary equivalent transformer when doing injection probe testing. The effect this has on induced current on the other lines can be seen in figure 11. It effectively reduces the induced current on the other I/O due to more reflected power going back to the amplifier caused by the shorted turn. Note that figure 11 has two sets of profiles to compare, the blue colored lines are when the other I/O is not tied to an artificial network (LISN), and the red lines are when they are tied to a LISN. In both cases however, when the line that is a shorted turn is included, it generally reduces the coupled current onto other I/O.

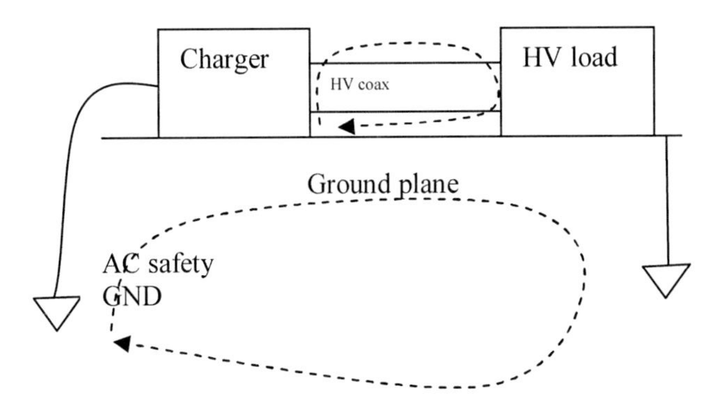

Figure 10. Ground paths in setup

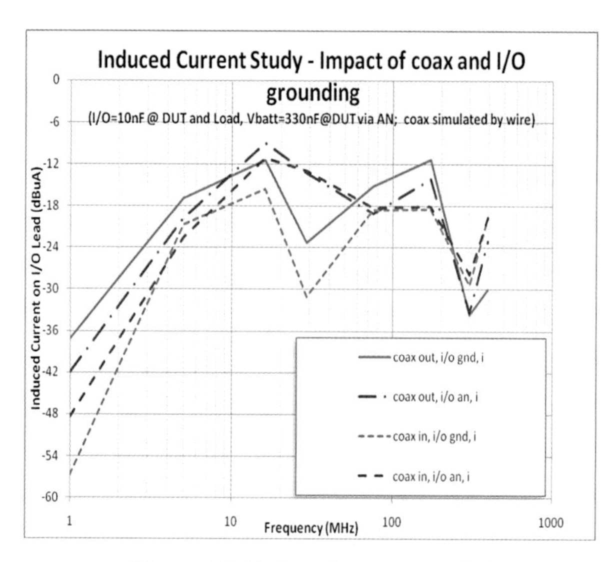

Figure 11. Induced current on I/O

Radiated Emissions

For radiated emissions also, there are two modes to be concerned with. In charging mode the both the FCC/EN conducted emissions for frequencies below 30Mhz needs to be performed as well as the radiated emissions needs to be done for frequencies above 30Mhz. In vehicle active mode, the OEM specific EMC standards need to be performed.

SUMMARY/CONCLUSIONS

Applicable EMC tests for both the EVSE and On-board charger were reviewed in this paper. Along with applicable tests, important aspects of the setup were also reviewed. Key things to consider are simulating loading characteristics when testing an individual component and also grounding to represent vehicle installation. An appropriate mix of IEC and OEM based EMC specs needs to be done to successfully test these charging system components.

REFERENCES

1. UL2231

2. SAE International Surface Vehicle Standard Series, SAE Standard Series J1113.

3. SAE International Surface Vehicle Recommended Practice, "SAE Electric Vehicle and Plug in Hybrid Electric Vehicle Conductive Charge Coupler," SAE Standard J1772™, Rev. Jan. 2010.

4. IEC 61000 EMC series

CONTACT INFORMATION

Vipul Patel (vipul.patel@gm.com) and Donald Seyerle (donald.r.seyerle@gm.com) both are in the Electromagnetic Compatibility Lab at GM Proving Grounds in Milford, MI.

DEFINITIONS/ABBREVIATIONS

EVSE

Electric Vehicle Supply Equipment

IEC

International Electro-technical Commission

The Engineering Meetings Board has approved this paper for publication. It has successfully completed SAE's peer review process under the supervision of the session organizer. This process requires a minimum of three (3) reviews by industry experts.

ISSN 0148-7191

Positions and opinions advanced in this paper are those of the author(s) and not necessarily those of SAE. The author is solely responsible for the content of the paper.

SAE Customer Service:
Tel: 877-606-7323 (inside USA and Canada)
Tel: 724-776-4970 (outside USA)
Fax: 724-776-0790
Email: CustomerService@sae.org
SAE Web Address: http://www.sae.org
Printed in USA

Development of the Chevrolet Volt Portable EVSE

2011-01-0878
Published
04/12/2011

Tony Argote and Gery Kissel
General Motors Company

doi:10.4271/2011-01-0878

ABSTRACT

The plug-in vehicles developed in the 1990's ushered in the first standards for electrified vehicles. These standards included requirements for Electric Vehicle Supply Equipment or EVSEs. EVSE is a general term for all the non vehicle components needed to charge a plug-in vehicle. These components include cabling, connectors and shock safety equipment. EVSEs are used to charge vehicles at home, work and in commercial settings.

Many people identify EVSEs with public charge stations. While public charge stations are iconic with plug-in vehicles, these are just one type of EVSE. Until public EVSEs become readily available, plug-in vehicle drivers will need to partially rely on portable versions of EVSE.

Portable EVSEs are required to provide the identical function and safety protection as their stationary cousins but their portability brings unique challenges and design considerations.

Thermal, vibration, ergonomic and vehicle storage are just a few of the challenges facing design of portable EVSEs that stationary EVSEs are completely or partially immune to.

This paper will explore the development of the portable EVSE for the Chevrolet Volt. Aspects of the portable EVSE design that will be explored include:

- Design requirements driven by portability
- Validation requirements and testing
- Ergonomics
- Supply base limitations

INTRODUCTION

One of the main aspects of the Chevrolet Volt is that the vehicle can be charged from any common AC receptacle in less than 10 hours. For this concept to be realized, it was decided that a portable EVSE would not only be included with the Chevrolet Volt, but a storage area under the rear cargo floor would be provided.

Storing the portable EVSE on the vehicle presented many challenges. The two obvious are fundamental to any vehicle component, physical size and mass. As the portable EVSE specification developed however, other challenges surfaced.

With the portable EVSE possibly stored on the vehicle the device would need to tolerate the vehicle environment. Factors including temperature and vibration needed to be comprehended. While standards such as SAE J1772™ Electric Vehicle Conductive Charge Coupler and industry safety documents from Underwriters Laboratory (UL) provide guidance on temperature and vibration requirements these were considered insufficient to tolerate the vehicle environment. In order to provide the Chevrolet Volt customer with a robust and reliable portable EVSE, it was decided to impose stricter requirements on the Volt's portable EVSE that were similar to other passenger compartment electronic modules.

The challenge of developing a portable EVSE robust enough to survive a vehicle passenger compartment is the focus of this paper.

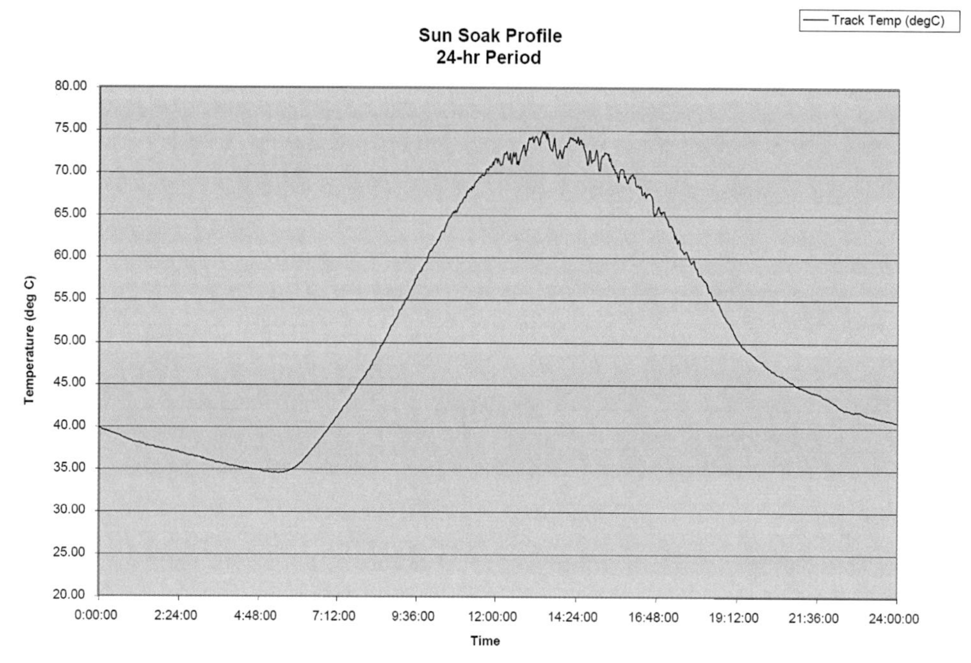

Figure 1. Typical Vehicle Sun Soak Profile

THERMAL

AMBIENT TEMPERATURE

SAE J1772™ requires the charge coupler to operate in an ambient temperature between −30 and +50 degrees Celsius. SAE J1772™ does not currently identify an operating ambient temperature range for the rest of the portable cordset including the control electronics. Temperature requirements referenced in safety standards documents such as UL2202 Electric Vehicle Charging System Equipment are intended to establish limits related to product safety and not for vehicle durability. For our operating ambient temperature requirements vehicle cold soak and sun soak data was referenced. An example of vehicle sun soak data appears in Figure 1.

Use cases for the portable EVSE were also considered. It was considered likely that a customer would consider charging their vehicle during extremes of vehicle cold or hot soak using the portable EVSE that was stored in the vehicle. After review of the vehicle data, the ambient operating temperature requirement of −40 to +80 degrees Celsius for the portable cordset was established. In addition to ambient operating temperature range, a non-operating storage temperature excursion requirement of +85 degrees Celsius was established to comprehend events such a paint baking during repairs after a collision.

HIGH TEMPERATURE DURABILITY

The Chevrolet Volt portable EVSE was subjected to sustained high temperature to evaluate material degradation and diffusion based failure mechanisms. These high temperature tests were conducted with the portable EVSE operating. All functional requirements had to be met during and after the test with all inputs/outputs operating normally. To complement the ambient temperature testing noted above, the portable cordset was tested operating at 60 degrees Celsius for 1000 hours.

THERMAL SHOCK

The portability of the EVSE allows for use cases where the EVSE may be subjected to sudden changes in ambient temperature. The test severities of Chevrolet Volt portable EVSE are based on the mounting location, internal use location, external use location, and in vehicle during storage. The use locations drive the minimum electrical, mechanical, thermal, and climatic requirements. Understanding the usage cases that would be characteristic of the portable EVSE, performance criteria was set to allow for cyclical indoor to outdoor temperature changes within the operating temperature requirement of the EVSE. Cyclical thermal testing of the unit within the operating temperature requirement extremes, or Thermal Shock testing, is a method of validating the robustness of the EVSE design against thermal fatigue.

At the molecular level, temperature changes produce stress between two bonding materials when those materials have different coefficients of thermal expansion (CTE). In thermal shock testing, the accelerated cyclical thermal exposure of the EVSE produces cyclical stress, which over time may lead to fatigue. By design, material selection for bonding material interfaces must comprehend CTE to avoid thermal fatigue.

Understanding failure modes driven by mismatches in subcomponent interface CTEs is essential in assuring long term durability of the EVSE.

For the development of the Chevrolet Volt portable EVSE, a thermal shock profile was developed, which specified temperature ranges within the units operating temperature specification, dwell times, and number of cycles, all correlated to the projected representative unit lifetime and usage.

As with ambient temperature, the Thermal Shock profile developed for the Chevrolet Volt portable EVSE exceeds the minimum temperature ranges required in SAE J1772™.

POWER TEMPERATURE CYCLING

Power Temperature Cycling (PTC) of the EVSE is the second step in quantifying the design's susceptibility to thermal fatigue, immediately following Thermal Shock testing. In contrast to Thermal Shock testing, in PTC the EVSE is powered and constantly monitored. Under these conditions, while the EVSE is subjected to the power and temperature cycling stresses, failure modes related to integrated circuit dies, solder creep, and mechanical attachments may all be observed.

Active monitoring of the EVSE electrical outputs during PTC allows for detection of variations in the unit's performance. The pilot signal is an example of one EVSE output that is held to a tolerance by specification. Shifts in component values during PTC may essentially stack up and impact the pilot signal output, which as a downstream affect changes the communicated current availability to the EV, and in some cases may interrupt charge. Attention to CTE values in the design of the EVSE control the variation and susceptibility of the impact of thermal cycling; PTC testing is used as a validation method against the EVSE specification.

VIBRATION

A portable cordset's control electronics will be comprised of a variety of electronic components. These electronic components will include many leaded components such as relays, capacitors, transformers, fuses, etc. In any electronic module located on a vehicle, leaded components require special attention to ensure they do not fail due to vibration fatigue.

The vibration requirements in SAE J1772™ are specified for the vehicle charge inlet, not for the portable cordset's control electronics. Vibration requirements referenced in safety standards documents such as UL2202 Electric Vehicle Charging System Equipment are intended to establish limits related to product safety and not for vehicle durability.

For the development of the Chevrolet Volt portable EVSE a vibration and thermal soak profile was developed, as well as 6 axes shock loads to represent pothole impacts and minor collisions.

SOLAR ULTRAVIOLET EXPOSURE

As EVs are expected to be charged using portable EVSEs in both garage (indoor) locations as well as outdoor locations, solar ultraviolet exposure is a factor taken into consideration in establishing design and performance specifications for the portable EVSE. When subjected to solar ultraviolet exposure, the portable EVSE must maintain performance requirements, safety requirements, and appearance requirements; the latter of which is a considerable challenge.

Ultraviolet impact on material performance has been studied sufficiently to provide material related specifications for safety and long term durability. As an example, within UL2251 Plugs, Receptacles and Couplers for Electric Vehicles, the vehicle plug is required to pass an ultraviolet light exposure test as referenced in UL746C Polymeric Materials Use in Electrical Equipment Evaluations, section 28. This particular specification is a standard for safety of polymeric materials used in electrical equipment. In this regard, the portable EVSE is tested to meet mechanical, thermal, and electrical performance requirements after accelerated solar ultraviolet exposure.

The portable EVSE may also have appearance requirements, as is the case for the Chevrolet Volt portable EVSE, which are impacted by ultraviolet exposure. While industry requirements for unattended household equipment are understood for common electrical equipment and materials, appearance properties, such as colorfast, are still an area under development. This difference between readily available materials with rated safety performance and materials with appearance durability is a challenge which often results in tradeoffs. For the portable EVSE, safety is an overriding priority, thus the selection of readily available compounds for materials are often limited in terms of long term appearance performance. Colorfast development, as an example, will undoubtedly be an area of development that will enable competitive portable EVSE offerings that meet automotive OEM appearance requirements.

ERGONOMICS

A portable EVSE must be designed for ergonomic ease of use. The portable EVSE vehicle plug is estimated to be the component most used by the consumer, and thus required development of ergonomic design criteria. Pictured below, the Chevrolet Volt portable EVSE vehicle plug incorporated criteria for a range of hand profiles, as well as gloved and ungloved hand criteria for users in any ambient. The vehicle plug also incorporates an ultra-bright LED flashlight feature, which maximizes outdoor use in dark or unlit locations when trying to plug in the EVSE while consuming minimal power.

Figure 2. Chevrolet Volt Portable EVSE Vehicle Plug

In addition to the vehicle plug, the portable EVSE main enclosure for the Chevrolet Volt, pictured bellow in Figure 2, included ergonomic design criteria. The key considerations in the development of human interface and usage requirements for the EVSE included portability, storage and simplicity. To support portability, the EVSE was designed to with a soft-touch carry handle and a cord reel styled main enclosure. The soft-touch carry handle was designed with the same design criteria used for the vehicle plug. The cord reel design of the EVSE was developed to allow the customer to wrap the 6.1 Meters of EV cable from the main enclosure to the vehicle plug, which aids in managing the EV cable as well as the storage of the EVSE into the rear compartment within the vehicle. In this respect, the human factors criteria driving the reel design assisted in integrating the EVSE design into the vehicle design. Undoubtedly portable EVSEs will offer a variety of features and methods of execution to integrate portability with storage, and functionality with ease of use. With time and feedback from the consumer, the design of portable EVSEs will evolve with ergonomics and human factors playing a key role.

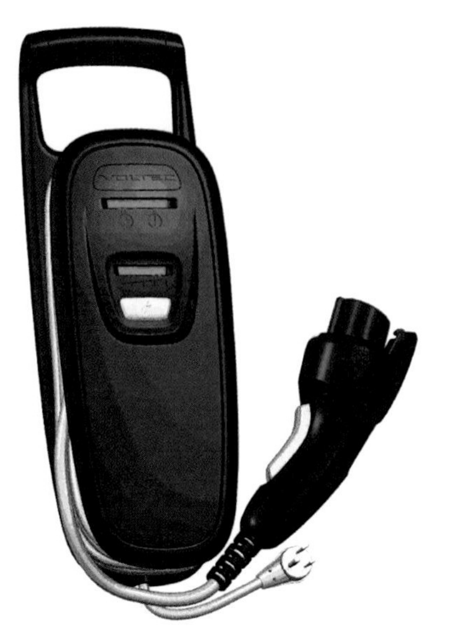

Figure 3. Chevrolet Volt Portable EVSE

SUPPLY BASE LIMITATION

With the requirements for the Chevrolet Volt portable cordset complete, the sourcing process began. It became apparent almost immediately that the supply base yielded its own set of challenges.

A couple of equipment suppliers that manufactured portable EVSE's in the 1990's and early 2000's were still in business. In fact, some of the public EVSEs installed by these companies in the 1990's were still in place and functional. However none had portable EVSEs designed to meet our requirements. Also none of these companies had ever been a full Tier 1 supplier to an automotive OEM and did not have the resources or experience to support a full production program.

Conversely, traditional automotive OEM electronic module Tier 1 suppliers had no experience with the challenges of designing a product to conform to consumer product safety standards.

General Motors solution to the supply base limitation was to encourage an automotive OEM electronic module Tier 1 supplier to develop a portable cordset with a partner company that already provided electrical distribution equipment and safety electronics to the consumer market. The rationale for this approach was that the automotive Tier 1 supplier comprehends the challenges of developing electronics for the automotive environment and can provide all the logistical support to launch a production program. The Tier 1's partner would provide the experience needed in safety electronics and certification of the portable EVSE..

SUMMARY/CONCLUSIONS

Storing the Chevrolet Volt portable EVSE on the vehicle presented many challenges. With the portable EVSE possibly stored on the vehicle the EVSE would need to tolerate the vehicle environment. The Chevrolet Volt portable EVSE was designed to requirements similar to other passenger compartment electronic modules.

Thermal and vibration performance requirements for the portable EVSE were developed. These performance requirements are more rigorous than the safety based performance requirements outlined in SAE J1772™ and other safety standards from Underwriters Laboratory.

Innovation from a broadening supply base should result in portable EVSEs with continued improvement in robustness, durability and ergonomics.

REFERENCES

1. SAE International Surface Vehicle Recommended Practice, "SAE Electric Vehicle and PLug in Hybrid Electric Vehicle Conductive Charge Coupler," SAE Standard J1772™, Rev. January 2010.

2. Underwriters Laboratory Standard, "Electric Vehicle Charging System Equipment," UL2202

3. Underwriters Laboratory Standard, "Plugs, Receptacles and Couplers for Electric Vehicles," UL2251

4. Underwriters Laboratory Standard, "Polymeric Materials Use in Electrical Equipment Evaluations," UL746C

CONTACT INFORMATION

Gery J. Kissel
gery.j.kissel@gm.com

Tony Argote Jr.
tony.argote@gm.com

DEFINITIONS/ABBREVIATIONS

CTE
Coefficient of Thermal Expansion

EVSE
Electric Vehicle Supply Equipment

TS
Thermal Shock

PTC
Power Temperature Cycling

UL
Underwriters Laboratory

The Engineering Meetings Board has approved this paper for publication. It has successfully completed SAE's peer review process under the supervision of the session organizer. This process requires a minimum of three (3) reviews by industry experts.

ISSN 0148-7191

Positions and opinions advanced in this paper are those of the author(s) and not necessarily those of SAE. The author is solely responsible for the content of the paper.

SAE Customer Service:
Tel: 877-606-7323 (inside USA and Canada)
Tel: 724-776-4970 (outside USA)
Fax: 724-776-0790
Email: CustomerService@sae.org
SAE Web Address: http://www.sae.org
Printed in USA

About the Author

Lindsay Brooke is Senior Editor of SAE International's *Automotive Engineering International* magazine. He has written extensively about automotive technology, manufacturing, business, and history for 30 years. Before joining SAE International, Brooke was senior auto industry analyst at CSM Worldwide, specializing in technology forecasting for various industry clients. As an analyst he was widely quoted by *The Wall St. Journal,* National Public Radio, *Forbes, Fortune,* and other media.Prior to that he was editor of *Automotive Industries* magazine, with a brief hiatus as Chrysler Corp.'s manager of Engineering and Technology public relations during the 1990s. Brooke's writing on automotive topics has appeared in *The New York Times, AutoWeek, Popular Science, Popular Mechanics Online, Cycle World* and other periodicals. He is the author of four books, covering the history of the iconic Ford Model T and Triumph motorcycles. His work has received numerous professional accolades, including the annual Jesse H. Neal Award presented by the American Business Press for outstanding journalism. Brooke is a juror on the North American Car and Truck of the Year awards. He holds Bachelor's and Master's degrees in journalism and communications from Shippensburg (PA) University, and is a member of SAE International and the Automotive Press Association.On weekends he can be found riding his Triumph Bonneville and wrenching on his 1947 Willys CJ2A Jeep.